W0018486

Quadruplex Nucleic Acids

Editorial Board

Professor Stephen Neidle (Chairman), *The School of Pharmacy, University of London, UK*
Dr Simon F Campbell FRS
Dr Marius Clore, *National Institutes of Health, USA*
Professor David M J Lilley FRS, *University of Dundee, UK*

This Series is devoted to coverage of the interface between the chemical and biological sciences, especially structural biology, chemical biology, bio- and chemo-informatics, drug discovery and development, chemical enzymology and biophysical chemistry.

Ideal as reference and state-of-the-art guides at the graduate and post-graduate level.

Other titles in the series:

Biophysical and Structural Aspects of Bioenergetics
Edited by Mårten Wikström, *University of Helsinki, Finland*

Structure-based Drug Discovery: An Overview
Edited by Roderick E. Hubbard, *University of York, UK and Vernalis (R&D) Ltd, Cambridge, UK*

Exploiting Chemical Diversity for Drug Discovery
Edited by Paul A. Bartlett, *Department of Chemistry, University of California, Berkeley and Michael Entzeroth, S*Bio Pte Ltd, Singapore*

Structural Biology of Membrane Proteins
Edited by Reinhard Grisshammer and Susan K. Buchanan, *Laboratory of Molecular Biology, National Institutes of Health, Bethesda, Maryland, USA*

Protein–Carbohydrate Interactions in Infectious Disease
Edited by Carole A. Bewley, *National Institutes of Health, Bethesda, Maryland, USA*

Sequence-Specific DNA Binding Agents
Edited by Michael Waring, *Department of Pharmacology, University of Cambridge, Cambridge, UK*

Visit our website on www.rsc.org/biomolecularsciences

For further information please contact:
Sales and Customer Services, Royal Society of Chemistry, Thomas Graham House, Science Park, Milton Road, Cambridge CB4 0WF, UK
Telephone: +44 (0)1223 432360, Fax: +44 (0)1223 426017, Email: sales@rsc.org

Quadruplex Nucleic Acids

Edited by

Stephen Neidle
The School of Pharmacy, University of London, London, UK

Shankar Balasubramanian
Department of Chemistry, University of Cambridge, Cambridge, UK

RSCPublishing

ISBN-10: 0-85404-374-8
ISBN-13: 978-0-85404-374-3

A catalogue record for this book is available from the British Library

© The Royal Society of Chemistry 2006

All rights reserved

Apart from fair dealing for the purposes of research for non-commercial purposes or for private study, criticism or review, as permitted under the Copyright, Designs and Patents Act 1988 and the Copyright and Related Rights Regulations 2003, this publication may not be reproduced, stored or transmitted, in any form or by any means, without the prior permission in writing of The Royal Society of Chemistry, or in the case of reproduction in accordance with the terms of licences issued by the Copyright Licensing Agency in the UK, or in accordance with the terms of the licences issued by the appropriate Reproduction Rights Organization outside the UK. Enquiries concerning reproduction outside the terms stated here should be sent to The Royal Society of Chemistry at the address printed on this page.

Published by The Royal Society of Chemistry,
Thomas Graham House, Science Park, Milton Road,
Cambridge CB4 0WF, UK

Registered Charity Number 207890

For further information see our web site at www.rsc.org

Typeset by Macmillan India Ltd, Bangalore, India
Printed by Henry Lings Ltd, Dorchester, Dorset, UK

Preface

The knowledge that guanine-rich DNA can form unusual structures has its origin in findings that precede the double-helix structure by 50 years. However, the nature of the guanine aggregation, at the polynucleotide helix level, was only revealed in the 1960s by fibre diffraction and other biophysical studies. These showed that guanine polynucleotides form four-stranded helices held together, not by the Watson–Crick base-pair motif, but by four guanines hydrogen-bonded in-plane G-quartets. This type of structure appeared initially to have little biological relevance until it was demonstrated that natural G-rich sequences in telomeric DNA at the ends of chromosomes could form such structures, at least *in vitro*. The study of quadruplex biophysical chemistry made steady progress in the 1990s, with NMR and to some extent X-ray crystallography defining some structural detail. The biology of quadruplexes remained a relatively unexplored area until it was demonstrated that small-molecule stabilisation of the single-stranded ends of telomeric DNA into quadruplexes, is an effective way of inhibiting telomerase activity, and could therefore lead to anti-telomerase anticancer therapy. Furthermore, a growing number of quadruplex-specific proteins are being discovered. There is now the increasing realisation that non-telomeric sequences in human and other genomes can form (or perhaps can be induced to form) quadruplexes, and that these quadruplexes play a role in the regulation of gene expression. The diversity in quadruplex architecture is also of increasing interest, even though much remains to be established, and rules relating sequence to fold cannot yet be defined.

We believe that it is now timely to review many of these themes, and so clarify where the quadruplex area is going. We have been fortunate in being able to bring together many of the leading experts in this exciting field, and are grateful to them for having delivered their manuscripts so promptly.

Shankar Balasubramanian, Cambridge
Stephen Neidle, London

Contents

Chapter 3   Structural Diversity of G-Quadruplex Scaffolds   81
Anh Tuân Phan, Vitaly Kuryavyi, Kim Ngoc Luu and
Dinshaw J. Patel

Fundamentals of Quadruplex Structures

GARY NIGEL PARKINSON

The School of Pharmacy, University of London, 29–39 Brunswick Square, Bloomsbury, WC1N 1AX, London, UK

1.1 Background and Introduction to Quadruplexes

Self-association of guanosine at millimolar concentrations has been observed in solution since the 19th century as characterized by the ready formation of polycrystalline gels. In the 1960s Gellert *et al.*[1] determined the associated guanine bases to be in a tetrameric arrangement by crystallographic methods, described simply as a G-quartet arrangement. The four guanine bases form a square co-planar array where each base is both a hydrogen bond donor and hydrogen bond acceptor. Utilization of both the N1 and N2 of one face with the O6 and N7 of the second face on guanosine yields eight hydrogen bonds per planar G-quartet [Figure 1(a-b)]. With the development of chemical synthesis of extended polyguanine oligonucleotides strands additional associations were observed in the laboratory environment. CD and IR spectroscopy confirmed the same self-assembly and association of the guanines into G-quartets, while X-ray fiber diffraction studies demonstrated a four-stranded motif with stacked tetrad planes, termed quadruplexes.[2–5] These stacked tetrads align themselves to give a similar appearance to that of duplex DNA [Figure 2(a)], characterized by a regular rise and twist between the tetrad planes and generating a right-handed helical twist [Figure 2(c)]. In this case the phosphate backbones, linking the nucleosides together, generate four grooves of variable width, instead of two, giving the quadruplex DNA motif a characteristic duplex DNA feel.

Interest in the structural arrangements of G-quadruplexes was ignited in the early 1990s by the identification of G-rich repetitive sequences located at the end of chromosomes and a protein, with a reverse transcriptase activity, involved in their maintenance. This ground-breaking work was carried out by Blackburn *et al.* in Joe Gall's research group. They applied sequencing techniques developed previously in Sanger's research laboratory in the late 1970s, a group that was conducting the first comprehensive sequencing experiments on genomic material. It was quickly realized that these guanine-rich

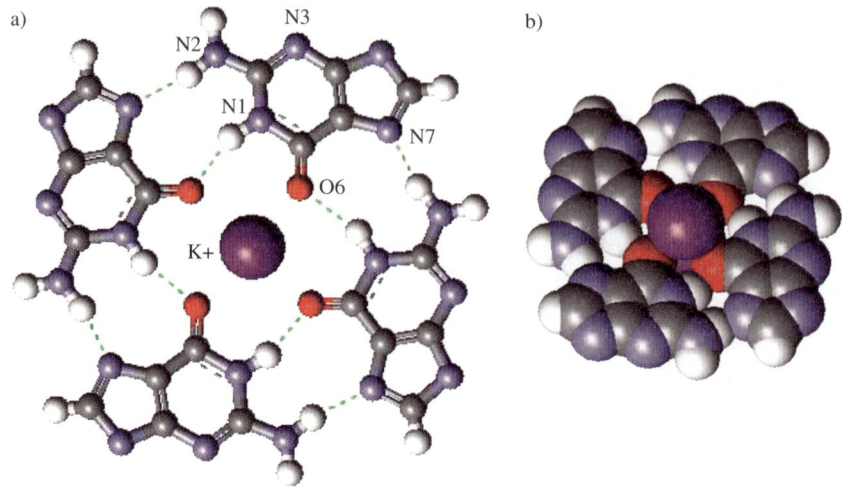

Figure 1   *(a) A guanine quartet highlighting the hydrogen bonding network between the Hoogsteen and Watson-Crick faces of the guanine bases, including the central potassium cation, deoxyribose sugars removed for clarity; (b) space filling model of a guanine tetrad with associated K^+ metal ions located above and below plane, deoxyribose sugars removed for clarity*

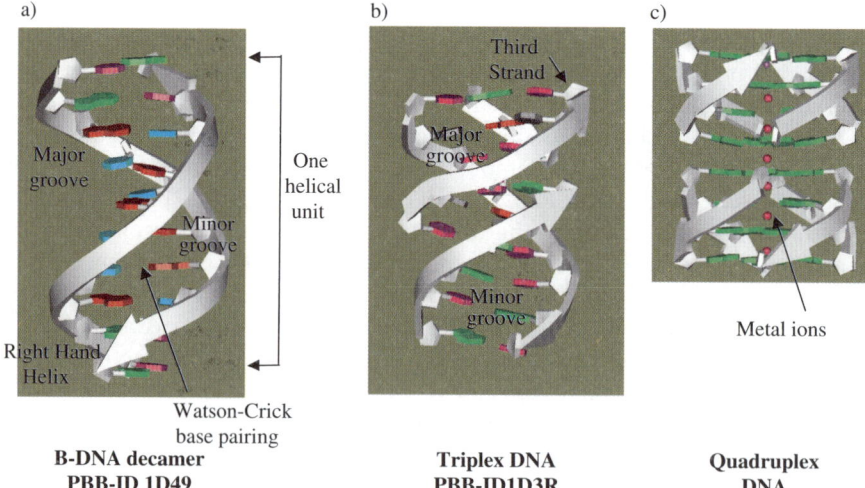

Figure 2   *(a) A B-DNA decamer[72] (PDB ID 1D49) showing one full helical turn, and major and minor grooves; (b) triplex DNA[73] (PDB ID 1D3R) showing in insertion of a third strand into the major groove of the duplex DNA; and (c) two stacked parallel stranded G-quadruplexes in a 5' to 5', head-to-head interaction*

repetitive telomeric DNA sequences could form higher order structures and were likely to be involved in chromosomal maintenance. Structural studies[6–14] on the telomeric sequences revealed both parallel and anti-parallel strand orientations, as well as mixed *anti* and *syn* glycosidic torsion angles with the specific features of the quadruplex structural motif dependent upon sequence. The structures determined for the telomeric 3' overhang were of particular interest in terms of chromosomal DNA packaging, and molecular self-assembly, particularly as these G-rich sequences can form compact, well-defined and stable structural motifs.

1.2 Fundamental Components of DNA and RNA

The following is a very basic overview of nucleic acids and their associated properties, focused in areas that relate to quadruplex structure. For a more comprehensive description references 15–19 provide excellent reading. Nucleic acids are polymers of nucleotide units. Each nucleotide unit is composed of three important building blocks: the bases, sugars, and phosphate groups. The nucleoside consists of a base attached to a pentose sugar ring [Figure 3(a)]. In RNA the sugar is a ribose, and in DNA a deoxyribose. The unmodified bases utilized in DNA comprise guanine, cytosine, adenine, and thymine, while for RNA the thymine base is substituted by uracil. The phosphate groups are attached at the 5' side of the nucleoside serving as the linking element between the nucleosides to form a nucleotide. Polymers formed from these three basic components have particular properties that make them ideal for the long-term storage of genetic information in living cells. The nucleotide units are chemically stable and the individual strands have the ability to associate together *via* complementary bases to form a stacked duplex structure. These polymers are then able to store complimentary copies of genetic information in a compact form that can both disassociate and re-associate.

1.2.1 Building Blocks

The bases are the key components that confer chemical variability to DNA/RNA. The bases have complementary hydrogen bond donors and acceptors that generate specific associations between bases. There are two faces involved in hydrogen bond formation, the Watson–Crick and the Hoogsteen face, as shown in Figure 3(b). Association *via* base pairing is normally seen between purines (Guanine, Adenine) and pyrimidines, (Thymine, Cytosine, and Uracil) bases utilizing the basic Watson–Crick base-pairing motif. The G:C base pairing, with its three hydrogen bonds is more stable than the A:T/A:U base pairing with only two hydrogen bonds. This is partially reflected in the higher melting temperatures for GC rich sequences. Other hydrogen bonding arrangements are possible between base pairs and these include a reversed Watson–Crick, a GT wobble pair as well as the use of the Hoogsteen face to give Hoogsteen pairings and reversed Hoogsteen pairings. It is the additional use of the Hoogsteen faces that is critical in the formation and stabilization of tetrads [Figure 3(b)]. The bases with their nitrogen atoms have the added capability to

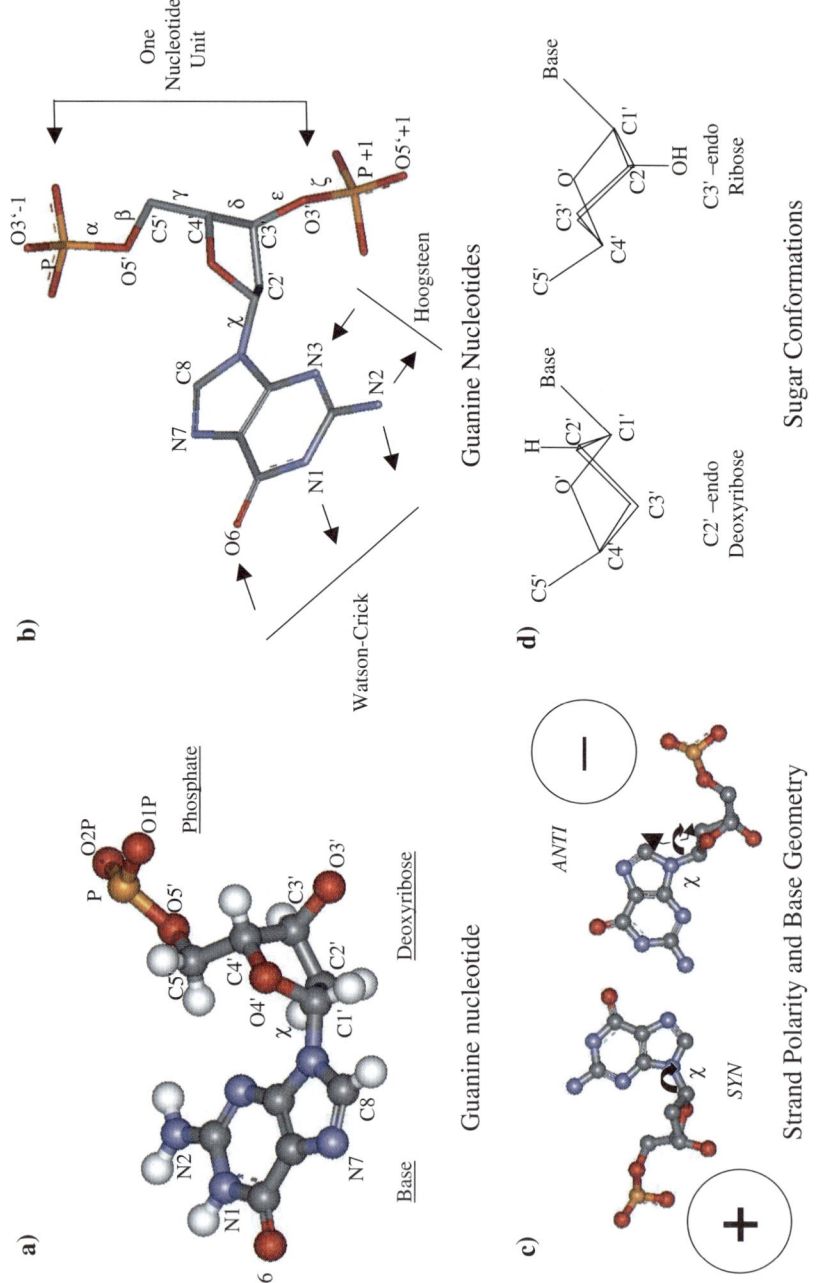

a) Guanine nucleotide

b) Guanine Nucleotides

c) Strand Polarity and Base Geometry

d) Sugar Conformations

change protonation states based on pH. At neutral pH standard base pairing occurs but at elevated (pH >9.2) or reduced pH (pH <4.2) additional base associations have been observed. The bases are covalently linked to the sugar *via* the glycosylic bond. The energy minima conformations available to this linkage, within an extended nucleic acid structure, are important in determining how DNA folds and the stability of these folded structures. This dihedral angle linkage χ defined as O4'-C1'-N9-C4 for purines and O4'-C1'-N1-C2 for pyrimidine bases, Figure 3(c). Glycosidic torsion angles in duplex DNA/ RNA fall into two tightly confined conformations. The related orientations adopted by the stacked bases have significant consequences for the overall shape and depth of the DNA/RNA grooves, altering the width, shape, and hydrogen bonding pattern. Two of the most common conformations observed in folded DNA structures are *syn* and *anti* shown in Figure 3(c), where *syn* (0< χ <90°) and *anti* (−120< χ <180°).

The pentose sugars also have a critical impact on the conformation of nucleic acid structures. They are non-planar with several energetically stable conformations. In duplex DNA/RNA structures the C2'-*endo* and C3'-*endo* conformation are the two most commonly observed [Figure 3(d)]. In an RNA duplex arrangement the sugar conformation is generally restricted to C3'-*endo*, partly due to steric hindrance, and partly due to hydrogen bonding associated with the O2' hydroxyl group on the pentose sugar. In the case of DNA (deoxyribose) the removal of the hydroxyl group at the O2' position allows additional energetically favorable sugar pucker conformations. The energy barrier between C2'-*endo* to C3'-*endo* sugar pucker is 8–21 kJ.[20] In B-form DNA for example, the sugar is primarily in the C2'-*endo* conformation. The same C2'-*endo* sugar pucker is commonly observed in G-quadruplex structures for nucleosides involved in tetrad formation. In addition a strong link exists between sugar pucker and the glycosidic torsion angle in extended nucleic acid structures, particularly as C3' and C4' atoms of the sugar also form part of the backbone structure. The sugars have no protonation sites however, the O4' oxygens are extensively utilized by waters forming a network of hydrogen bonding interactions that contribute to the stability of nucleic acid structures.

Figure 3 *(a) A ball and stick representation of a guanine nucleotide highlighting the three building blocks of DNA, the base, sugar (pentose ring), and phosphate groups. (b) The two hydrogen bonding faces of guanine nucleotide known as the Watson–Crick and Hoogsteen edges, both of which are utilized in guanine tetrad formation. Arrows indicate hydrogen bond donors and acceptors. (c) Nucleotides showing both syn and anti glycosylic torsion angles χ [O1-C1'- N9(1)-C8(6)] with backbone strand polarity shown as (+) for the 5'–3' direction and as (−) for 3'–5' the direction. The anti conformation is the lower energy arrangement and the preferred orientation in duplex B form DNA. (d) Two major sugar pucker conformations as found in duplex and quadruplex DNA. The C2'-endo conformation for a deoxyribose ring commonly found in B- form and quadruplex DNA and C3'-endo conformation for a ribose sugar associated with A-form DNA and RNA*

In nucleic acid structures the phosphate groups are the most exposed and susceptible elements to the effects of solvent. Salt concentrations and levels of hydration have a profound effect on nucleic acid structure, as demonstrated by the salt dependent transition from B to A form DNA. The linking phosphodiester groups have a formal minus one charge resulting in a repulsive electrostatic interaction with other phosphate groups along its own chain and with other associated chains. In addition, the phosphomonoester group at the 5′ terminal end of the DNA/RNA strand introduces one extra charge. All these charged oxygen atoms on the phosphate groups are commonly counter balanced by water molecules and metal ions, stabilizing grooves and loop motifs. The consequence of the repulsive negative charges along the phosphate backbone is also apparent within quadruplex structures in modulating folding and the arrangement of loops that link adjacent strands together.

The extended phosphate backbone also introduces a great potential for conformational variability in nucleic acids structures. The DNA backbone consists of six atoms compared to the three in the protein backbone; consequently the backbone conformation is defined by six dihedral angles [Figure 3(b)]. This apparent variability however, is greatly limited by steric constraints. For example several of the backbone torsion angles (α, β, γ, ζ) are restricted to three energetically favorable orientations, albeit with low energy barriers. This results in three broad energy minimas, -60, 60, $180°$. It would be still expected that a large number of backbone conformations would be observed in nucleic structures. This indeed is seen for single nucleotide structures, however in extended polynucleotides bound to complementary strands, other factors take over in the efficient compaction of the units into higher order helical structures. Within these duplex arrangements it is observed that certain dihedral angles associate together into sets of low-energy conformations.[18] One such arrangement identified as BI/BII comes from the linked interplay of two dihedrals (ζ and ε). Several other relationships have also been identified, involving for example, α and γ. In addition some dihedral angles are tightly constrained to the sugar pucker conformation such as δ and ε. Other factors affecting backbone conformation come from packing together discrete DNA or RNA structures, either by association in solution or by the tight packing within a crystal lattice. Visually the distribution of backbone dihedral angles is well represented by the use of a wheel plot. An example is show in Figure 4(a) for B-form DNA where the lines represent individual dihedral angles for each nucleotide unit. The clustering and tight distribution of dihedrals can be seen. For comparison, wheel plots are shown for an anti-parallel quadruplex structure,[21] Figure 4(b) and a parallel quadruplexes arrangement[22] Figure 4(c). The similarity in dihedral angle distribution and clustering in these two plots, when compared to duplex B-form DNA, is striking. The obvious difference seen in the anti-parallel quadruplex is the distribution of both *syn* and *anti* conformations for the glycosylic torsion angles. Quadruplex structures, in common with RNA structures and unlike helical DNA have single stranded nucleotides unconstrained by normal Watson–Crick base pairing. These single stranded runs have a great deal of flexibility, and are structurally more versatile, typically

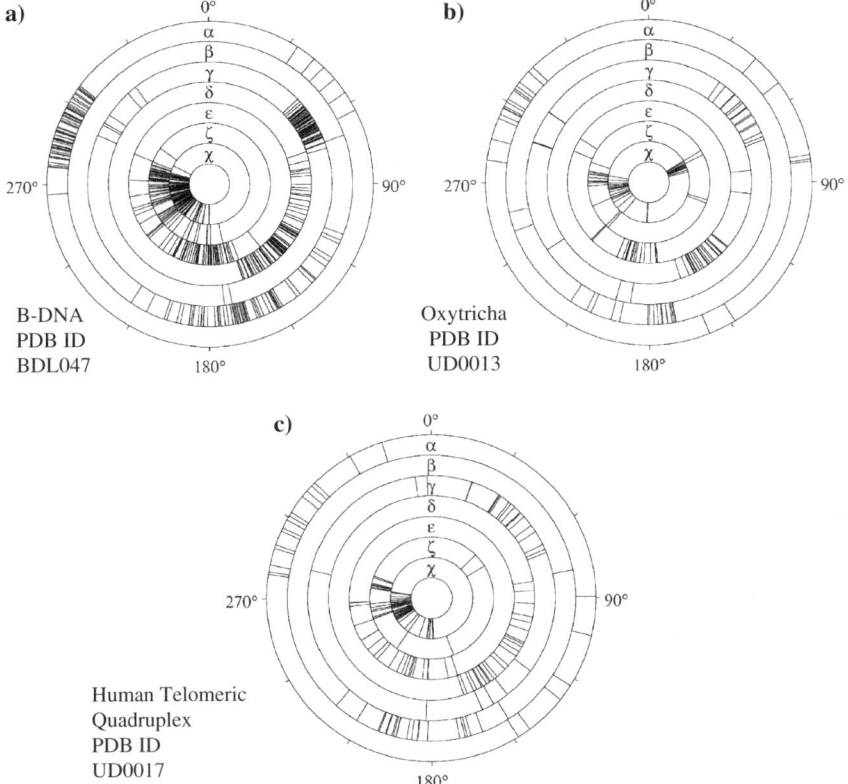

Figure 4 *Wheel plots of major torsional angles of various DNA crystal structures including backbone and glycosidic angles derived from the NDB (www.ndb-server.org). (a) B-DNA dodecamer (PDB ID BDL047); (b) quadruplex crystal structure with four tetrads containing the oxytricha⁺ telomeric sequence, a G_4T_4 repeat (PDB ID UD0013); and (c) parallel stranded human telomeric structure of sequence TTAGGG (PDB ID UD0017)*

involved in inter and intra molecular contacts. The conformational flexibility available to the backbone is a key factor in how quadruplexes fold and pack together.

1.2.2 DNA/RNA as a Helical Structure

One view of double stranded DNA was as a featureless linear rod for the storage of genetic information. This view was quickly revised to one of a standard, double stranded B-form DNA, associated with the chromosome bent around chromatin and packaged into the nucleus. Its passive role was revised again as it soon became apparent that structural flexibility was required for DNA to function. In fact, this structural heterogeneity observed in DNA has long been explored and for the case of duplex DNA, classified into several diverse structural groupings A, B, C, D. These motifs and their role in a

Table 1   *A comparative summary of key structural parameters for duplex and quadruplex folded DNA*

Structural type	B-DNA	A-DNA	Z-DNA	Quad-parallel	Quad-anti-parallel
Rise (Å)	3.40	2.90	3.7	3.13	3.30
Twist (°)	36.7	32.7	−10/−50	30.0	30.0
Groove width (Å)	11.7/5.7	2.7/11	8.5	10.2	12
					8.9/12.2
Strand polarity	+−	+−	+−	++++	+−+−, +−−−
					++−−
Helix	RH	RH	LH	RH	RH
No. of bases per turn	10.5	11.0	12.0	12.0	12.0
Width (Å)	18	26		23	21–23
C1′-C1′	10			16	16
Sugar pucker	C2′	C3′	C2′/C3′	C2′	C2′

Groove width: Backbone phosphate i and the $i+3$ phosphate on the opposing strand.

biological setting has long been sort. A common theme for all these forms of DNA is in a stacked duplex arrangement, with anti-parallel polynucleotide strands (the 5′–3′-phosphodiester backbone linkages running in opposite directions) forming a right-handed stacked helix. A schematic of B-form DNA, the most common motif found in a biological setting is shown in Figure 2(a),[23] many of its most important characteristics are highlighted. Key structural features for B-form DNA and other forms are laid out in Table 1, tabulated from X-ray structural data. There is great structural diversity available to DNA and one of the more unusual polymorph of duplex DNA is the Z-form. This motif has a left handed helical rise, alternating *syn-anti* glycosidic angles with variable groove widths. Structural descriptors for the DNA are included in Table 1 for comparison. A structure of a B to Z-form DNA transition has recently been determined by X-ray crystallography.[24] In addition to single and double stranded DNA, the polynucleotide strands can associate into a three-stranded arrangement termed a triplex [Figure 2(b)],[25] by the inclusion of a third strand into the major groove of duplex DNA. This is accomplished by utilizing additional hydrogen bonding and protonation states of the bases. The association of four polynucleotide strands termed a quadruplex only adds to the structural diversity seen in nucleic acid structure. Even though quadruplexes with their four strands might be considered a more diverse polymorph of DNA, they still retain characteristics that are similar to their double stranded cousins. The torsion angles adopt a range of values that are characteristic of duplex DNA, the bases also stack with a helical twist, while the base-pairs utilize standard Watson–Crick and Hoogsteen hydrogen bonding faces. In addition the dihedrals and sugar pucker conformations adopt similar low energy conformations to those observed in other DNA structural motifs. Characteristic structural descriptors of G-quadruplexes are tabulated in Table 1, highlighting their similarity to other forms of DNA, particularly to the B-form.

1.2.3 Stabilizing Factors in Quadruplexes

The same stabilizing factors found in duplex DNA structures such as base stacking, hydrogen bonding, hydration structure, and electrostatic interactions are associated with quadruplex DNA structures, although in G-tetrads only guanine (purine) base stacking needs to be accommodated in G-quadruplex formation. In contrast to these stabilizing factors, an important component of quadruplex instability comes from the arrangement of the guanine O6 carbonyl groups central to the G-quartet. The O6 atoms form a square planar arrangement for each tetrad, with a twist of 30° and rise of 3.3 Å between each tetrad step forming a bipyramidal antiprismatic arrangement for the eight O6 atoms. These negatively charged cavities located between the G-tetrads need to be stabilized by the coordination of cations [Figure 5(a–b)]. The selection of a suitable cation, based on size and charge, dramatically determines the overall stability of the final folded quadruplex. For all types of DNA/RNA, where strands associate, grooves are formed from the linked phosphate backbones, creating an extended buried channel of hydrogen bond donors and acceptors. These channels are populated by water molecules, ordered into hydration shells linking together the bases, sugars to the charged phosphate atoms, located on the outer surface, stabilizing the folded DNA structures.

1.2.3.1 Base-Stacking

The bases in DNA are non-polar in nature and have unfavorable interactions with polar solvents. Paired bases will then associate and stack on each other in order to reduce the area exposed to the solvent. Stability is then provided to the

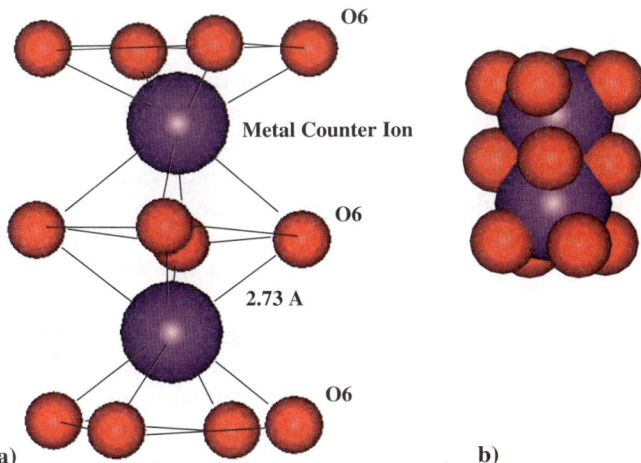

Figure 5 *Counter ion coordination between tetrad bases shown with twist between bases. (a) Potassium metal ion is shown coordinated between eight carbonyl oxygens with an average 2.73 coordination distance. (b) A space filling model with potassium counter ions*

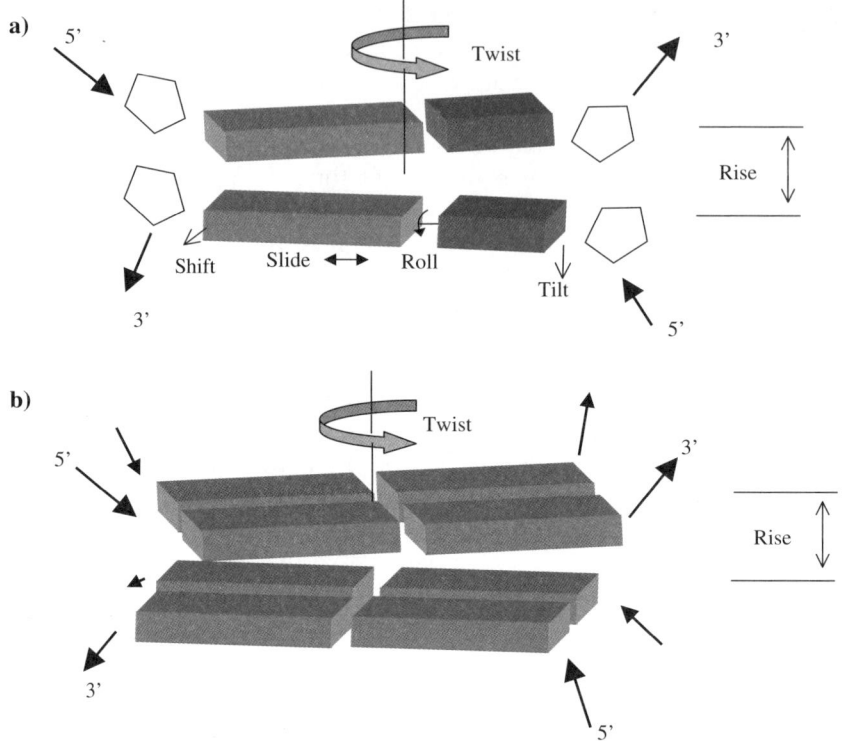

Figure 6 *Example of base stacking between base pairs. Main geometric descriptors of twist and rise are shown in relation to helical axis and backbone strand polarity. (a) Duplex DNA and a GC base pair; (b) a stacked tetrad consisting exclusively of guanines bases*

DNA helices by the stacking of the bases through a combination of hydrophobic, electrostatic, and van der Waals forces. In addition a small negative charge is also associated with the stacked bases providing a slight tendency for the bases to repel each other. The stacked bases are polarized resulting in the planar face of an aromatic molecule becoming electron rich, while the hydrogens around the edge of the ring become electron poor. In canonical B-DNA stacking energies have been estimated to be between -9.5 and -13.2 kcal mol^{-1} for GC base-pair steps, whereas an AT base-pair steps have a lower stacking energy of about -5.4 kcal mol^{-1}. These stacked bases cannot sit directly on each other due to steric constraints but are twisted about the helical axis. The distance between the bases pairs is defined as the rise while the rotation about the helical axis is defined as the twist [Figure 6(a)]. Table 1 has a list of typical values of rise and twist for the different DNA forms. Other parameters such as roll, slide, and tilt are used to provide a detailed description of the bases relative to one another and to the global helical axis. Variations in roll and tilt are critical in changing DNA structural motifs say from B to A forms, and to give

DNA a global buckle and bend required in the packaging of DNA around the histone, or other protein scaffolds.

In quadruplex structures, just as in duplex DNA, the stacked guanines are neither exactly symmetric or planar. This variation in geometry in the G-tetrad plane can be described in the same geometric terms as duplex DNA with rise, twist, roll tilt, and slide [Figure 6(b)] for both tetrad steps and individual guanine bases. Crystal structures of quadruplexes reveal an unexpected heterogeneity in guanine stacking along the length of the quadruplex. Distortion of the planar tetrads can come from the crystal packing forces and also from linkers of reduced length that impose distance constraints on the G-tetrads. Guanines forming the top G-quartet planes as well as to those stacked below tilt and/or buckle to accommodate the induced strain. As expected, away from the influence of the linking nucleotides the bases involved in tetrad formation at the center of the stacked quadruplex are more planar. An additional variation in tetrad geometry, in part, comes from the stacking of guanines bases in the alternating *syn, anti, syn* glycosidic arrangements. This alternating pattern is found in the anti-parallel backbone arrangements where the bases need to be flip in order to retain their Watson–Crick and Hoogsteen base pairing. In parallel stranded structures only a single stacked arrangement is observed as all the bases have *anti* glycosidic torsion angles [Figure 7(a)]. This variation in base stacking affects the arrangement of the carboxyl O6 oxygen atoms forming the cavity, located central to the quadruplex core, shown in Figure 7(a–c), where the position of the O6 atoms are rotated relative to one another dependent upon the type of stacking. This may play a role in metal ion selectivity in the central core.

1.2.3.2 Hydrogen Bonding and Other Energies

Hydrogen bonding is a major mode of interaction in DNA and RNA, controlling both secondary and tertiary conformations. Watson–Crick base pairing predominates in the double helix where as the utilization of the Hoogsteen face in guanine quartet increases the potential number of hydrogen bond to eight. Studies were undertaken to understand the apparent increases in energies associated with quadruplex formation. Calculations showed the energy for an average hydrogen bond increased from 0.22 to 0.42 eV, when involved in the G-quartet formation. This significant increase in binding energies was attributed to resonance-assisted hydrogen bonding.[25] Additional binding energies associated with the G-quadruplex formation were assigned to metal ions associated with the O6 carboxyl atoms located in the central G-tetrad stack.

1.2.3.3 Hydration Structure

The quadruplexes, similar to duplex DNA have grooves contain a network of water molecules ordered about hydrogen bonding donors and acceptors, the exocyclic amino groups N2, and heterocyclic N3 atoms for quadruplexes. The 4-fold symmetry seen in the parallel stranded quadruplexes is reflected in the symmetric hydration structure seen in the four grooves formed by the stacked

a)

b)

c)

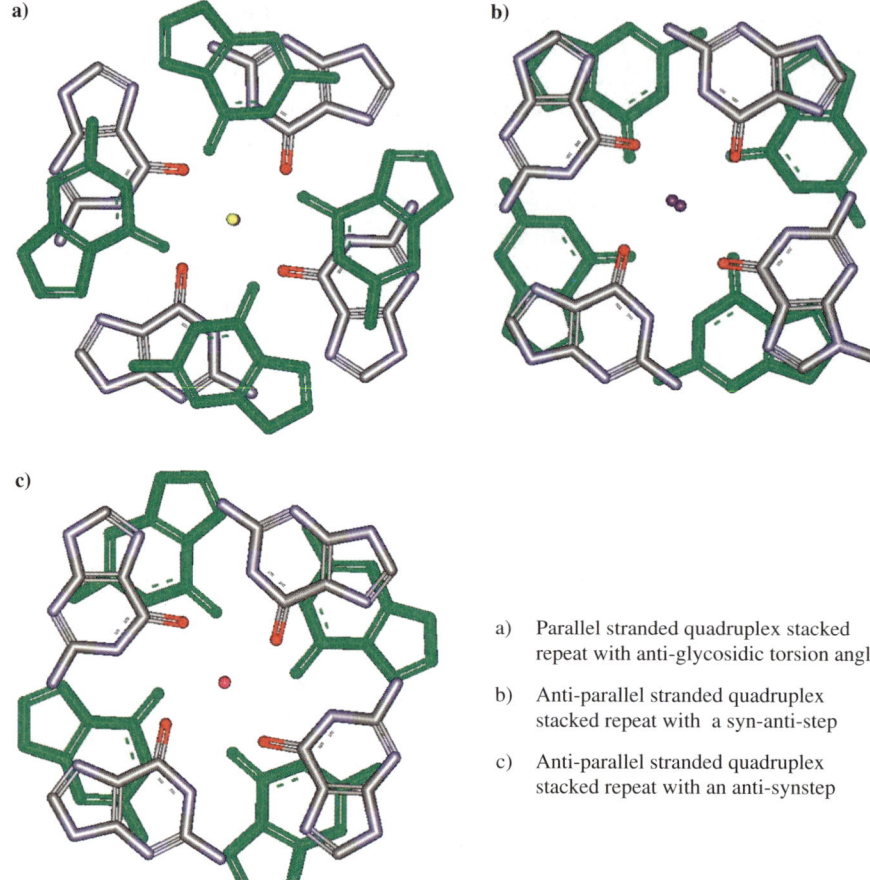

a) Parallel stranded quadruplex stacked repeat with anti-glycosidic torsion angles

b) Anti-parallel stranded quadruplex stacked repeat with a syn-anti-step

c) Anti-parallel stranded quadruplex stacked repeat with an anti-synstep

Figure 7 *A series of stacked guanine tetrads with various strand arrangements syn and anti relationships. (a) Parallel stranded quadruplex stacked repeat where all bases have anti-glycosidic torsion angles stacked over a second tetrad containing only anti-glycosidic torsion angles. (b) First step in an anti-parallel stranded quadruplex with a syn-anti-syn-antitetrad stacked arrangement over an adjacent anti-syn-anti-syn tetrad. (c) Second step in an anti-parallel stranded quadruplex with anti-syn-anti-syn tetrad stacked over a syn-anti-syn-anti-tetrad*

tetrads. In anti-parallel quadruplexes with alternating glycosidic torsion angles a more complicated and less ordered hydration structure is observed. Figure 8(a) shows a guanine quartet with pseudo-electron density derived from the averaged water positions of three overlain parallel stranded G-quartets. The labeled N2 acceptor and N3 donor atoms highlight the ordered nature of the waters, in particular the 4-fold symmetry around the tetrad. The sugar O4' oxygens, commonly associated as hydrogen bond acceptors, are also extensively utilized to create a second tier of ordered water molecules in the grooves. A network of ordered water molecules from the human telomeric crystal

a)

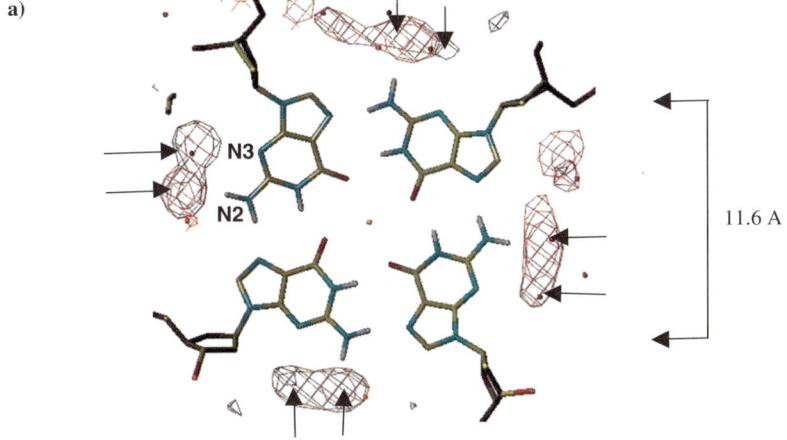

Guanine tetrad and associated waters

b)

Groove2

Figure 8 *(a) A guanine quartet with pseudo-electron density, averaged water occupancy derived from three overlain G-quartets. The N2 acceptor and N3 donor atoms are labeled. (b) Hydration structure associated with grooves highlighting hydrogen donor and acceptors (N2, N3, 0P2, O4')*

structure is shown in Figure 8(b), highlighting the interaction of the water molecules with N2, N3, O4' base sugar atoms to the external phosphate oxygens.[26]

1.2.3.4　Metal Ion Coordination

The dependence of quadruplex formation on metal ion coordination has been known since quadruplexes were first identified, while it was Arnott *et al.*[3] first suggested that the O6 carboxyl oxygen would coordinate cations. In fact a range of cations both monovalent and divalent are able to stabilize quadruplex formation to varying degrees. These are listed in Table 2 with their various ionic radii. The dependence of ions on stability can be seen from the arrangement of the carboxyl oxygen O6 atoms forming the cavity central to the stacked G-tetrad, [Figures 7(a–c) and 1.5 a and b]. Ions that coordinate effectively enhance stability. This is most clearly seen with the folding of quadruplexes in both Na^+ and K^+ containing solutions.[27] Increasing the monovalent cation concentration added little to the dissociation of the intermoleculer quadruplexes but contributed significantly to their rate of refolding. It was observed that a 200-fold increase in the rate of folding for sequence $d(TG_4T)$ was achieved by the addition of only a 10-fold increase of Na^+ ions. This effect was enhanced with the addition of potassium ions resulting in a significant increase in the overall stability of the quadruplex compared to sodium ions. A shift of $+32°C$ in the melting temperature was achieved by the substitution of K^+ for Na^+ for the $d(TG_3T)_4$ quadruplex forming sequence. It appears that the potassium ions have ideal characteristics to effectively coordinate the carboxyl oxygen atoms positioned 2.73 A from each O6 atom. Sodium ions, with a slightly smaller radii, have been observed in the crystal structures to be positioned either slightly above or below the central position, closer to the tetrad planes. A general trend in alkali ions from the most stable to least is as follows $K^+ > Na^+ > Rb^+ > NH_4^+ > Cs^+ > Li^+$. The idea of the cations stabilizing one folded state over another was first described by Sen and Gilbert[28] as a Na^+-K^+ switch. This switching has biological relevance as the intracellular ion concentration for K^+ is high compared to Na^+, while the situation is reversed for the extracellular region. This cation dependence on folding can be seen clearly between linear and folded unimolecular quadruplexes. It is also clear that the effects of cation type on the folded form is a great deal more complicated for the intramolecular quadruplexes, comprising four or more G-runs in a single strand, that the intermolecular quadruplexes. The addition of Mg^{2+} ions and other multivalent cations such as spermine assists in the association of the quadruplexes but can destabilize the resulting folded quadruplex, reducing the overall stability.

Table 2　*Ionic radii of quadruplex stabilizing cations*

Element	K^+	Na^+	NH_4^+	Rb^+	Cs^+	Li^+	Ca^{2+}
Ionic radius	1.52 1.33^{100}	1.16 (0.97)	1.43^{71}	1.66	1.81	0.9	0.99

1.3 Folding and Topology of Quadruplexes

The formation of a quadruplex simply requires four guanine repeats to self-associate. The simplest topology is formed by the self-association of four DNA strands in solution containing short guanine runs, for example

Xn Gp Xn

where Xn is any nucleotide of length n and Gp is any number of guanines involved in tetrad formation of length p.

Structures containing these short guanine runs termed intermolecular or tetramolecular quadruplexes, such as d(TGGGGT)$_4$ have been determined by crystallographic and NMR methods to reveal a parallel stranded arrangement,[29,30] where the phosphate backbone runs in the same direction, and all the bases are in an *anti* glycosidic orientation, as shown schematically in Figure 2(c). There are four possible ways that strands can self associate, as shown in Figure 9(a–d), however in monomeric solutions containing single G-runs, only parallel arrangements have so far been observed [Figure 9(a)]. In all the structures forming this parallel arrangement the 5′ and 3′ ends of the runs are capped by non-guanine bases. In the case of d(GGGT) where the guanine runs are not capped at both ends by alterative bases, a more complex arrangement is observed.[31] Here interlocked quadruplexes form from eight associated strands. It would appear then, that even single strands might form complex structural arrangements in solution, while still utilizing the guanine tetrads as fundamental building blocks. RNA also has the ability to form a quadruplex structure as first identified by NMR techniques on the sequence r(UGGG-GU).[32] In K$^+$ the sequence forms a parallel stranded quadruplex with an additional U-quartet stacked above the G-tetrad plane.

More complex structures and topologies can form from strands containing two guanine repeats separated by non-guanine nucleotides, for example

Xn Go Xp Go Xn

where Xn is any non-guanine nucleotide of length n, Go is any number of guanines involved in tetrad formation of length o, and Xp is any nucleotide of length p involved in loop formation.

Association between two of these strands result in assembly of four stranded G-quadruplexes, termed dimeric quadruplexes. Numerous examples of dimeric quadruplexes have been reported by both NMR,[33–38] and crystallographic methods[39,40] for both DNA and RNA on sequences containing two G runs and short nucleotide linkers. The topology of these quadruplex depends on which of the strands are connected together. Figure 9(a) and (c–e) shows the four possible ways that strands can associate. The element linking the strands together can be, diagonal, lateral (edgewise), or external (propeller, strand exchange, double-chain-reversal) to the quadruplex [Figure 10(a–e)]. These linking nucleotides are critical in determining quadruplex stability both in terms of length and sequence[41,42] as discussed below.

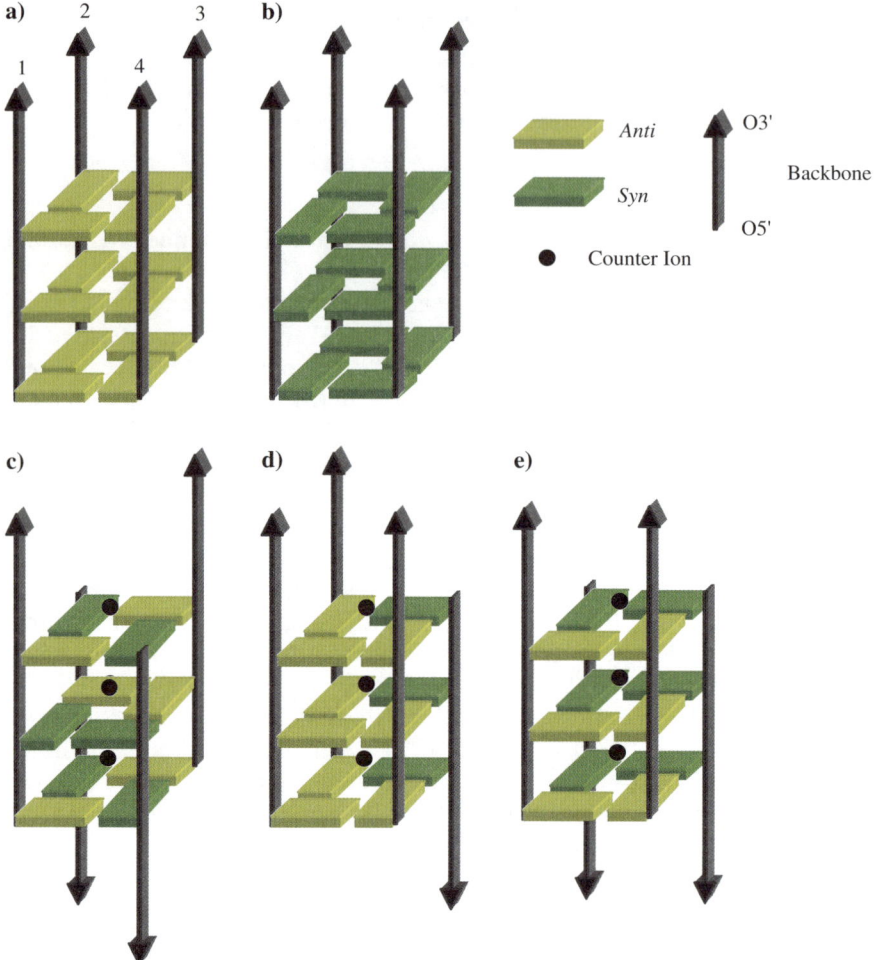

Figure 9 *Schematic diagrams showing potential strand polarity and related glycosidic torsional angles for intermoleculer quadruplexes. (a) Parallel with all anti glycosidic torsional angles; (b) parallel all syn glycosidic torsional angles; (c–e) alternating strands arrangements with the corresponding mixed syn anti relationships*

In the forming of dimeric quadruplexes we first need two strands to be linked together. If we link two of the strands by a lateral loop the strands will be constrained to be anti-parallel and require an inversion of one of the paired bases and the glycosidic torsion angle, in order to retain an appropriate hydrogen bonding arrangement. This is usually accommodated in one of the two ways, either each strand has a run of alternating *syn, anti* glycosidic angles for the guanine bases, or one strand has exclusively *syn* and the other strand exclusively *anti* glycosidic angles. The two remaining strands can now be linked together and take any strand polarity, but the glycosidic torsion angles of the guanines are constrained such that the tetrads formed have normal hydrogen bonding between the Hoogsteen and Watson–Crick faces, as shown in Figure 10(a–c). The link can take the form of a lateral loop adjacent or opposite to the first loop or by an external loop bringing three of the four strands parallel to one another. Many such structural arrangements have been identified by NMR and X-ray crystallographic methods. Two other arrangements for dimeric quadruplexes are possible, first where the four strands are linked by two external loops (double chain reversal), as seen in the human telomeric structure of sequence d(TAGGGTTAGGGT) [Figure 10(d)] or by the use of loops connecting across the diagonal [Figure 10(e)] face typified by the *Oxytricha nova* sequence d(GGGGTTTTGGGG). In these last two cases the first connecting linkage defines the second linkage and its symmetry.

Intramolecular quadruplexes associate from the following general sequence.

$$\text{Xn Go Xp Go Xp Go Xp Go Xn}$$

where Xn is any non-guanine nucleotide of length *n*, Go is any number of guanines involved in tetrad formation of length *o*, and Xp is any nucleotide of length *p* involved in loop formation.

The folding topologies available to intramolecular quadruplexes are more varied and complex than those available to the intermolecular or bimolecular quadruplexes due to the additional linking nucleotides. Structural examples from X-ray[22,43] and NMR[14,44-48] show just a few of the possible structures arrangements. Several examples of theoretical folding topologies are shown in Figure 11, where L = Lateral loop, E = External (double chain reversal) loop, and D = Diagonal loop, *=any type of loop, the models drawn comprise of lateral, external, and mixed loop forms. Those not shown are ELL and LDE, which are just inverted loop arrangements differing in the 5′ to 3′ direction. The only linking arrangement not available to these four guanine repeats motifs are those starting or ending with a diagonal loop. Again the same rules apply for the selection of the *syn, anti* glycosidic dihedral angles and their relationship to the backbone strand polarity.

Sequences containing four or more extended G-rich repeats are mainly associated with telomeric regions located at the ends of the chromosome. Organisms with linear chromosomes protect their DNA with G-rich repetitive sequences contained within duplex DNA that ends in a single stranded 3′ overhang. These 3′ extended single stranded G-repeats are then available for quadruplex formation without interference from a complimentary 5′ C-rich strand. If the G-rich repeat sequence is associated with its complimentary C-rich strand then the double stranded strands will have to be unzipped and stabilized in order allow the folding and stabilization of the G-rich strand into a quadruplex. Repetitive G-rich sequences can also be found distributed more generally throughout the genome, associated with promoter regions, as well as coding and non-coding regions.

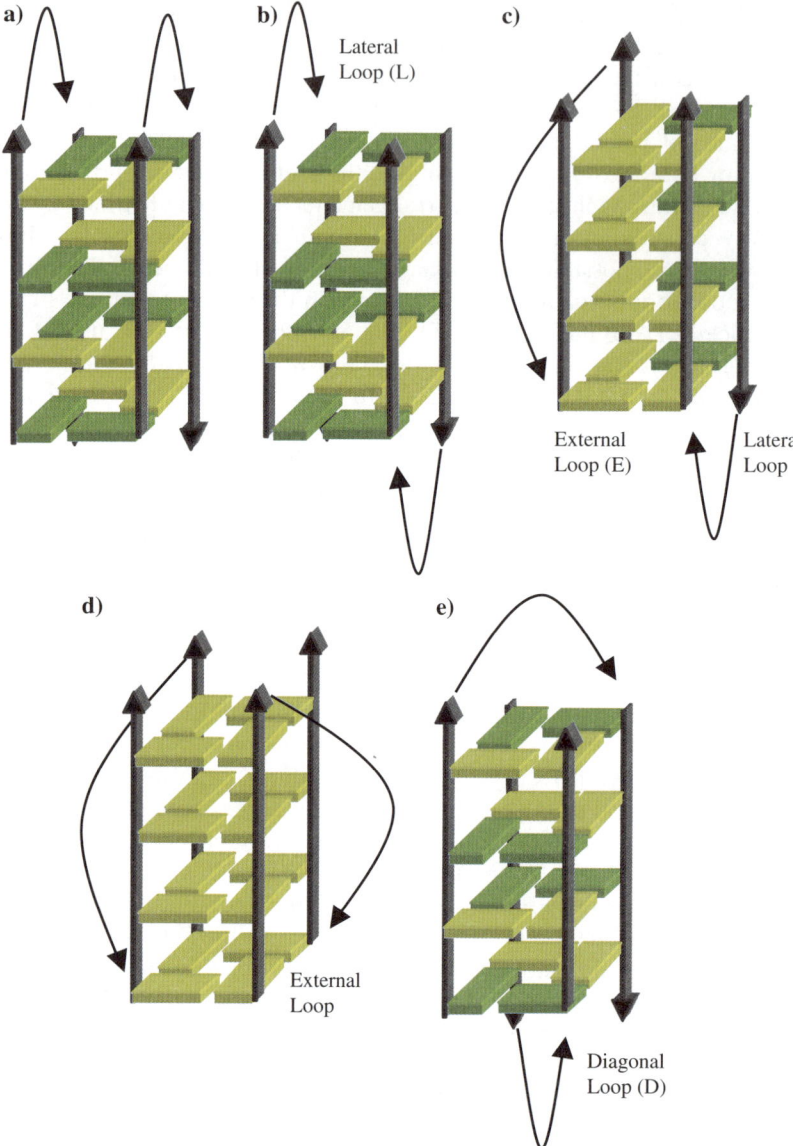

Figure 10   *Schematic diagrams showing potential linking loop arrangements and resulting topologies for bimolecular quadruplexes. (a) Lateral loops connecting anti-parallel strands arranged on the same face; (b) lateral loops connecting anti-parallel strands arranged on either end. (c) Diagonal loops connecting opposite anti-parallel strands. (d) External chain reversal loops connecting parallel strands together; (e) Diagonal loops connecting opposite anti-parallel strands*

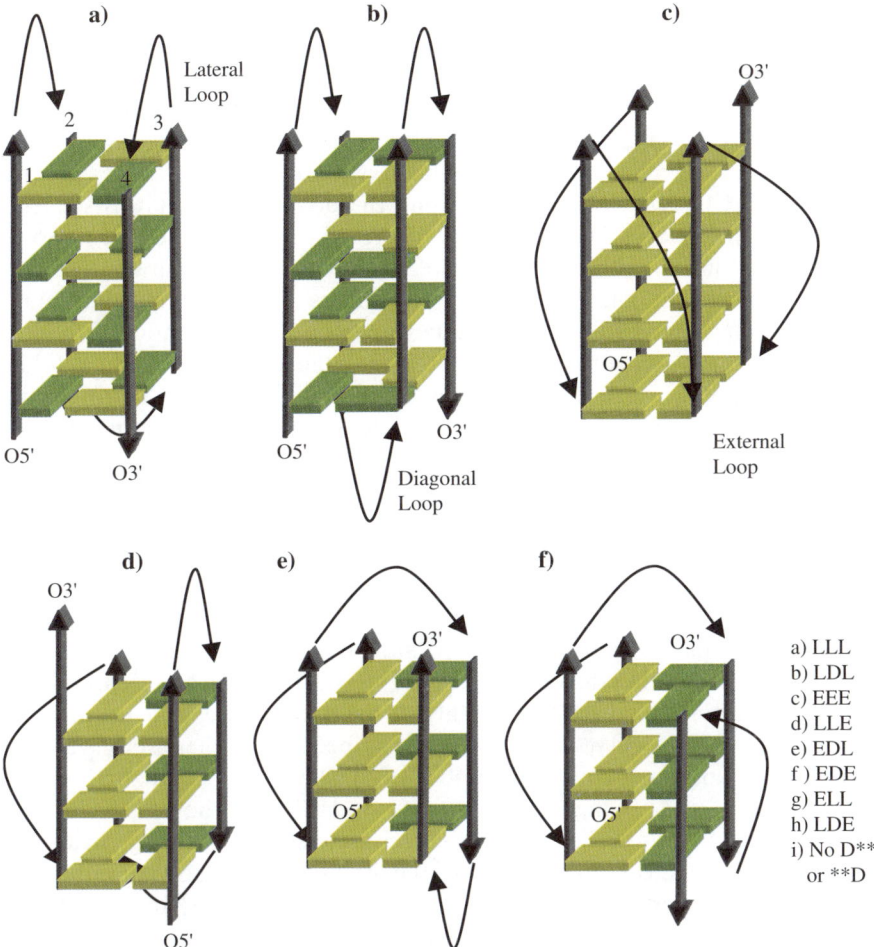

Figure 11 *Schematic diagrams showing potential linking loop arrangements and resulting topologies for intramolecular quadruplexes. (a) Lateral loops connecting anti-parallel strands arranged on the same face; (b) mixed lateral and diagonal loops connecting anti-parallel strands. (c) External chain reversal loops connecting parallel strands together. (d) Mixed external and lateral loops connecting strands; (e) Mixed external, diagonal, and lateral loops connecting strands together; (f) Mixed external and diagonal loops connecting strands together*

1.3.1 Loop Length, Sequence, and Stability

An additional constraint for the folding topologies for bimolecular and inter-molecular quadruplexes relates to the number (length) of the linking nucleotides. Short linker lengths, two or less, will prevent diagonal loops from forming due to the distance to be spanned across G tetrad. However, the short sequences can accommodate both lateral and external loops. Indeed even a single connecting nucleotide, forming an external connecting loop, can accommodate a stack of up to three guanine tetrads, while a stack of four guanines tetrads requires at least two nucleotides in the linker. Short linker lengths then impose certain topologies constraints on the folded quadruplexes resulting in a propensity for them to fold into parallel stranded structures.[49] This was identified by CD gel electrophoresis using the d(GGGGTGGGG)$_2$ sequence containing one thymine as nucleotide linker.[50] The introduction of two thymines into the linker resulted in quadruplexes having a mixed parallel and anti-parallel arrangement, while lengthening the linkers to three or four nucleotides resulted in only anti-parallel folded quadruplexes. Linker lengths of three or more impose little constraint on possible loop arrangements or the final folded quadruplex motif. Longer linkers are involved in all three types of connecting loops, diagonal, lateral, and external. To date, no one simple relationship has been identified linking quadruplex stability to loop length.

An increase in melting temperature is associated with the overall stabilizing effect of additional stacked G-tetrad on a quadruplex structure.[51] For intermolecular quadruplexes it was determined that for each guanine tetrad added, the association rate constant was increased by a factor of ten. By extending DNA and RNA guanine runs by one, from G$_3$ to G$_4$, their melting temperatures (T1/2) increased from 16 to 54.5°C and 50 to 89°C. However, it was also observed that increasing the G-runs from G$_3$ to G$_4$, in an intramolecular folded quadruplex did not increase the stability of the quadruplex. This is unlike the case for the intermolecular quadruplexes described above where stabilization is enhanced by an increase in the number of stacked tetrads. In this case the critical factor is a change in both G's and T's in the sequence from d((G$_4$T$_2$)$_3$T$_2$G$_4$) to d((G$_3$T$_3$) $_3$T$_3$G$_3$). The consequence is a change in folding topology and an increase the stability of the modified structure even though the number of guanines is reduced. It would then appear that the stability of quadruplex structures depends on the types of fold formed. In addition, alterations to telomeric sequences that are known to fold into stable quadruplex structures, for example by the addition or removal of nucleotides to the loops has, in general, a detrimental effect on overall stability of the resulting folded structure.[52]

Nucleotides located in the linker regions for quadruplex forming sequences are typified by thymines and adenines. The selection of thymines over adenines for connecting loops may be an important factor in quadruplex stability. The replacement of the TTA loop sequence in the human telomeric sequence, a well represented loop sequence in many telomeric sequences, with a run of AAA's results in the complete destabilization of the quadruplex structure.[42] In telomeric sequences the linkers are between two and four nucleotides in length, and comprised primarily of thymines with at most one adenine, while the guanines repeats are between two to four in length. In non-telomeric sequences guanines can be accommodated in the loops especially where the repeating guanine pattern is not of a uniform length. Here the unmatched guanines if located in a run of guanines can result in structural heterogeneity as the unmatched guanine can be accommodated at either end of the stacked tetrad. Cytosines are not commonly observed in quadruplex forming sequences as they will tend to base pair with guanines and disrupt tetrad formation. They are however seen in fission yeasts d(TTAC(A)(C)G(1-8)) and some budding yeast telomeric sequences but comprise only one nucleotide within quite long sequences of up to 25 nucleotides per repeat. (Figure 12)

1.3.2 Polymorphism in Sequence and in Topology

1.3.2.1 Sequence (Variable Number of G-Repeats, Same Topology)

Not only four G-repeats but any number of G-repeats can be included into a run of guanines and fold into an intramolecular quadruplex. Indeed in the case of five G-repeats, one of the central three G-rich runs can be accommodated into an extended loop. The *c-myc* oncogene promoter sequence found in the nuclease hypersensitivity element III(1) (NHE III(1)) of sequence d(TG$_4$AG$_3$T-G$_4$AG$_3$TG$_4$A$_2$G$_2$) containing six G runs is one such example that has been shown to form a quadruplex structure.[53] The full 27nt length sequence shows a propensity to form multiple G-quadruplex structures[46,47,54,55] while a truncated 24nt version containing five G-repeats forms a stable parallel stranded motif in potassium with a fold back of one guanine on the 3' end into the three guanine tetrad stack.[56] This is an unusual arrangement as the G-runs are normally contagious within a stack of G-tetrads, while here two of the guanines in a three stacked tetrad come from one strands while the third tetrad in the stack is completed by a single guanine from a different G-run. In another example, a sequence containing three human telomeric repeats, folded in Na$^+$, forms a four stranded quadruplex by utilizing a strand from an unassociated DNA fragment, irrespective as to whether this fragment contains the same full sequence or just a short G-run of three guanines.[57] The folded topology has two lateral loops, with three parallel strands and a mixed population of syn-anti glycosidic torsion angles. These strand arrangements complicate and increase the complexity of the possible folding topologies we might expect from certain linear G-rich sequences.

1.3.2.2 Topology (Different Topology, Same Sequence)

In solution, the presence of multiple folding topologies has been observed for many quadruplex forming sequences. An example given is the human telomeric sequence repeat sequence d(TTAGGG). Here the NMR and crystal structure determinations have provided us with two quite diverse folded conformations for the intramolecular quadruplex of sequence d(AGGG(TTAGGG)$_3$). In addition when folded as a bimolecular quadruplex containing two G-runs the human sequence adopts both the parallel and an alternative anti-parallel conformation.[55] The *Tetrahymena* telomeric sequence also folds into two different topologies depending on the number of G-repeats. Again the d(TG$_4$T$_2$G$_4$T) sequence folding as a bimolecular quadruplex with two edgewise loops, folds differently from the intramolecular sequence d((T$_2$G$_4$)$_4$). In a

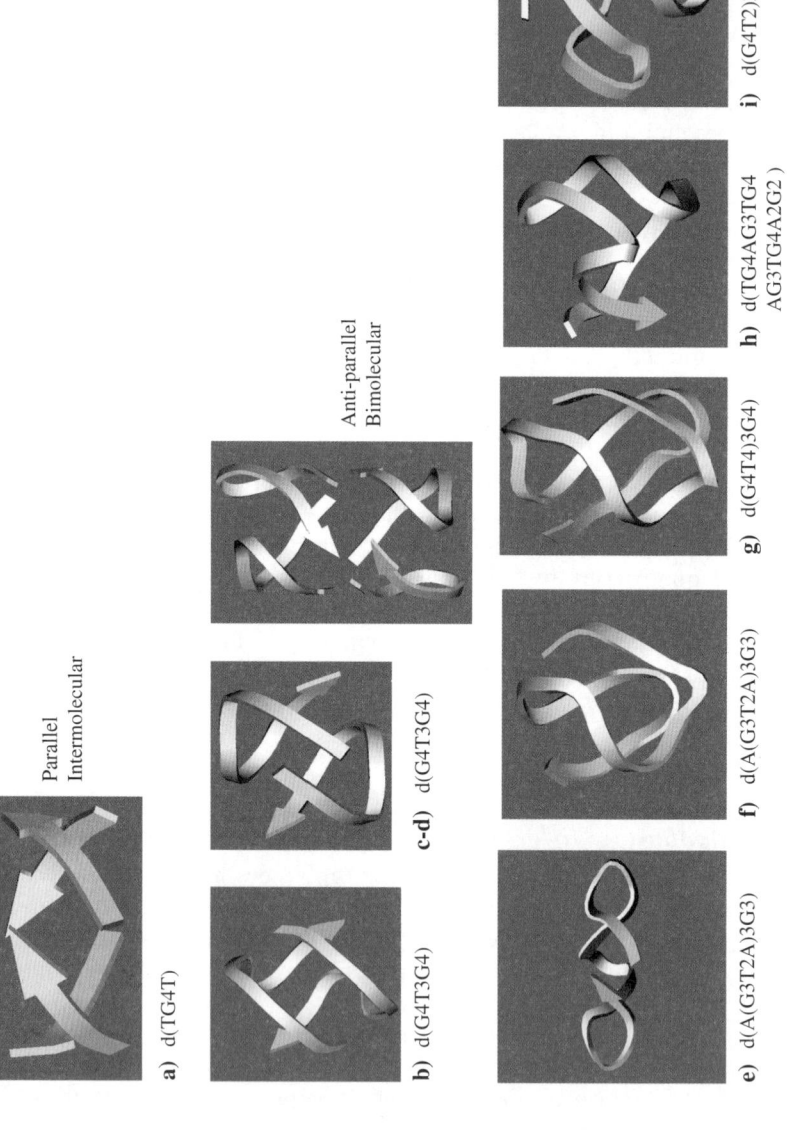

Figure 12 *Folding topologies with backbones drawn as arrows 5' to 3', all taken from known crystal structures. (a) A parallel intermolecular structure; (b - d) Bimolecular anti-parallel structural motifs; (e - i) Intra molecular parallel and anti-parallel structures*

recently submitted crystal structure to the PDB by Hazel *et al.* (PDB ID, 2AVJ) we find that within one crystal asymmetric unit there are two independent and diverse folding topologies for the sequence d($G_4T_3G_4$), folding as a bimolecular quadruplexes. In one arrangement we see lateral loops positioned at either end of the stacked G-tetrads while in the second folded arrangement there are two stacked quadruplexes with the lateral loops positioned at the same end of the G-tetrads. The second arrangement is unexpected as the negatively charged phosphate backbones are in close proximity to one another. It was expected that the repulsive negative charges on the backbones would be sufficient to restrict such an arrangement.

The switching between various conformations can be induced by the presence of different cations and temperature. The switching induced by ions has been exemplified by the human telomeric sequence where the presence of either Na^+ or K^+ metal ions induces the stabilization of different structural forms.[58,59] In all cases the human telomeric sequence, when presented by its complementary C-rich strand will refold into duplex DNA. This is not the case for the *Tetrahymena* telomeric sequence or the *c-myc* promoter sequence where in high K^+ salt conditions they will preferentially remain in a quadruplex folded state, even at high concentrations of their own complimentary C-rich strands. However, in sodium these two folded sequences, when presented by there complimentary strands unfold into duplex DNA. There are also several examples of where temperature can be used to select different folded motifs. By using the human telomeric sequence d($UAG_3T^{br}UAG_3T$) it was identified that an anti-parallel structure formed at below 50°C can be converted to a parallel form by an increase in temperature.[58] In this case the structures are in equilibrium between folded and unfolded states where the relative folding kinetics between these states and thermodynamics determined the final conformation observed in solution.

So whether the nucleotide sequence is determined in solution by NMR techniques or packed into crystalline lattices and resolved by crystallographic techniques, we observe a structural diversity dependent upon a variety of factors, and not only on the sequence used. In summary, whether these guanine repeats come in sets of two, three, four, or five the final topology depends upon many factors such as the type of nucleotide linker between the guanines, their length, sequence, temperature, concentration, and type of counter ion associated with the O6 oxygen atoms of the guanine nucleotides, making structural predictions very challenging.

1.3.3 Other Quadruplex Forming Sequences

1.3.3.1 The i-Motif

The i-motif can be formed from a cytosine-rich strand with protonation of the intercalated $C \cdot C^+$ base pairs [Figure 13(a)]. In the case of duplex DNA the disassociation of the cytosine and guanine-rich strands is required before being refolded into a C-rich i-motif and a G-rich G-quadruplex motif. Under normal physiological cellular conditions the guanine and its complimentary sequences

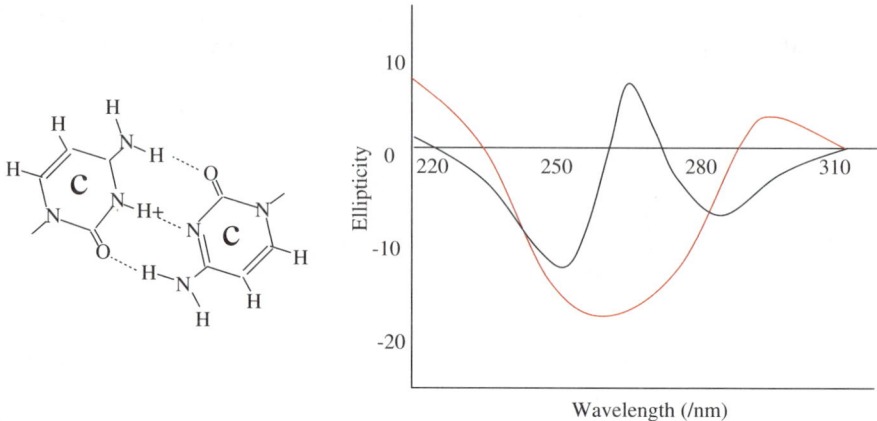

Figure 13 *(a) Protonation state of the intercalated $C \cdot C^+$ base pairs particular to the i-motif; (b) characteristic CD spectrum with ellipticity plotted against wavelength, for intramolecular anti-parallel and parallel quadruplexes*

will form duplex DNA. However, there are conditions at low pH, less than 4.5, or at higher temperatures, that the cytosine-rich and guanine-rich strands can disassociate and refold into both the i-motif, for the C-rich strand and a G-quadruplex structure for the G-rich strand. This i-motif has been shown to form from the C-rich strand of the human telomeric repeat sequence (5′TTAGGG):(5′CCCTAA).[60] In this structure the strands in the i-motif run anti-parallel to one another forming an intramolecular four-stranded quadruplex, with cytosine-protonated cytosine base-pairing across the diagonal strands. The cytosine bases hydrogen bond to parallel strands forming two wide grooves and two narrow grooves.

1.3.3.2 Self Association of Watson–Crick Duplex Sequences to Form Tetraplexes

Self association of Watson–Crick duplex sequences can come in two varieties, either the base-pairs will associate through the major grooves or through the minor grooves. Association of two sets of base-paired DNA will give a familiar tetraplex arrangement of four nucleotide bases. The same *syn* and *anti* glycosidic constraint for self associating telomeric guanine rich sequences is applicable to the duplex associated tetraplexes.

1.3.3.3 G C Tetrads

Interest in G:C tetrad forming base-pairs comes from the discovery of the fragile X syndrome and the associated d(CGG)n:d(CCG) repeats. Initial structural prediction for the CG rich repeat sequences identified a parallel stranded motif stabilized by the methylation of the cytosine residues. Subsequent structural analyses by NMR methods on a sequence that contained CG_2 repeats, linked by thymine bases, revealed a head-to-tail hairpin arrangement *via* a lateral loop with *syn-anti-syn-anti* glycosidic torsion angles,[61] exactly the same arrangement seen previously in telomeric G-rich bimolecular quadruplex structures.

Another interesting DNA sequence containing GC rich regions is a G_3C repeat, identified in chromosome 19 as a site of viral integration as a specific site of entry by the adeno-associated virus. An NMR structure solution containing this model G_3C site, $d(G_3CT_4G_3C)$[62] in a Na^+ solution was determined. It was found to fold into a hairpin arrangement with head-to-tail dimerization with three stacked G:C:G:C tetrads, a motif seen previously observed in other bimolecular quadruplexes containing two G rich repeats.

1.3.3.4 Other Tetrad Forming Sequences

Sequences other than CG-containing runs can also associate together to form tetrads through the major groove. The human quadruplex structure contains one such example. Formed from two $d(TAGGGTTAGGGT)_2$ repeats the thymine and adenine nucleotides at the 5′end, hydrogen bond *via* Watson–Crick base pairing to adjacent anti-parallel strands. These in turn hydrogen bond across their Hoogsteen edge, O6(thymine) to N6(Adenine), and are capped by a Na^+ counter ion. Another example comes from an NMR study on the sequence d(GAGCAGGT). This forms a head-to-head dimeric quadruplex containing G-C-G-C, G-G-G-G and slipped A-T-A-T tetrads.[63] This is truly a diverse structure with A-T-A-T and G-C-G-C tetrads formed through the major and minor grooves, chain reversal and hairpin turns all contributing to a complex DNA structure. The way the different DNA strands can association together has implications for recombination.

There are also several examples of G:C Watson–Crick base pairs associating through their minor grooves edges to form G:C:G:C tetrads. One such structure is a very interesting four stranded quadruplex structure formed from a heptanucleotide sequence d(GCATGCT).[64] Here the individual strands form an intra-molecular hairpin and are stabilized by complimentary Watson–Crick base pairing. The strands are linked together through a series of G(N2)-C(O2) intra-strand H-bonds to form a physiologically relevant structure. Further stabilization comes from H-bonds to the phosphate oxygens and hydration structure. Even though the structure has a core of two GCGC tetrads they are non-planar with a 30° tilt. In spite of the tilted bases the four-stranded anti-parallel backbone generates a recognizable quadruplex topology.

1.4 Experimental Methods Used in Quadruplex Characterization

1.4.1 UV Melting Experiments

The melting of DNA can be easily monitored by a hyperchromic shift resulting from the unstacking of base bases and dissociation of the duplex DNA. This quick and simple method can also be applied to quadruplex DNA utilizing the same principles. Measuring the spectral absorbance at 295 nm, a maximum for

G quadruplex motifs, is carried out on pre-annealed samples which would be heated to 90°C at 0.1°C min^{-1}. The maximum in the first derivative will provide a suitable melting temperature, T_m, or the use of a two state model for the melting curve fitted by van't Hoff analysis, would provide additional thermodynamic data only if the samples show minimal hysteresis during several heating and cooling cycles. Intramolecular quadruplexes associate fast enough in most cases for equilibrium to be established, allowing thermodynamic measurements to be successfully performed. Most bimolecular quadruplexes also associate fast enough but are more likely to show hysteresis. However linear four-stranded quadruplexes exhibit very slow folding kinetics restricting accurate measurements.

1.4.2 Circular Dichroism Spectroscopy

Circular dichroism (CD) is an indirect method for the characterization of quadruplex topology reliant on known structural motifs in solution for comparison. There are a few characteristic spectral properties that can be assigned, however these can only be applied in a general manner. In the simple case of parallel quadruplexes, generated by the association of four independent G-rich strands (intermolecular), they are characterized by a positive ellipticity maximum at 264 nm and a negative minimum at 240 nm. The anti-parallel (intramolecular) complexes usually possess a positive maximum at 295 nm and a negative minimum at 265 nm.[65–68] An example is shown in Figure 13(b) where through the extension of the linker loop length, from one to three thymines, between the guanine repeats, results in a dramatic change on ellipticity. This can be representative of a change from a parallel to anti-parallel arrangement of the connecting backbone nucleotides.

1.4.3 Polyacrylamide Gel Electrophoresis

The rapid identification of folded motifs against unfolded single stranded DNA states can quickly performed under strict native conditions. Additional analysis of possible structures present in the bands requires careful consideration, particularly in respect of the speed (voltage) of the run and stability of the quadruplex motif. Quadruplex stability can be sensitive to temperature and buffering conditions used during the experiment. Additional analysis on the observed bands is required to determine the folded forms present. Visualization of the DNA can be carried out by radioactive labeling or EtBr staining or other dyes.

1.4.4 Fluorescence Resonance Energy Transfer (FRET)

A range useful of assays can be rapidly carried out using fluorescent labeled DNA. DNA can be synthesized with its 3′ and 5′ ends tagged with fluorescent dyes, at the 5′ end 6-carboxy fluorescein (FAM) the donor, and tetramethyl-6-carboxyrhodamine (TAM) at the 3′ end as the acceptor. When the two dyes are in close proximity the fluorescence is quenched, but upon separation the fluorescence is enhanced. Simple melting experiments can be designed to utilize this property where pre-folded DNA oligonucleotides, that have the labels in close proximity, can be heated and their melting temperature determined under a range of conditions. This technique is most suited to long sequences that fold as intramolecular quadruplexes, and where a folded state brings the two fluorophores into close proximity.

1.4.5 Sedimentation Velocity Analysis and Sedimentation Equilibrium

Density gradient ultra-centrifugation can be used to determine the buoyant density of DNA and quadruplexes. Sedimentation experiments can also be used to measure sedimentation coefficients and molecular weights. Bead modeling on atomic-level model structures can be used to calculate theoretical hydrodynamic properties of DNA the structures. The calculated hydrodynamic properties (translational diffusion, sedimentation coefficients (S) and correlation times) in conjunction with the program HYDROPRO[69] has been used to help differentiate between known or inferred structural conformations measured by analytical ultra-centrifugation.

1.4.6 Cross-Linking DNA and Topology

Cross-linking of DNA bases by the use of difunctional platinum complexes has been successfully used to identify guanine nucleotides that are in close proximity to one another, and so provide distance constraints for the construction of G-quadruplex folded motifs.[70] Compounds suitable for cross-linking, such as *cis*- and *trans*-Pt(NH$_3$)$_2$ or *cis*- and *trans*-diamminediaquaplatinum(II) are able to covalently link the N7 of purine bases together. The N7 of guanines involved in G-tetrad formation will not be able to cross-link if they are involved in tetrad formation. In sequences that contain adenine additional cross-linking can occur, but with a reduced efficiency as they are 10 times less nucleophilic. Once the products are formed the folded structures can be investigated by mass spectrometry or by 3'-exonuclease digestion and subsequent analysis by gel electrophoresis.

References

1. M. Gellert, M.N. Lipsett and D.R. Davies, *Proc. Natl. Acad. Sci. USA.*, 1962, **48**, 2013.
2. P. Tougard, J.F. Chantot and W. Guschlbauer, *Biochim. Biophys. Acta.*, 1973, **308**(7), 9.
3. S. Arnott, R. Chandrasekaran and C.M. Marttila, *Biochem. J*, 1974, **141**(2), 537.
4. S.B. Zimmerman, G.H. Cohen and D.R. Davies, *J. Mol. Biol.*, 1975, **92**(2), 181.
5. S.B. Zimmerman, *Lett. Biopolym.*, 1975, **14**(4), 889.
6. F. Aboul-ela, A.I. Murchie and D.M. Lilley, *Nature*, 1992, **360**(6401), 280.
7. F. Aboul-ela, A.I. Murchie, D.G. Norman and D.M. Lilley, *J. Mol. Biol.*, 1994, **243**(3), 458.

8. C. Cheong and P.B. Moore, *Biochemistry*, 1992, **31**(36), 8406–8414.
9. C. Kang, X. Zhang, R. Ratliff, R. Moyzis and A. Rich, *Nature*, 1992, **356**(6365), 126.
10. G. Laughlan, A.I. Murchie, D.G. Norman, M.H. Moore, P.C. Moody, D.M. Lilley and B. Luisi, *Science*, 1994, **265**(5171), 520.
11. P. Schultze, F.W. Smith and J. Feigon, *Structure*, 1994, **2**(3), 221.
12. F.W. Smith and J. Feigon, *Biochemistry*, 1993, **32**(33), 8682.
13. Y. Wang and D.J. Patel, *Biochemistry*, 1992, **31**(35), 8112–8119.
14. Y. Wang and D.J. Patel, *Structure*, 1993, **1**(4), 263–282.
15. S. Neidle, *DNA Structure and Recognition*, Oxford University Press, Oxford, 1984.
16. S. Neidle, *Nucleic Acid Structure and Recognition*, Oxford University Press, Oxford, 2002.
17. S. Neidle (ed), *Oxford Handbook of Nucleic Acid Structure*, Oxford University Press, Oxford, 1999.
18. S. Neidle (ed), *Topics in Nucleic Acid Structure*, Macmillan Press Ltd., London and Basingstoke, 1981.
19. W. Saenger, *Principles of Nucleic Acid Structure*, Springer-Verlag, New York, 1984.
20. S. Arnott, R. Chandrasekharan, D.L. Birdsall, A.G. Leslie and R.L. Ratliffe, *Nature*, 1980, **283**, 743.
21. F.W. Smith and J. Feigon, *Nature*, 1992, **356**(6365), 164.
22. G.N. Parkinson, M.P. Lee and S. Neidle, *Nature*, 2002, **417**(6891), 876.
23. J.R. Quintan, K. Grzeskowiak, K. Yanagi and R.E. Dickerson, *J. Mol. Biol.*, 1992, **225**, 379.
24. S.C. Ha, K. Lowenhaupt, A. Rich, Y.G. Kim and K.K. Kim, *Nature*, 2005, **437**(7062), 1183.
25. S. Rhee, Z-J. Han, K. Liu, H.T. Miles and D.R. Davies, *Biochemistry*, 1999, **38**, 16810.
26. R. Otero, M. Schock, L.M. Molina, E. Laegsgaard, I. Stensgaard, B. Hammer and F. Besenbacher, *Angew. Chem. Int. Ed. Engl.*, 2005, **44**(15), 2270.
27. J.L. Mergny, A. De Cian, A. Ghelab, B. Sacca and L. Lacroix, *Nucleic Acids Res.*, 2005, **33**(1), 81.
28. D. Sen and W.A. Gilbert, *Nature*, 1990, **344**(6265), 410.
29. G. Laughlan, A.I. Murchie, D.G. Norman, M.H. Moore, P.C. Moody, D.M. Lilley and B. Luisi, *Science*, 1994, **265**(5171), 520.
30. K. Phillips, Z. Dauter, A.I. Murchie, D.M. Lilley and B. Luisi, *J. Mol. Biol.*, 1997, **273**(1), 171.
31. Y. Krishnan-Ghosh, D. Liu and S. Balasubramanian, *J. Am. Chem. Soc.*, 2004, **126**(35), 11009.
32. C. Cheong and P.B. Moore, *Biochemistry*, 1992, **31**(36), 8406.
33. A.T. Phan and D.J. Patel, *J. Am. Chem. Soc.*, 2003, **125**(49), 15021.
34. A.T. Phan, Y.S. Modi and D.J. Patel, *J. Mol. Biol.*, 2004, **338**(1), 93.
35. M. Crnugelj, N.V. Hud and J. Plavec, *J. Mol. Biol.*, 2002, **320**(5), 911.

36. F.W. Smith, F.W. Lau and J. Feigon, *Proc. Natl. Acad. Sci. U S A.*, 1994, **91**(22), 10546.
37. M. Crnugelj, N.V. Hud and J. Plavec, *J. Am. Chem. Soc.*, 2003, **125**(26), 7866.
38. H. Liu, A. Kugimiya, A. Matsugami, M. Katahira and S. Uesugi, *Nucleic Acids Res. Suppl.*, 2002, **2**, 177.
39. S. Haider, G.N. Parkinson and S. Neidle, *J. Mol. Biol.*, 2003, **326**(1), 117.
40. M.P. Horvath and S.C. Schultz, *J. Mol. Biol.*, 2001, **310**(2), 367.
41. I. Smirnov and R.H. Shafer, *Biochemistry*, 2000, **39**, 1462.
42. A. Risitano and K.R. Fox, *Biochemistry*, 2003, **42**, 6507.
43. K. Padmanabhan, K.P. Padmanabhan, J.D. Ferrara, J.E. Sadler and A. Tulinsky, *J. Biol. Chem.*, 1993, **268**(24), 17651.
44. Y. Wang and D.J. Patel, *Structure*, 1994, **2**(12), 1141.
45. V. Kuryavyi, A. Majumdar, A. Shallop, N. Chernichenko, E. Skripkin, R. Jones and D.J. Patel, *J. Mol. Biol.*, 2001, **310**(1), 181.
46. A.T. Phan, Y.S. Modi and D.J. Patel, *J. Am. Chem. Soc.*, 2004, **126**(28), 8710.
47. A. Ambrus, D. Chen, J. Dai, R.A. Jones and D. Yang, *Biochemistry*, 2005, **44**(6), 2048.
48. P. Schultze, R.F. Macaya and J. Feigon, *J. Mol. Biol.*, 1994, **235**(5), 1532.
49. P. Hazel, J. Huppert, S. Balasubramanian and S. Neidle, *J. Am. Chem. Soc.*, 2004, **126**(50), 16405.
50. P. Balagurumoorthy, S.K. Brahmachari, D. Mohanty, M. Bansal and V. Sasisekharan, *Nucleic Acids Res.*, 1992, **20**(15), 4061.
51. L. Petraccone, E. Erra, I. Duro, V. Esposito, A. Randazzo, L. Mayol, C.A. Mattia, G. Barone and C. Giancola, *Nucleosides Nucleotides Nucleic Acids*, 2005, **24**(5–7), 757.
52. A. Risitano and K.R. Fox, *Nucleic Acids Res.*, 2004, **32**(8), 2598.
53. T. Simonsson, P. Pecinka and M. Kubista, *Nucleic Acids Res.*, 1998, **26**(5), 1167.
54. A. Siddiqui-Jain, C.L. Grand, D.J. Bearss and L.H. Hurley, *Proc. Natl. Acad. Sci. USA.*, 2002, **99**(18), 11593.
55. J. Seenisamy, E.M. Rezler, T.J. Powell, D. Tye, V. Gokhale, C.S. Joshi, A. Siddiqui-Jain and L.H. Hurley, *J. Am. Chem. Soc.*, 2004, **126**(28), 8702.
56. A.T. Phan, V. Kuryavyi, H.Y. Gaw and D.J. Patel, *Nat. Chem. Biol.*, 2005, **1**, 167.
57. N. Zhang, A.T. Phan and D.J. Patel, *J. Am. Chem. Soc.*, 2005, **127**(49), 17277.
58. A.T. Phan and D.J. Patel, *J. Am. Chem. Soc.*, 2003, **125**(49), 15021.
59. I.N. Rujan, J.C. Meleney and P.H. Bolton, *Nucleic Acids Res.*, 2005, **33**(6), 2022.
60. A.T. Phan and J.L. Mergny, *Nucleic Acids Res.*, 2002, **30**(21), 4618.
61. A. Kettani, R.A. Kumar and D.J. Patel, *J. Mol. Biol.*, 1995, **254**(4), 638.
62. A. Kettani, S. Bouaziz, A. Gorin, H. Zhao, R.A. Jones and D.J. Patel, *J. Mol. Biol.*, 1998, **282**(3), 619.

63. N. Zhang, A. Gorin, A. Majumdar, A. Kettani, N. Chernichenko, E. Skripkin and D.J. Patel, *J. Mol. Biol.*, 2001, **312**(5), 1073.
64. J.H. Thorpe, S.C. Teixeira, B.C. Gale and C.J. Cardin, *Nucleic Acids Res.*, 2003, **31**(3), 844.
65. M. Lu, Q. Guo and N.R. Kallenback, *Biochemistry*, 1992, **31**, 2455.
66. P. Balagurumoorthy, S.K. Brahmachari, D. Mohanty, M. Bansal and V. Sasisekharan, *Nucleic Acids Res.*, 1992, **20**, 4061.
67. P. Balagurumoorthy and S.K. Brahmachari, *J. Biol. Chem.*, 1994, **269**, 21858.
68. M. Lu, Q. Guo and N.R. Kallenbach, *Biochemistry*, 1993, **32**, 598.
69. B. Carrasco and J. García de la Torre, *Biophys. J.*, 1999, **76**, 3044.
70. S. Redon, S. Bombard, M.A. Elizondo-Riojas and J.C. Chottard, *Biochemistry*, 2001, **40**, 8463.
71. N.V. Hud, P. Schultze, V. Sklenar and J. Feigon, *J. Mol. Biol.*, 1999, **285**, 233.
72. J.R. Quintana, K. Grzeskowiak, K. Yanagi and R.E. Dickerson, *J. Mol. Biol.*, 1992, **225**(2), 379.
73. S. Rhee, Z. Han, K. Liu, H.T. Miles and D.R. Davies, *Biochemistry*, 1999, **38**, 16810.

CHAPTER 2

Energetics, Kinetics and Dynamics of Quadruplex Folding

JEAN-LOUIS MERGNY, JULIEN GROS, ANNE DE CIAN, ANNE BOURDONCLE, FRÉDÉRIC ROSU, BARBARA SACCÀ, LIONEL GUITTAT,† SAMIR AMRANE, MARTIN MILLS,‡ PATRIZIA ALBERTI, MASASHI TAKASUGI AND LAURENT LACROIX

Laboratoire de Biophysique, Muséum National d'Histoire Naturelle USM503, INSERM U565, CNRS UMR 5153, 43 rue Cuvier, 75231 Paris, Cedex 05 (France)

2.1 Introduction

The inclination of guanosine monophosphate (GMP) or guanine-rich poly- and oligonucleotides to self-assemble has been recognized for several decades.[1-4] G-quadruplexes result from the hydrophobic stacking of several quartets; each quartet being a planar association of four guanines held together by eight hydrogen bonds.[5] A cation (typically Na^+ or K^+) is located between two quartets forming cation–dipole interactions with eight guanines. This reduces the electronic repulsion of the 2×4 central oxygen atoms, thus enhancing hydrogen bond strength and stabilizing quartet stacking.

It is only in the past decade that the level of interest in these peculiar structures has increased, due to the hypothesis of a relevant role of G-quadruplex structures in key biological processes[6-11] and recent demonstrations of their existence *in vivo*.[12-14] G-quartets may now have applications in areas ranging from supramolecular chemistry to medicinal chemistry (for a recent review see ref. 15). Quadruplex DNA is an excellent module for the design of devices for nanotechnology and guanine-rich sequences are also likely to form higher order structures such as synapsable DNA[16,17] or G-wires.[18,19]

---

†Current address: Department of Cell Biology and Physiology, Washington University School of Medicine, Saint Louis, MO (USA)
‡Current address: Department of Molecular and Cellular Biology, University of Cape-Town, South Africa

It is therefore important to understand the rules that govern the formation of these complexes and determine their stability and folding kinetics. In this review, we will address this problem, with an emphasis on the thermodynamics and kinetics of quadruplexes formed by oligodeoxyribonucleotides (short DNA strands: in this manuscript, the d-prefix will be omitted for most DNA sequences, which are always provided in the 5′ to 3′ direction). We will treat the intra-, bi- and tetramolecular quadruplexes separately (Figure 1). Note that even the kinetics of association of GMP alone and isolated guanine bases are complex; in fact, the initial studies on helix formation by guanylic acid[2] clearly showed hysteresis of the melting process, with a "considerable time dependence of the optical density". However, a detailed review on the kinetics of these aggregates is beyond the scope of this manuscript and additional information on quartet formation by isolated bases can be found in the chapter by J. Davis and co-workers (also see ref. 15).

2.2   Thermodynamics

Although this review focuses on the kinetics and dynamics of quadruplexes, one cannot avoid a preliminary discussion on the thermodynamics of these structures.

2.2.1   Enthalpy and Entropy of Formation

The formation of quadruplex structures, whatever their type, is clearly enthalpy driven, with an enthalpy per quartet of -15 to -25 kcal mol^{-1}. For example, concerning intramolecular quadruplexes, the van't Hoff analysis of the melting process of unmodified DNA samples reveals a $\Delta H°$ of -19.8 and -19.0 kcal mol^{-1} for the $G_2T_2G_2TGTG_2T_2G_2$ (thrombin aptamer) and $AG_2(T_2AG_2)_3$ samples, respectively.[20,21] For the $G_3(T_2AG_3)_3$ sample, we obtain a slightly less favourable enthalpy (-15 kcal mol^{-1}) in agreement with Vorlickova *et al.* (-16 kcal mol^{-1}).[22] Addition of a 5′ adenine residue (giving the $AG_3(T_2AG_3)_3$ sequence) stabilizes the quadruplex and increases the $\Delta H°$/quartet to -18 kcal mol^{-1} (ref. 20) or -24 (in Na$^+$) and -26 (in K$^+$) kcal mol^{-1}.[23] Note that longer repeats, such as $G_3(T_2AG_3)_7$ have relatively similar melting temperatures, but significantly lower enthalpies per quartet.[22] The $(T_4G_4)_4$ repeat has an enthalpy of -26 kcal mol^{-1} per quartet.[24] Relatively similar values have been found with model-independent techniques such as differential scanning calorimetry (DSC).

Bi- and tetramolecular structures also give quite comparable enthalpies. However, the validity of the van't Hoff analysis for some of these quadruplexes is questionable, as the melting curve used to determine thermodynamic parameters rarely corresponds to a true equilibrium curve. We will address this point later in the manuscript. An enthalpy of formation of -21 to -26 kcal mol^{-1} per quartet is found for the $G_2T_2NT_2G_2$ sequence[25] in comparison to -22.5 kcal mol^{-1} per quartet measured for the $G_4T_4G_4$ sequence.[26] Concerning parallel quadruplexes, analysis of the calorimetric or thermokinetic data gives a value of -72 kcal mol^{-1} (ref. 27) and -76.5 kcal mol^{-1} (ref. 28) for the TG$_4$T sample

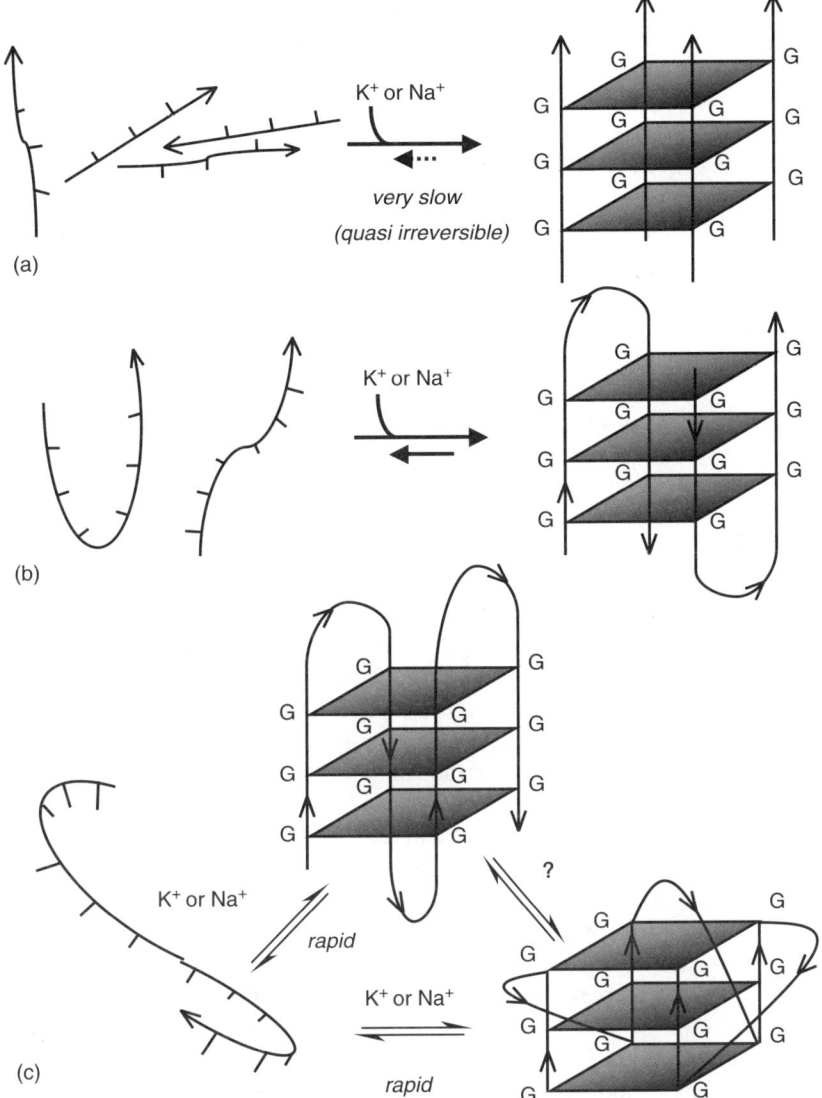

Figure 1 *Summary of the different equilibria involving G-quadruplexes. We wish to emphasize the differences between tetramolecular (panel a), bimolecular (panel b) and intramolecular (panel c) complexes. In the latter case, two possible quadruplex folds are presented; a similar situation may occur for bimolecular structures.*

(*i.e.*, −18 kcal mol^{-1} per quartet) and −61 kcal mol^{-1} for the TTAG$_3$T sample[29] (*i.e.*, −20 kcal mol^{-1} per quartet) in good agreement with enthalpies found for intramolecular quadruplexes. Overall, the enthalpy per quartet is (unsurprisingly) more negative than the enthalpy per base pair in a double-helix.[30]

This very negative (*i.e.*, favourable) enthalpy is partially compensated by a negative (and unfavourable) entropy of formation. The range of values found for $\Delta S°$ is wider than for enthalpy and depends on the stoichiometry of strand association and on the nature and length of the loops. Therefore, G-quadruplex formation shows the classical hallmarks of many nucleic acids structures, with $\Delta H < 0$ and $\Delta S < 0$. This effect arises (i) from the enthalpic gain of stacking large aromatic surfaces containing polarizable atoms (base stacking itself is driven by electrostatic and van der Waals interactions) and (ii) cation–dipole interactions (see below).

Despite the negative contribution of entropy to stability, most quadruplex structures are stable under physiological conditions. For example, most intra-molecular quadruplexes have a $\Delta G° < 0$ at 37°C in a buffer that mimics intracellular conditions (near neutral pH, high K^+ concentration, with or without Mg^{2+}). This simply reflects the fact that at physiological temperature the predominant, and sometimes only, species will be the quadruplex structure rather than the single-stranded form. This observation has been well documented for the human telomeric motif.

The last issue concerns the determination of heat capacity difference ($\Delta Cp°$). The most convenient method to determine this parameter is probably the isothermal titration calorimetry (ITC). However, this approach is not well suited for self-pairing association, as in quadruplexes. The relative paucity of $\Delta Cp°$ values for quadruplex formation is unfortunate, as this parameter can provide valuable information on the hydrophobic interactions involved in the stabilization of the structure (an empirical correlation has been found between the change in accessible solvent surface area and $\Delta Cp°$). $\Delta Cp°$ may be extracted from DSC data, using the difference between pre- and post-transition baselines. No baseline difference is observed by Petraccone *et al.*,[31] suggesting that $\Delta Cp°$ for parallel quadruplex formation is close to zero. Additional experiments will be required to confirm this observation for other sequences. Related to the heat capacity change is the interaction between the solvent (this case, water) and the quadruplex. Although some crystal structures provide a detailed hydration pattern of G-quadruplexes,[32] our understanding of solvent contribution to quadruplex stability is still limited. Water molecules make well-ordered hydra-tion spines in the grooves of the G-quadruplex through extensive H-bonds with the guanine bases and with the sugar-phosphate backbone.[33] G-quadruplex loops are widely hydrated, and water bridges connect the loops to a central K^+ ion and to the hydration spines running along the grooves[32] Unfortunately, these observations only give limited insight into the quantitative contribution of such interactions for quadruplex stability. Analysis of molecular crowd-ing[34,35] or studies under high pressure might provide additional information on such effects.

2.2.2 Loop Contribution to Folding and Dynamics

The previous paragraph addressed the impact of quartets on G-quadruplex stability. What about the other bases present in the sequence? Do they participate in the stability? A possible answer to this question comes from the analysis of many NMR solutions structures, where the thymine residues at the loops[36] or at the end of tetramolecular quadruplexes[37,38] are often under-defined; they are present in different interchangeable conformations, all NMR-visible. In fact, whereas adenines stack on top of adjacent G-tetrads,[36] thymidines often experience important rotational freedom, as shown for example by the thymine loops of $G_3T_4G_3$ in K^+ solution which appear to interconvert between two preferred conformations.[39] Additionally, molecular dynamics (MD) simulations of the quadruplexes formed by G_4 and $G_4T_4G_4$ with sodium cations in the central ion channel yield to exceptionally stable trajectories of the quadruplex stem, indicating a very rigid and stable structure, associated with a significant geometrical plasticity of the thymine residues.[40] Altogether, these data suggest the limited contribution of thymine bases to the overall stability of the quadruplex structure.

In contrast, $G_4T_4G_4$ (Oxy-1.5 from the *Oxytricha nova* telomeric motif) forms a well-defined, symmetrical structure with ordered thymine loops.[41] In the case of $G_3T_4G_3$, in Na^+ solution, the loop structure is compact and incorporates many of the features found in duplex hairpin loops including base stacking, intraloop hydrogen bonding and extensive van der Waals' interactions. The first and third thymine loop stack over the outermost G-quartet and are associated by hydrogen bonding.[42] The latter appears to contribute to the stability and final conformation of the quadruplex.[43] These observations, together with thermodynamic data, support instead the important role played by the nature and length of the loops in quadruplex stability.[44-47]

2.2.3 Stability under Physiological Conditions

Quadruplexes may find applications in diverse fields, such as biotechnology, nanotechnology, electronics, and last but not least, biology. In the latter case, it is of course essential that such structures are stable inside a cellular environment. Although a number of other unusual DNA structures have been discovered (for a review see refs, 48 and 49), their formation often requires non-physiological concentrations of divalent cations or relatively acidic conditions. In absence of stabilizing factors, their formation in the intracellular medium is therefore questionable. In contrast, quadruplexes are relatively stable – and sometimes even extremely stable – under near-physiological conditions. Intramolecular quadruplexes often have a melting temperature above 37°C in a near physiological buffer[20,46] and are therefore stable under physiological conditions.[20,50] Stability depends on loop length, but stable quadruplexes may be formed with a central loop of 3–7 bases[46] or probably more (Bourdoncle *et al.*, in preparation). In a recent article, Vorlickova *et al.* analysed the stability of human telomeric fragments of variable length (9–99 base-long).[22] T_m and $\Delta H°$ depend on the number of T_2AG_3 repeats: stability surprisingly decreases with increasing repeat number.

2.2.4 Which Quadruplexes are the Most Stable: Intra-, Bi- or Tetramolecular? Parallel or Antiparallel?

This important question has been addressed by several teams. Lu *et al.*[26] concluded that the parallel strand structure is thermodynamically more stable than the antiparallel one. Petraccone *et al.*[51] reached a similar conclusion: the tetramolecular structure is thermodynamically more stable than the bimolecular one, which, in turn, is more stable than the unimolecular one.

Nevertheless, one should note that it is not straightforward to compare the stability of structures with different molecularities. For example, the comparison of equilibrium association constants for unimolecular, bi- or tetramolecular quadruplexes is meaningless, as these constants are not expressed in the same units (unit-less, M^{-1} and M^{-3}, respectively). Similarly, a comparison of the ΔG° might be deceptive. For example, if one compares an intramolecular structure with a bimolecular one (self-complementary) which have the same $T_m = 60^{\circ}C$ at 1 μM strand concentration, one will determine a $\Delta G^{\circ}{}_{(Tm)}$ of 0 and -9.2 kcal mol^{-1}, respectively.[52] Does this mean that the intermolecular complex is more stable? Clearly not! One faces a similar problem when comparing association rate constants (k_{on}), which are expressed in s^{-1}, M^{-1} s^{-1} and M^{-3} s^{-1} for unimolecular, bi- or tetramolecular quadruplexes, respectively.

This caveat does not mean that one cannot compare other parameters. For example, it is legitimate to compare dissociation rate constants (k_{off}) that are expressed in s^{-1} for all the quadruplexes. One may analyse the enthalpies of quadruplex formation; as discussed earlier, the enthalpy per quartet is probably comparable for the three types of quadruplexes. A further level of complexity is added by the very different kinetics of formation. As a result, the thermodynamically most stable structure is not necessarily the kinetically favoured. Venczel and Sen[53] investigated the formation of higher order structures in the TGTG$_3$TGTGTGTG$_3$ sequence. Dramatic switches in the formation of G4 *vs.* G'2 structures occur in solution; rapidly forming G'2 bimolecular structures accumulate in highly stabilizing potassium (and strontium) solutions at the expense of the thermodynamically more stable G4 parallel structures.[54]

One may wonder what will be the predominant conformation for a given sequence. Phase diagrams may be the best way to define the relative abundances of species as a function of ionic strength, pH, dications and of course, strand concentration. The last parameter will of course play a crucial role when dealing with structures of different molecularities. Finally, another important parameter is molecular crowding: polyethylene glycol induces a structural transition from the antiparallel to the parallel G quadruplex in G$_4$T$_4$G$_4$.[34] Both the volume excluded by polyethylene glycol and the chemical interactions with polycations such as putrescine are critical for quadruplex conformation.[34] Note, however, that molecular crowding induces drastically different structures in the G-quadruplexes formed by *Tetrahymena* (T$_2$G$_4$)$_n$ and human (T$_2$AG$_3$)$_n$ telomere sequences: the latter seems to be unaffected by crowding agents[55] (but Li *et al.* recently reached a different conclusion[56]). The two motifs differ by only one base, which indicates that the single A → G mutation in the telomere sequence is crucial for the regulation of its polymorphic nature and, therefore, for its potential biological functions and physical properties.

In order to determine what would be the most favourable strand orientation in a quadruplex, several laboratories designed quadruplexes with forced geometries. Lu *et al.*[26] constructed two G-tetraplex structures containing identical G-tetrad base pairs from oligonucleotides. One has the truncated telomeric sequence from *Oxytricha*, $G_4T_4G_4$, which forms an antiparallel G-quartet structure; the second is constrained to form a parallel G-strand arrangement by insertion of a 5'-p-5' linkage between two T_2G_4 sequences. Each oligomer forms a defined G-tetraplex dimeric structure in the presence of Na^+. The standard-state enthalpies, entropies and free energy for formation of these tetraplexes have been determined. The parallel strand structure is thermodynamically more stable than the antiparallel one, primarily because of both the greater enthalpy and entropy of formation. A more recent approach involves the so-called "bunch oligonucleotides" in which orientation of the strands is predetermined by a covalent link between the four strands.[57] Again, the parallel orientation is favoured over the antiparallel.[58]

If one is interested in the thermal stability of the preformed complexes, it is clear that tetramolecular complexes are the most heat stable. For example, a small oligonucleotide such as TG_4T may form a parallel quadruplex with four quartets; its thermal stability in K^+ is so high that it can resist boiling at least for a few minutes whereas one may observe the thermal denaturation of $G_4T_4G_4$ and $(G_4T_4)_4$ under identical conditions.

2.2.5 Competition with Duplexes

This thermal stability prompted several laboratories to test whether quadruplex formation could compete with duplex formation. Duplex–quadruplex interconversion has been studied for a number of different sequences.[23,59–63] For example, using an equimolar mixture of the telomeric oligonucleotides AG_3(T-TAG_3)$_3$ and $(C_3TAA)_3C_3T$, we defined which structures exist and which is the predominant species under a variety of experimental conditions.[50] Under near-physiological conditions of pH, temperature and salt concentration, telomeric DNA is predominantly in a double-helical form. However, at lower pH values or higher temperatures, the G-quadruplex and/or the i-motif efficiently compete with the duplex.[23,50] We then demonstrated that, in the μM range, the duplex is the thermodynamically favoured species and that folding into a quadruplex delays, but does not prevent, formation of a Watson–Crick duplex.[64,65] The increased stability of the duplex as compared to the quadruplex has been confirmed by UV-melting analysis of the oligonucleotides under a variety of conditions. For example, the $\Delta G°$ for quadruplex formation at 37°C in 0.1 M KCl is -3.8 kcal mol^{-1}, and the $\Delta G°$ for duplex formation under identical conditions is significantly more negative (<-5 kcal mol^{-1}). Invasion of the thrombin aptamer quadruplex fails if one uses a complementary DNA strand; whereas a short PNA probe actively disrupts the quadruplex.[61]

These results suggest that human telomeric quadruplexes are marginally less stable than duplexes. In the absence of external factors, such as supercoiling or structure-specific proteins, a genomic region with a strong asymmetry between guanine-rich and a cytosine-rich strand is likely to remain double-stranded. However, the condition in a living cell is inherently molecularly crowded. This molecular crowding affects the structure and stability of the telomeric G-rich and C-rich strands and prevents any duplex formation.[66] Ciliate telomeric motifs such as $(G_4T_4)_n$ (*Oxytricha*) prefer quadruplexes over duplexes under certain conditions. The $G_4T_4G_4$ sequence forms a duplex in the presence of its complementary strand in 150 mM Na^+, but retains a quadruplex conformation in K^+.[60] Higher sodium concentration (> 550 mM) are required to induce duplex disproportionation.

This duplex–quadruplex competition is of special interest in the case of gene promoters. As discussed in the chapter by L. Hurley, guanine rich sequences are present in the promoter region of key oncogenes, and one can observe an induction of duplex to G-quadruplex transition in the presence of a quadruplex ligand and/or KCl.[67,68]

2.3   Cations Play a Central Role!

A detailed study of the dynamics of ions in a quadruplex will be found in the chapter by N. Hud and co-workers; we will therefore limit our discussion to the role of ions on the kinetics and thermodynamics of quadruplexes.

2.3.1   Modes of Interaction

One should first distinguish two very different types of monocation–quadruplex interaction: the first is based on central ions which are specifically sandwiched between quartets while the second is based on external ions which contribute to partial charge screening of the quadruplex. The central ions interact with a quadruplex in a specific fashion. X-ray crystallography provided definitive evidence of dehydrated cation coordination by G-quartets along the central axis.[33,69] Xu *et al.*[70] demonstrated that K^+ preferentially binds to a small number of specific sites on a tetramolecular quadruplex. Quadruplex specific stabilization by cations has been evaluated for a long time. Hardin *et al.*[59] defined the following order $K^+ > Ca^{2+} > Na^+ > Mg^{2+} > Li^+$ and $K^+ > Rb^+ > Cs^+$. A recent study on the human intramolecular quadruplex determined that $Sr^{2+} > K^+ > Na^+ \geq Rb^+ > Li^+ > Cs^+$.[71] Venczel and Sen analysed type Ia and IIa cations and found the following order $Sr^{2+} > Ba^{2+} > Ca^{2+} > Mg^{2+}$ and $K^+ > Rb^+ > Na^+ > Li^+ = Cs^+$.[4,53] The effect of Sr^{2+} ions on the formation of quadruplex structures by the human telomere sequence $(T_2AG_3)_4$ is actually stronger and different from that of the other ions tested. The T_m in 0.2 mM Sr^{2+} is equivalent to the T_m in ≈ 120 mM K^+ but the reaction is much less enthalpy-driven in the former case.[71] Lead (Pb^{2+})[72,73] and ammonium (NH_4^+)[4,74] are also effective to promote quadruplex formation. NH_4^+ ions are especially important for quadruplexes studied by mass spectrometry.[75,76]

Cations are also important for the self-assembly of nucleotides. Wong and Wu[77] have found that the cation-induced stability of a 5′-GMP structure is determined by the affinity of monovalent cations for the channel site (with the order $K^+ > NH_4^+ > Rb^+ > Na^+ > Cs^+ > Li^+$). Polycations such as spermine also interact with quadruplex structures,[78] but their effects on the thermodynamics of the quadruplex are still not completely understood.

Some cations may actively disrupt quadruplexes: stabilization of quadruplex structure by K^+ is specifically counteracted by low concentrations of Mn^{2+}, Co^{2+} or Ni^{2+}.[79] When the quadruplex and monovalent cation concentrations are low enough, or the temperature is sufficiently high, several divalent cations, for example, Ca^{2+}, Co^{2+}, Mn^{2+}, Zn^{2+}, Ni^{2+} and Mg^{2+} can induce quadruplex dissociation.[80] Divalent cations destabilize the $G_4T_4G_4$ antiparallel dimeric G-quartet according to the following order: $Zn^{2+} > Co^{2+} > Mn^{2+} > Mg^{2+} > Ca^{2+}$ (ref. 81) and induce a transition to a parallel quadruplex structure. More controversial is the supposed strong inhibitory effect on G-quadruplex formation played by Li^+ ions.[82] In our hands, the addition of various amounts of LiCl to an intramolecular quadruplex in a sodium or potassium buffer does not significantly affect its thermal stability (Lacroix *et al.*, unpublished results), arguing that Li^+ ions "ignore" rather than destabilize G-quadruplexes.

The two best-studied ions are Na^+ and K^+. The preference of quadruplex central cavity for potassium over sodium ions is the result of two opposite effects: from one side the free energy of Na^+ binding to a quadruplex is more favourable than that of K^+, but from the other side this effect is more than compensated by the much greater cost of Na^+ dehydration.[83,84] The net result is a free energy change in favour of the potassium form. The number of released ions upon melting of the human telomeric quadruplex can be estimated by melting experiments as 5 in the presence of NaCl and 6 in the presence of KCl.[71] Note that this value should not be interpreted as the number of ions bound in the quadruplex channel, as the total number of ions released on thermal denaturation includes the ions bound in the quadruplex channel, and the difference between the number of ions condensed on the quadruplex and on the random coil. In our hands, the differences in thermal stability between the sodium and potassium forms of a quadruplex are very sequence dependent: ΔT_m values ($T_{m(K+)} - T_{m(Na+)}$) are 8 and 30 °C for the human telomeric motif and the thrombin aptamer, respectively.[20] Takenaka *et al.*[85] found a 43,000-fold preference of the human telomeric intramolecular quadruplex for K^+ over Na^+. This value is surprisingly large, given the relatively small difference in thermal stability between the sodium and potassium forms of this quadruplex,[20] corresponding to a $\Delta\Delta G°$ of 2 kcal mol^{-1} (L. Lacroix, unpublished).

2.3.2 Location of Ions

The smaller Na^+ ion is able to reside within the plane of a quartet, whereas ammonium and potassium ions coordinate between two G-quartets. On the other hand, "external" ions interact with the quadruplex in a much less defined manner, with little, if any, preference for cation nature. Na^+ and K^+ compete on the same footing for atmospheric binding. For most other nucleic acids structures, there are no "internal" ions, and cations only interact in a relative passive way ("atmospheric binding") with no preference for cation nature.

It is clear that, if available, monocations such as potassium and sodium will be incorporated into a quadruplex. The potassium and ammonium ions are too large to be coordinated by a single G-quartet in a coplanar fashion. As a result, coordination of these ions occurs between two G-quartets planes. Each quadruplex involving n quartets will then accommodate $(n − 1)$ of these specific ions. For example, quantitative determination of ammonium peak intensity revealed that three NH_4^+ ions are placed between four quartets.[86] In contrast, the smaller Na^+ ion (ionic radius of 1.18 Å) allows for in-plane coordination. Multiple Na^+ ions are therefore not restricted to the spacing between G-quartets, and can move further away form each other to reduce electrostatic repulsions.[69] In any case, empty sites between quartets are probably very rare. In fact, although vacant coordination sites are likely to exist (as ions move between sites, see below), their lifetime must be very short[86] as demonstrated, for example, by Federiconi *et al.*[87] who determined a site occupancy of 0.97 K^+ ions per tetramer in GMP quadruplexes prepared in 0.5 M KCl.

Is it possible to obtain an "empty" quadruplex, whose central cavity is devoid of cations? Without a bound cation, the cyclic arrangement of four oxygen atoms clustered in the centre of the G-quartet would be electronically unfavourable and MD simulations without cations in the channel show destabilization of the G-DNA structure, underscoring the central role of these ions for the structural integrity of the molecule.[40] As noted by Borzo *et al.*, self-assembly of 5′-GMP, unlike a normal stacking interaction is not determined predominantly by hydrophobic forces. Cation binding plays an important role, together with and reinforcing the hydrogen bonding of the guanines into planar tetramers.[88] *Ab initio* calculations show that carbonyl-cation enthalpies provide more stabilization than either hydrogen bonding or stacking interactions.[89] Ions therefore play a strong stabilizing role on the quadruplex, but this does not exclude that a relatively unstable quadruplex stem could be formed in the absence of ions. A recent article demonstrated that intramolecular quadruplexes with non-nucleosidic loops could form a stable complex in the presence of lithium only.[47] Formation of a G-quartet in the absence of a templating metal cation has been obtained in a few examples (i) in the case of the multimerization of a 8-(N,N-dimethyl-aniline)guanosine derivative,[90] (ii) in the presence of a highly specific quadruplex ligand, telomestatin,[91,92] (iii) for the thrombin aptamer at low temperature and (iv) for very long guanine sequences (poly dG). In this case, the results indicate that the unstable quadruplex structure formed in the absence of small cations is presumably different from that existing in the presence of cations.[93]

2.3.3 Dynamics of Ions

^{23}Na NMR relaxation measurements, performed as a function of temperature, allowed to determine the kinetics of sodium ion complexation. The lifetime of specifically bound sodium ions is estimated to be 180 μs at 20°C.[94] Cations bound by G-quartets are exchanging with ions in the bulk solution at a rate greater than 10^3 s^{-1} (ref. 83) (*i.e.*, at a rate much faster than quadruplex opening, no disruption of the quadruplex is required to permit cation release/uptake). A similar observation was made for nucleotide-based G-quartets: Na$^+$ ions moved in and out of the GMP$_8 \cdot$Na$^+$ octamer 10^4–10^8 times per second, which is several orders of magnitude faster than nucleotide exchange.[95] The NH$_4^+$ ion has a much longer binding lifetime (250 ms or more), probably as a result of its larger ionic radius:[86] the G-quartet must assume an altered or partially denatured state to allow K$^+$ or NH$_4^+$ movement between coordination states, whereas sodium ions may move freely along the central axis.[96,97] NH$_4^+$ ion movement inside the G$_4$T$_4$G$_4$ G-quadruplex is accelerated in the presence of smaller Na$^+$ ions.[98] The first indication that a quadruplex may accommodate different ions simultaneously has been provided from a K$^+$ titration of a Na$^+$ G$_3$T$_4$G$_3$ dimeric quadruplex:[83] an additional mixture of the coordinated ionic species was observed. Interestingly, there is no evidence for direct movement of ammonium from the bulk to a central inner site (located between two internal quartets).

2.3.4 Proton Exchange, Transient Opening of the Quartet

A number of NMR or Raman studies indicate that amino and imino protons involved in quartet H-bonds may exchange without disruption of the quartet. Raman studies have been applied to the (T$_4$G$_4$)$_4$ and G$_{12}$ quadruplexes. The two N2 amino protons are not equivalent, as they exchange with the deuterated solvent at distinct rates: this demonstrates that the amino group rotation around the C2–N2 bond is highly restricted in the G$_{12}$ quadruplex.[99] The non H-bonded, exposed N2 proton exchanges much faster than its H-bonded neighbour. Nevertheless, full deuteration of the N2 amino groups is complete within 24 h at 10°C. In contrast, the N1 imino proton does not exchange at all at 10°C, and the exchange rate remains slow even at 95°C, illustrating again the extraordinary thermal stability of the G-quadruplex. However, the imino protons of peripheral or terminal quartets exchange more easily. NMR studies of telomeric quadruplexes show indeed that the N1 imino proton exchange in the outer two quartets proceeds more quickly than that of the inner quartets.[100–102] These results indicate that the guanines in the terminal layers of stacked quartets are more distorted or more flexible than those of the inner layers. Even in the case of the thrombin aptamer, which involves only two quartets (and therefore no inner quartet), the eight imino residues in the quadruplex showed different proton exchange rates.[103] A recent detailed analysis of individual H-bonds couplings of the Oxy-1.5 guanine quadruplex revealed that the 5′ strand end is the most thermolabile region of the DNA quadruplex.[104] On the basis of these differences, one can propose a preferential unfolding pathway for a quadruplex, with initial disruption of the less stable outer G-quartets. Surprisingly, the idea that compact DNA structures exchange H for D slower than unfolded ones might be a misconception in the gas-phase, as fast hydrogen/deuterium exchange is indeed observed for a DNA G-quadruplex by ESI–MS.[105]

2.4 Multiple Quadruplex Conformations

Most quadruplexes rely on the formation of a single building block, the G-quartet. However, variation in the loop geometry and strand orientation may lead to distinct conformations. Several distinct conformations have often been reported for a single guanine-rich sequence, depending on the incubation conditions and the experimental approach. Although some of these differences are artefacts, it is still clear that at least a few biologically relevant sequences are indeed able to adopt very different folding schemes. Such polymorphism will complicate kinetic and thermodynamic analyses and rational design of quadruplex ligands.

2.4.1 Initial Observations

The first indication that several conformations could be observed came from the comparison of the NMR solution and an earlier X-ray structure of the *Oxytricha* bimolecular quadruplex. The initial X-ray structure contained many of the features found in the solution structure[106] but differed in the orientation of the thymine loops: edge-loops were observed by X-ray as compared to diagonal-loops in the NMR structure.[107] However, a recent analysis corrected this earlier crystallographic study and demonstrated that this quadruplex adopts a structure with two strands forming an antiparallel diagonal arrangement,[32] indicating that the native structure is the same in solution and in the crystalline state.

A similar discrepancy between NMR and crystallographic studies was found for the thrombin-binding aptamer $G_2T_2G_2TGTG_2T_2G_2$.[108,109] In all cases, the thrombin-binding aptamer was found to fold into a structure containing two planar guanine quartets as its core. The NMR and crystal structures, however, have fundamentally different folding patterns owing to differences in the way these central bases are connected. Also in this case, it was possible to reconciliate the two structures and to show that the NMR model was consistent with the X-ray diffraction data.[110]

2.4.2 Telomeric Repeats

The most striking and recent example of structural complexity is the intramolecular human telomeric motif. In sodium, the $AG_3(T_2AG_3)_3$ sequence adopts an intramolecular fold with a central diagonal loop. As a result, each strand has one parallel and one antiparallel neighbour.[36] In potassium, the situation is different. Crystallographic studies indicate that the very same sequence adopts a completely different folding scheme: all four DNA strands are parallel, with the three linking trinucleotide loops positioned on the exterior of the quadruplex core, in a propeller-like arrangement.[111,112] Whether this folding pattern is relevant in solution is the subject of recent controversies.[56] Crosslinking studies show that the parallel structure probably exists in solution whatever the cation and confirm the existence of the antiparallel structure in presence of both sodium and potassium cations.[113,114] Inosine substitution[115] and CD experiments[116] confirm that the parallel intramolecular quadruplex structure may be observed in solution. On the other hand, the crystal structure predicts a sedimentation coefficient value that differs dramatically from that experimentally observed in K^+ solution[56] and circularization experiments suggest that the antiparallel form is present.[117] Systematic single-substitutions of adenine bases with 2-aminopurines also yields to fluorescence results at odds with quantitative predictions from the reported crystal structure. Based on these observations, the authors conclude that the predominant quadruplex in potassium is not the propeller-type. Molecular crowding agents also affect the conformation of the quadruplex.[55,56] As intracellular conditions indeed correspond to a crowded environment, even if the diluted solution structure disagrees with the crystal conformation, the latter could well be biologically relevant in the nucleus!

This polymorphism is not limited to the human telomeric intramolecular quadruplex (*i.e.*, with four guanine repeats). A shorter version of the human telomeric motif, with only two repeats, may also adopt several conformations.[22,111,118,119] $TAG_3T_2AG_3T$ can form both parallel and antiparallel G-quadruplex structures in K^+-containing solution and the two structures can coexist and interconvert one into the other. Sequence variants preferentially adopt the parallel or antiparallel folding scheme: TAG_3UTAG_3T, which contains a single thymine-to-uracil substitution at position 6, forms a predominantly parallel dimeric G-quadruplex with double-chain-reversal loops and all guanines in *anti* conformation. Another modified sequence, UAG_3BTAG_3T (with B = 5-BromoUracyl), forms a predominantly antiparallel dimeric G-quadruplex with edgewise loops; the structure is asymmetric with six *syn* guanines and six *anti* guanines. On the other hand, a longer variant of the human telomeric motif, with five telomeric repeats seems to adopt a single strictly parallel arrangement[22] as shown by CD and PAGE. Its denaturation is a two state process with isoelliptic points, suggesting a single melting structure.

Telomeric repeats from other organisms are also prone to quadruplex polymorphism. The *Tetrahymena* $TG_4T_2G_4T$ telomeric sequence, for example, may adopt several conformations.[120] This sequence forms two novel G-quadruplex structures in Na^+-containing solution. In the first structure (head-to-head), the two loops are at one end of the G-tetrad core; in the second structure (head-to-tail), the two loops are located on opposite ends of the G-tetrad core. In contrast to the human telomere sequence, the proportions of the two forms are similar for a wide range of temperatures; their unfolding rates are also similar, with an activation enthalpy of 37 kcal mol^{-1}. The $(G_4T_4)_n$ sequence may also interconvert between parallel and antiparallel structures.[54,60,99,121–123] Low concentrations of either Na^+ or K^+ facilitate formation of the antiparallel quadruplex, while high cation concentration promotes formation of the parallel complex.[60] However, the midpoint of quadruplex conversion is 240 mM for Na^+, as compared to 20 mM for K^+.

2.4.3 Consequences of Polymorphism

This structural polymorphism has important implications for thermodynamic and kinetic analyses. In some cases, the two-state hypothesis (which is required for a number of calculations) is obviously inaccurate. Such polymorphism may also explain why complex kinetic behaviours are observed in situations where simpler models are expected. It also complicates molecular modelling, as well as drug-quadruplex analysis. In the latter case, a quadruplex ligand may have a different affinity for the possible folding configurations. Depending on the interconversion rate between these conformations, a shift in their relative abundancies may – or may not – be obtained when a quadruplex ligand is present.[124] NMR is probably the best-suited technique to analyse these effects.

2.5 Kinetic Analysis: Technical Considerations

2.5.1 Elementary Precautions

One of the problems one may face when analysing quadruplexes, is that, once formed, some structures are so difficult to disrupt that even incubating for 10–30 min at 100°C does not lead to significant quadruplex unfolding, even in the presence of 7 M urea.[125] When undesired quadruplex formation is observed before a kinetic experiment, the addition of limited amounts of sodium hydroxide followed by HCl neutralization is a convenient and fast method to disrupt these complexes[1,126,127] but results in a net increase in salt concentration. As a consequence, we suggest using a 0.1 M LiOH solution to unfold these oligomers.

Several controls may be performed to confirm that the folded structure indeed corresponds to a quadruplex. The cation dependency of its stability (which follows a $K^+ > Na^+ > Li^+$ preference order) is a good indication. One may also demonstrate that the thermal difference spectra or circular dichroism spectra of all these structures are in agreement with the formation of quadruplexes.[20] Furthermore, non-denaturing gel electrophoresis may confirm that a single, major retarded (for bi- and tetramolecular) band is observed when the oligonucleotides are pre-incubated at high concentration in the presence of NaCl or KCl.

2.5.2 General Considerations

Many techniques besides NMR and crystallography may be used to study G-quadruplex formation, for example: circular dichroism, Raman spectroscopy, electrophoresis, nuclease sensitivity, chemical probing and calorimetry. However, tetra, bi- and intramolecular quadruplexes behave quite differently, and their folding kinetics require specific experimental constraints. To summarize:

(i) Intramolecular structures (G4' DNA) fold and unfold (relatively) quickly, and fast mixing experiments (*i.e.*, using a stopped flow accessory) are often required to measure their association and dissociation rates. On the other hand, it is straightforward to obtain equilibrium-melting curves, and this facilitates the determination of thermodynamic parameters (*i.e.*, $\Delta H°$, $\Delta S°$, $\Delta G°$ at 37°C, equilibrium constant). These melting curves may be obtained by UV absorbance (at 240 or 295 nm), circular dichroism or fluorescence provided that a suitable fluorescent reporter system is attached to the oligonucleotide.

(ii) Tetramolecular quadruplexes (G4 DNA) exhibit a totally different behaviour. Most melting curves recorded by heating a preformed tetramolecular quadruplex do not correspond to equilibrium melting curves; the "T_m" deduced from these experiments (which depends on the heating rate but not on strand concentration[27,128]) is therefore inaccurate. It is still informative, though, as it reflects the temperature dependency of the dissociation process.[27,128] In order to distinguish this apparent melting temperature from the true thermodynamic T_m, we prefer to call this value $T_{1/2}$. Additionally, k_{off} values extracted from a melting curve may be confirmed by "T-jump experiments" in which a preformed quadruplex (at low temperature) is suddenly transferred to high temperature, and the time-course of isothermal quadruplex dissociation is recorded.[129,130] In order to determine the association rate constant (k_{on}), one can perform an isothermal kinetic analysis. In general, association experiments are carried out at relatively high strand concentrations (up to 500 μM). These experiments are usually performed at a relatively low temperature, for which dissociation is very slow (once formed, the complex has a very long lifetime). From the Arrhenius representations of association and dissociation processes, it is possible, at least in theory, to recalculate the equilibrium constant at every temperature. However, one should note that the experimental points are not determined in the same temperature range (low T for association, high T for dissociation): an extrapolation of these data is therefore required. The slow kinetics of association and dissociation of tetramolecular G-quadruplexes often make equilibrium measurement impractical.[27,130]

(iii) Bimolecular quadruplexes (G2′ DNA) exhibit an intermediate behaviour. A small but significant hysteresis is often present in all denaturation/renaturation experiments.

2.5.3 Limitations of the Current Methods

Many of the results presented in the next pages have been obtained with methods that either require (i) covalent labelling of a quadruplex-forming oligonucleotide with a fluorescent tag (FRET experiments) or (ii) attachment of the oligonucleotide to a surface (surface plasmon resonance). However, one must keep in mind that such modifications may significantly affect (by an order of magnitude or more) the kinetics of folding and unfolding. For example, the dR-FAM-$^{5'}$G$_4$T tetramolecular quadruplex has a lifetime of 183 h at 37°C, as compared to 3 min for the G$_4$T$^{3'}$-dR-FAM.[128] Which one reflects the true kinetics of the unmodified G$_4$T sequence? Working with an unmodified quadruplex is therefore desirable; but even the simple 5' terminal radioactive labelling with a phosphate (required for most non-denaturing electrophoresis experiments) may affect the stability of a quadruplex. Whenever possible, methods that do not require modification, labelling or attachment should be preferred (CD and UV spectroscopy are good examples). Unfortunately, these methods are not always applicable.

2.6 Kinetics of Intramolecular Quadruplexes

2.6.1 Presentation

Intramolecular quadruplexes are often referred to as G4' DNA. Four guanine blocks must be present on the same oligonucleotide sequence and folding of this oligonucleotide into a quadruplex will create three loops. As discussed before, intramolecular quadruplexes are very polymorphic; a single sequence may adopt distinct conformations, which will complicate kinetic and thermodynamic analyses.

2.6.2 Kinetic Analysis

Thermal denaturation/renaturation experiments performed on intramolecular quadruplexes often give reversible melting curves, which simplifies the determination of thermodynamic parameters, but does not give much information on the kinetics of folding and unfolding. However, when very fast temperature gradients are chosen on a real-time PCR apparatus,[131] a hysteresis phenomenon is clearly visible on the emission signal of a tagged quadruplex-forming oligonucleotide.[44,47] Such a hysteresis should, in principle, allow for the determination of kinetic parameters. However, the hysteresis of the $(G_3T_2A)_3G_3$ quadruplex is probably too small for a reliable deconvolution of the kinetic parameters, even with a $0.1^\circ C \ s^{-1}$ temperature gradient.[47-49]

A convenient method to study the folding of an intramolecular quadruplex should be to perform fast-mixing experiments with a stopped flow accessory coupled to an UV, CD spectrometer (or fluorimeter, ensured that the oligonucleotides are covalently labelled to fluorescent groups). Unfortunately, we found no report on the kinetics of G4' DNA with this technique. We are currently performing such experiments on two different quadruplexes with a 3–12 base central loop (Bourdoncle *et al.*, in preparation). Preliminary results suggest that folding at 25°C requires at least a few seconds.

In principle, by using an SPR apparatus with a G-rich strand attached to the matrix and different strand concentrations of the complementary oligonucleotide in solution, one can determine the k_{on} and k_{off} values for the intramolecular quadruplex and the bimolecular duplex.[132-134] One of the major conclusions of these experiments is that both the folded and unfolded forms of the intramolecular quadruplex have relatively short lifetimes (<90 s).[133] However, this requires fitting the SPR sensorgrams with four unknown kinetic parameters and considerable precautions are required to assess the validity of these values. At least two different intramolecular quadruplexes have been tested with this method: the human telomeric repeat and the nuclease hypersensitive element (NHE III$_1$) in the *c-myc* promoter. For (TTAG$_3$)$_4$, Zhao *et al.* found $k_{on} = 1.2 \times 10^{-2}$ s^{-1} and $k_{off} = 1.3 \times 10^{-3}$ s^{-1} in K$^+$ at 25°C. Increasing the temperature to 37°C has a very limited impact on association ($k_{on} = 1.6 \times 10^{-2}$ s^{-1}) and slightly increased the dissociation ($k_{off} = 3.8 \times 10^{-3}$ s^{-1}). Identical measurements in a sodium buffer revealed a $k_{on} = 7 \times 10^{-3}$ s^{-1} and a $k_{off} = 4.6 \times 10^{-3}$ s^{-1} at 25°C. From these numerical values, one can determine the equilibrium constant ($K_a = k_{on}/k_{off}$), obtaining a $K_a = 1.5$ and 0.8 at 25 and 37°C, respectively, in Na$^+$. Unfortunately, as noted by Zhao *et al.*,[132] these values are in contradiction with van't Hoff and calorimetric data for the same sequence. SPR values suggest that (i) the melting temperature (for which $K_a = 1$) must be between 25 and 37°C and (ii) the ΔH°, deduced from the temperature dependency of the equilibrium constant is extremely low in absolute terms. Both results are incompatible with solution studies ($T_m = 52–58$°C for the unlabeled telomeric strand,[20,132] ΔH° = −45–48 kcal mol^{-1}). A similar caveat may be found in the data obtained for the *c-myc* NHE sequence (G$_4$AG$_3$TG$_4$AG$_3$TG$_4$A$_2$G$_2$TG$_4$). A value of $K_a = 0.54$ was found ($k_{on} = 8.3 \times 10^{-3}$ s^{-1} and $k_{off} = 1.6 \times 10^{-2}$ s^{-1}) in 150 mM Na$^+$ at 25°C, indicating that the oligonucleotide is mainly *unfolded* at this temperature.[133] The K_a value in potassium was marginally higher. Both results are extremely unlikely, given the very high thermal stability of the *c-myc* quadruplex.[135] It is not simple to explain this discrepancy. A first hypothesis is that the linking of the G-rich strand to the matrix strongly destabilizes the quadruplex (perhaps by two orders of magnitude[132]). In agreement with this hypothesis, kinetic parameters deduced from the reverse experiment (a C-strand attached to the matrix) often gave different results. A second hypothesis is that one of the assumptions made is wrong. Many authors assumes that, in order to bind to its complementary strand, the G-quadruplex must be completely unfolded: no interaction is supposed to occur between the C-strand and the quadruplex (*via* its loops or *via* a partially unfolded region). It would be interesting to demonstrate unambiguously the validity of this hypothesis, as it has profound general implications for the duplex–quadruplex equilibrium. A last possibility is that additional secondary structures of the G-rich oligonucleotide that do not participate in the hybridization might be present.[133]

Our (unsupported) opinion is that SPR data accurately predicts k_{on} but somewhat overestimates k_{off} (therefore leading to an underestimation of K_a). In line with this hypothesis are the k_{off} values found with other techniques. One of the first articles analysing the kinetics of quadruplex folding and unfolding used electrophoresis to analyse the (T$_4$G$_4$)$_4$ intramolecular quadruplex in 50 mM K$^+$ or Na$^+$.[136] The authors concluded that the association constant at physiological temperature is $> 2 \times 10^{-2}$ s^{-1} in K$^+$, as compared to 1.7×10^{-3} s^{-1} in Na$^+$ (*i.e.*, in reasonable agreement with SPR data). In contrast, k_{off} values are much lower (in the 10^{-5} s^{-1} range) leading to very slow unfolding of the quadruplex ($t_{1/2} = 4$ and 18 h at 37°C in Na$^+$ and K$^+$, respectively).[136] In order to determine k_{off} and k_{on}, different laboratories chose to study the hybridization of a complementary oligonucleotide to a preformed quadruplex.[76,118,137,138]

The kinetics of opening of the DNA quadruplex formed by the human telomeric repeat have been investigated using real-time fluorescence resonance energy transfer (FRET) measurements with a peptide nucleic acid (PNA) trap.[138] It has been found that quadruplex opening is zero-order with respect to PNA, indicating that the initial step is a rate-limiting internal rearrangement of the quadruplex. k_{off} values for a dual-labelled $(G_3T_2A)_3G_3$ strand were 5×10^{-4}, 6×10^{-3} and 1.6×10^{-2} s^{-1} at 20, 37 and 50°C in a 100 mM Na^+ solution, respectively.[138] Kinetic runs in 100 mM K^+ give even lower dissociation rates, leading to calculated time constants of 40 h at 20°C and 4 h at 37°C.[138] The activation energy of dissociation (E_{off}) deduced from the temperature-dependency of k_{off} is +25 kcal mol^{-1}. If one assumes a similar $\Delta H°$ (around -45 kcal mol^{-1}) for the labelled and unlabeled $(G_3T_2A)_3G_3$ quadruplexes, a significant portion of this negative enthalpy must then come from a negative activation energy of association E_{on} ($\Delta H° = E_{on} - E_{off}$), meaning that quadruplex folding is faster at lower temperature, in agreement with data on tetramolecular quadruplexes. Green *et al.* indeed found values of 2.4×10^{-2} and 1.1×10^{-2} s^{-1} for k_{on} at 37 and 50°C, respectively. On the other hand, association rates measured by SPR were relatively temperature insensitive.[132] Mass spectrometry is another useful method to observe quadruplex opening in an ammonium acetate buffer.

Analysis of the data is further complicated by the existence of several quadruplex conformations in solution. This is one possible reason to explain why a double-exponential fit is required to analyse the data.[138] The first-order kinetics observed by Green *et al.* suggest a slow rearrangement of the quadruplex prior to trapping. Finally, one should note that the double-labelling procedure leads to a 10°C decrease in melting temperature and we do not know whether this reflects a decrease in association and/or an increase in dissociation as compared to the unlabeled $(G_3T_2A)_3G_3$ quadruplex.

2.6.3 Quadruplex Interconversion

It is indeed likely that, under near physiological conditions, the human telomeric motif may adopt several different conformations, which may interconvert one into the other. It is possible to determine the number and distribution of quadruplex conformations by operating at a single molecule level. Using single molecule FRET, Ying *et al.*[139] demonstrated the existence of two stable folded conformations, in both sodium- and potassium-containing buffers. The free energy difference between these structures is small (<1 kcal mol^{-1}) in the range of temperatures studied (12–37°C). Molecular modelling suggests that the two species are the parallel and antiparallel quadruplex structures. Very interestingly, the opening of such a mixture of conformations follows a single exponential law (with time constants of 12–13 min). It is possible to put upper and lower limits on the rate of interconversion between the two conformations: it is slower than 1 ms, as the two species may be resolved in the fluorescence correlation spectroscopy (FCS) apparatus, but occurs within a few minutes.

The human telomeric sequence is not the only one forming different intramolecular quadruplexes. The *c-myc* quadruplex ("Pu27") may form at least two different structures in equilibrium in K^+-containing solution.[11] Phan *et al.*[140] revealed that they both correspond to distinct intramolecular propeller-type parallel-stranded G-quadruplexes. Very high T_m (75°C or more) are observed in a 90 mM K^+ solution. Even if the authors did not determine the kinetics of these quadruplexes, they performed an analysis of the exchange times of the guanine imino protons in the central G-tetrad. The exchange times of these four guanine imino protons range from 12 to 50 h at 25°C, indicating a long lifetime of the quadruplexes at room temperature.

2.7 Kinetics of Bimolecular Quadruplexes

2.7.1 Presentation

Oligonucleotides having two guanine blocks may dimerize to form a bimolecular quadruplex often referred to as G'2 DNA. At least two thymine residues are required to allow formation of antiparallel folded-back hairpin dimers from two-copy oligomers of sequence $(T_nG_4)_2$ in Na^+; additional thymines destabilize this structure[141] (in potassium the oligonucleotides rather adopt a parallel structure). One of the best-known bimolecular quadruplexes results from the dimerization of the $G_4T_4G_4$ oligomer, often referred to as Oxy-1.5 as it corresponds to 1.5 copies of the *Oxytricha* telomeric motif.[26,41,101,106,107,121] Oxy-1.5 forms a symmetrical bimolecular G-quadruplex with four G-quartets and thymine loops at opposite ends of the G-quartets. A diagonally looped quadruplex is formed in the presence of both Na^+ and K^+ counterions. A number of Oxy-1.5 variants have been investigated. $G_3T_4G_3$ forms a dimeric quadruplex in Na^+ with the thymine loops across the diagonal of the end quartets.[83,142] However, unlike Oxy-1.5, the dimer is not symmetric. $G_3T_3G_4$ forms unusual dimeric G-quadruplex structure with the same general fold in the presence of K^+, Na^+ or NH_4^+.[143]

2.7.2 Kinetic Analysis

Data on the kinetics of bimolecular quadruplexes are scarce. Attachment of a long single-stranded tail to an oligonucleotide bearing two guanine blocks destabilizes and delays bimolecular quadruplex formation. This property has been used to determine the effect of quadruplex ligands on quadruplex formation[144] (see below). The second order rate constant for the formation of the bimolecular structure was estimated to be 3×10^{-2} M^{-1} s^{-1} in the absence of ligand.

A relatively simple method to determine the kinetic parameters of a bimolecular quadruplex relies on the analysis of non-equilibrium melting profiles, based on the deconvolution we developed for triplexes[145] and i-DNA.[146] However, in order to be accurate, this analysis requires a relatively important difference between the heating and cooling curves. Several bimolecular quadruplexes actually melt with a small hysteresis, such as $G_3T_2AG_3$ (our unpublished information) or GRO29A.[147] In order to analyse such oligonucleotides, one has to artificially increase the difference between the heating and cooling profiles, by using faster temperature gradients or lower strand concentrations. On the other hand, as shown in Figure 2(a), one obtains a significant hysteresis for the $G_4T_4G_4$ Oxy 1.5 sequence in Na^+.[21,148] This allows the determination of k_{on} and k_{off} in a relative wide range of temperatures, as presented in the Arrhenius representation [Figure 2(b)]. Cevec and Plavec[148] found activation energies for

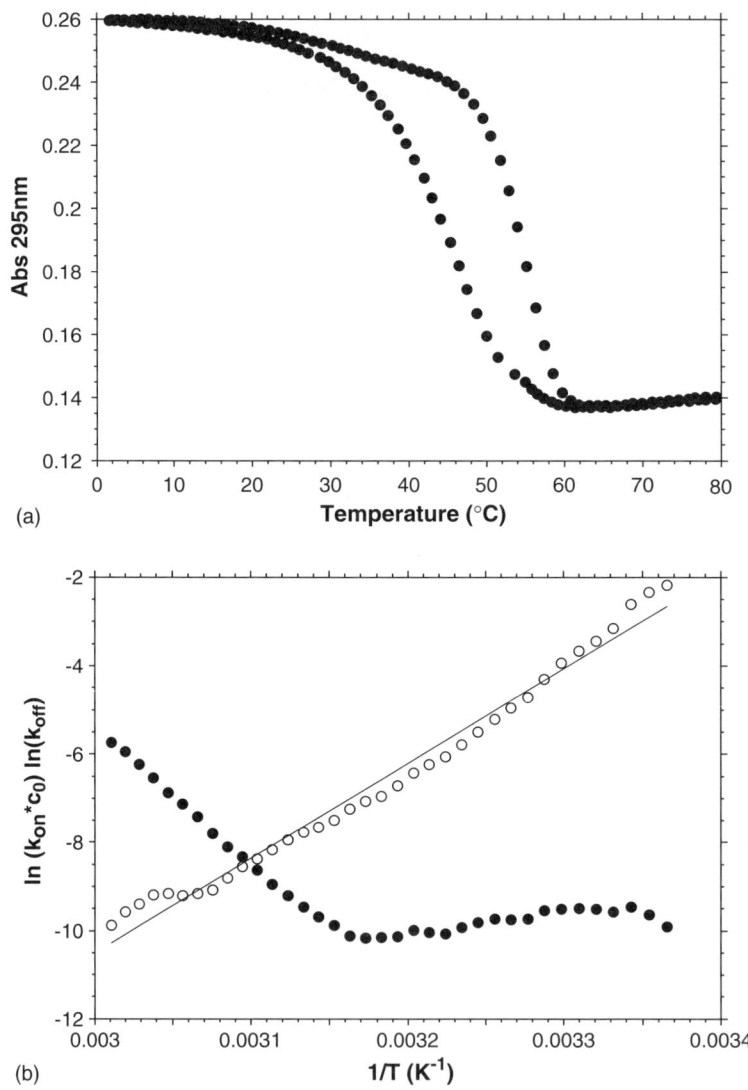

(a)

(b)

Figure 2 *Bimolecular quadruplex analysis. (a) Melting and cooling profiles of the Oxy 1.5 sequence ($G_4T_4G_4$) in 0.11 M Na^+ with a temperature gradient of 0.2°C min^{-1}. (b) Arrhenius representation of the association (In (K_{on} C_0);open circles) and dissociation (In (K_{off});filled circles) rates of the bimolecular quadruplex formed by $G_4T_4G_4$*

association (E_{on}) and dissociation (E_{off}) of -29 and 64 kcal mol^{-1}, respectively, indicating a enthalpy of -93 kcal mol^{-1} for quadruplex formation in 50 mM NaCl. The negative value found for the activation energy for association (E_{on}) again indicates that quadruplex formation is faster at low temperatures. We determined k_{off} values of 3×10^{-3} s^{-1} at 59°C and 5×10^{-5} at 44°C in 0.11 M NaCl, in fair agreement with the values found at a slightly lower ionic strength.[148] Surprisingly, the ln (k_{off}) *vs.* $1/T$ curve is not linear, and k_{off} values become relatively temperature independent below 44°C. Such behaviour could be an artefact resulting from the presence of several folded species in solution. Cevec and Plavec[148] also found a two-step dissociation process, which suggests premelting transitions in the loop region. Additional experiments are required to confirm this observation. Interestingly, oligonucleotides in which the T4 loop has been replaced with non-nucleosidic residues fold faster, exhibit no hysteresis, but are less stable than the $G_4T_4G_4$ Oxy-1.5 sequence.[148]

2.7.3 Quadruplex Interconversion

A simplified version of the human telomeric motif $G_3T_2AG_3$ adopts an anti-parallel quadruplex conformation, as demonstrated by a positive CD band at 290 nm.[22] Interestingly, this quadruplex is unstable in 150 mM KCl (but not NaCl), as it slowly (within days) isomerizes into a parallel guanine quadruplex.[22] This isomerization is facilitated at higher temperatures. The resulting structure may be similar to the crystal structure determined in potassium for the $TAG_3T_2AG_3T$ sequence.[111]

Other sequences interconvert between two bimolecular quadruplexes as shown by NMR studies. Phan *et al.*[120] demonstrated that a similar interconversion may occur between two asymmetric quadruplexes formed by the $TG_4T_2G_4T$ in Na$^+$. These two quadruplexes have very similar thermodynamic and kinetic properties. Their relative proportions and rate of unfolding are similar in a wide range of temperatures. The activation energy of dissociation is, in that case, $+37$ kcal mol^{-1}. At 37°C, the lifetime of both quadruplexes is approximately three times longer than for the human repeat in K$^+$. Phan and Patel[118] analysed the K$^+$ solution structure of a related sequence, $UAG_3T^{Br}UAG_3T$. This sequence may form two structures, a parallel and an antiparallel bimolecular quadruplex, which coexist and interconvert in solution. These two quadruplexes have different thermodynamic properties and different kinetics of folding and unfolding. The proportion of the two forms depends on the temperature (the antiparallel structure predominates at low temperature).

To study the $UAG_3T^{Br}UAG_3T$ quadruplexes kinetics, the authors performed different types of experiments. Addition of an excess of the complementary strand induces a time-dependent formation of the duplex, whose kinetics are related to quadruplex unfolding. A single exponential function perfectly fits the data for each form. At temperature below 40°C, the parallel quadruplex unfolds faster than the antiparallel one. As the antiparallel quadruplex has slightly higher activation energy of dissociation ($E_{off} = +43$ kcal mol^{-1} *vs.* $+34$ kcal mol^{-1} for the parallel form), kinetic stability is reversed above 40°C. At 37°C, the lifetimes of the two quadruplexes are very close, around 2 h.[118] To study their kinetics of folding, the authors performed a rapid cooling of the unfolded strands, which leads to a time-dependent formation of quadruplexes. The antiparallel quadruplex folds slightly faster than the parallel form.

2.8 Kinetics of Tetramolecular Quadruplexes

2.8.1 Presentation

All oligonucleotides considered in this paragraph contain only one block of guanines and form tetramolecular species. In the tetramolecular quadruplex configuration (also called G4-DNA), all strands are parallel, and all guanines are in the *anti* conformation with their sugars generally in the C2′ endo conformation (for DNA). These tetramolecular quadruplexes offer an interesting paradox: on one hand, their conformation is very well known, and a number of high[149] or very high (0.61 Å; ref. 150) resolution X-ray or NMR structures are available. This structural wealth might be explained in part by the extraordinary stiffness of this motif, as demonstrated by MD simulations.[40,151] On the other hand, one cannot but notice the paucity of thermodynamic and particularly kinetic data (with one notable exception[130]) on these structures.

2.8.2 Thermal Stability

Once formed, parallel quadruplexes are extremely inert, as the ring protons of the guanines involved in central quartets require days to exchange even at 60°C.[102] This is consistent with the exceptional thermal stability of this structure, which is still partially formed at 95–100°C in the presence of potassium.[24,54] In our hands, it is difficult to observe the melting of a preformed DNA quadruplex that involves four or more guanines (in 0.11 M K$^+$) or five or more guanines (in 0.11 M Na$^+$). Hence, thermal denaturation data may only be collected for a subset of sequences corresponding to oligomers containing relatively short runs of guanines.

As mentioned before, thermal denaturation of short parallel quadruplexes is often irreversible[27,31,152] (illustrated in Figure 3(a)). One possibility to circumvent this problem is to record the denaturation profile after an extremely long equilibrium time at each temperature.[31] However, this hysteresis phenomenon is sometimes extreme; no renaturation is observed at low strand concentration or unacceptable incubation times are required. Even at a rate as slow as 0.025°C min^{-1} (1.5° C h^{-1}!) the T_m for the $T_2G_4T_2$ quadruplex is overestimated by 40°C.[130] We actually took advantage of this irreversibility to determine dissociation rate constants (k_{off}) as a function of temperature[27] [Figure 3(b)]. An alternative approach to measure the lifetime of the quadruplex is to perform T-jump experiments of a preformed quadruplex.[128] Experiments with the $T_2G_4T_2$ sequence revealed a lifetime of 3 days at 55°C and 4.6 h at 65°C.[130] Using different unfolding temperatures, it is possible to calculate the activation energy of dissociation, E_{off}, which is around +50 kcal mol^{-1},[128] in fair agreement with our data (+43 kcal mol^{-1} for TG$_4$T[27]). Accurate estimation of the lifetime at physiological temperature requires an extrapolation of the

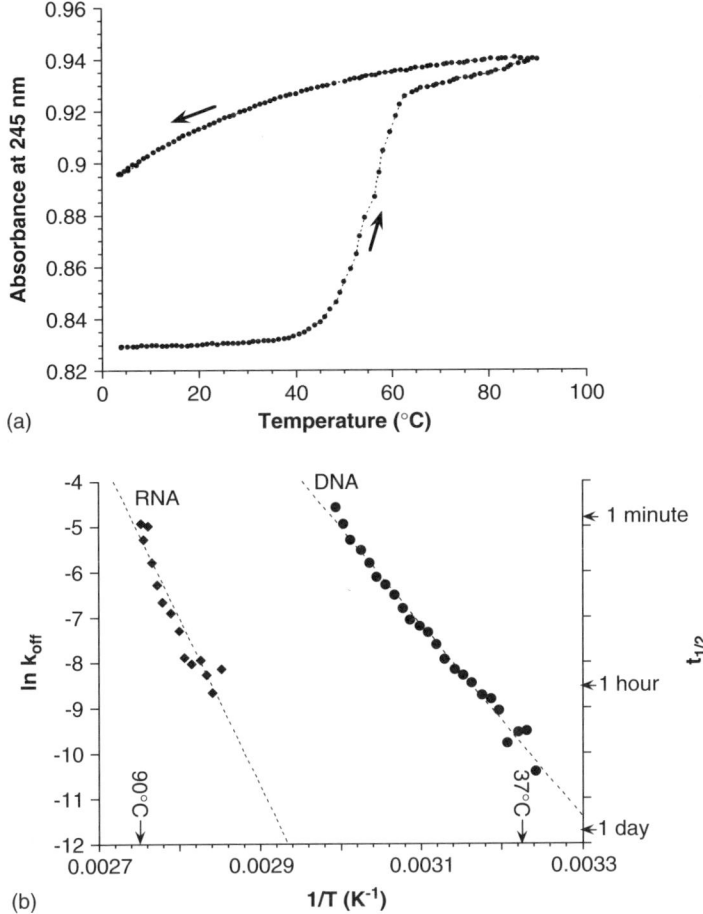

Figure 3 *Tetramolecular quadruplex dissociation. (a) Example of an irreversible melting curve (TG_4T) in 0.11 M Na^+ recorded at 245 nm with a temperature gradient of $0.2°C$ min^{-1}. Directions of temperature changes are indicated by arrows. The small difference observed at high temperature between the heating and cooling profiles results from a partial evaporation of the sample. (b) Arrhenius representation of the dissociation rate (ln (k_{off}) shown on the left Y-scale) and lifetime (right Y-scale) of the TG_4T (DNA) and UG_4U (RNA) quadruplexes in 0.11 M Na^+ (ref. 27)*

$\ln(k_{off})$ *vs.* $1/T$ plots, which are fortunately linear.[27,128] The slow unfolding ($k_{off} \approx 10^{-5}$–10^{-6} s^{-1} at 37°C) of some sequences[27,128,136] suggests that such structures would be long-living – days, weeks or even more – if they formed *in vivo*, unless they can be actively unfolded.

Another important issue is to determine whether the thermal melting occurs as an all-or-none phenomenon or involves partially unfolded species. Using single value determination analysis of CD melting spectra, Petraccone *et al.*[31] convincingly demonstrated that the quadruplex to single strands transition of

TG$_4$T involved only two significant spectral species, in full agreement with a simple dissociation pathway.

2.8.3 Association Process

It has been known for over 40 years that the renaturation of short guanine rich oligomers such as G$_3$ is very slow.[1] Absorbance and fluorescence spectroscopy were used to measure kinetics of association. In agreement with the seminal paper by Wyatt *et al.*,[130] who used size exclusion chromatography to study this process, we[27] and others[31] have found a fourth-order association rate with respect to strand concentration. We initially concentrated our efforts on well-known tetramolecular quadruplexes d-(TG$_4$T)$_4$[12,33,38,51,69,147,149,153] and r-(UG$_4$U)$_4$ formed by the tetramerization of 6-base long oligonucleotides. Isothermal renaturation experiments have been used to study the formation of the quadruplexes. Starting from the unfolded species, a time-dependent increase of absorbance at 295 nm may be observed [Figure 4(a)]. An opposite trend was seen at 240 nm, with a time-dependent decrease of absorbance. These variations reflect the spectral differences between the initial single-strand and the quadruplex. The order of the reaction *n* may be experimentally estimated by analysing the concentration dependency of the association process. Assuming that at $t = 0$; $\alpha = 1$, one can demonstrate that:[130]

$$\alpha = (1 + C_o^{n-1}.(n - 1).k_{on}.t)^{1/(1-n)} \tag{1}$$

where C_o is the initial strand concentration, α the fraction of unfolded strand and k_{on} the association rate constant. In most cases, this equation gave excellent fits, with $n = 4$ or between 3 and 4.[27,31,130] This strong dependency on strand concentration has important practical implications: a 10-fold increase in concentration may well mean switching from a reaction too slow to be followed to a reaction too fast to be measured![130]

The observed reactions seem to obey a simple kinetic pathway. Careful analysis of isothermal CD data by SVD analysis[31] or by "Walliman plots"[27,154] revealed only two species for the TG$_4$T DNA oligomer, in agreement with the absence of significantly populated intermediate states. However, the situation was more complex for DNA-PNA chimera, with three significant spectra species.[31] One should also note that oligomers ending with a terminal 3′ guanine such as TG$_{3-5}$ are more likely to form complex or higher order molecular species, as pointed out by the CD studies of Lieberman and Hardin.[155]

Generally, association processes for all these sequences are temperature-dependent, with large *negative* activation energy of association (E_{on}), both in sodium and potassium solutions [Figure 4(b)]. A value of −33 kcal mol^{-1} was found by Wyatt *et al.*[130] for T$_2$G$_4$T$_2$ as compared to −29 kcal mol^{-1} for TG$_4$T.[27] In another words, E_{on} accounts for one-third to one-half of the negative enthalpy of quadruplex formation ($\Delta H°_{ass}$ for the TG$_4$T quadruplex is −72 kcal mol^{-1}).

We investigated the effects of replacing sodium by potassium or ammonium.[156] As previously described, K$^+$ stabilizes quadruplexes, as demonstrated by an increase in apparent melting temperature ($T_{1/2}$), which in turn

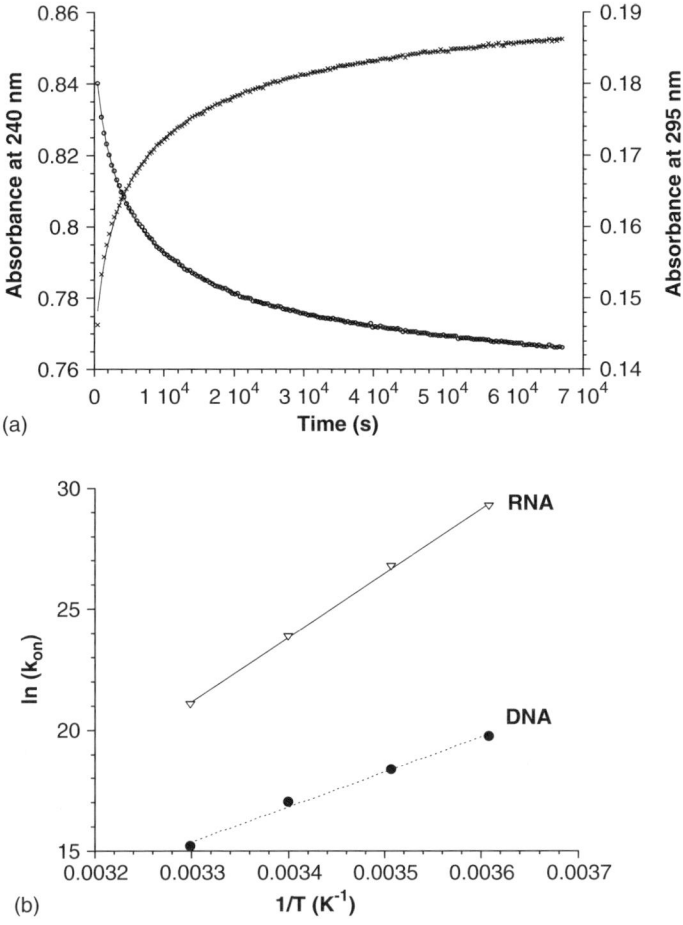

Figure 4 *Tetramolecular quadruplex association. (a) Example of an isothermal exper-*
iment. d-TG$_6$T oligonucleotide at 15 μM strand concentration in a 0.11 M Na$^+$
pH 7.2 buffer at 12°C. Two different wavelengths are presented: 240 nm
(circles; Y-scale shown on the left) and 295 nm (crosses; Y-scale shown on
the right). (b) Arrhenius representation of the association rate (ln (k$_{on}$)) of the
TG$_4$T (DNA) and r-UG$_4$U (RNA) quadruplexes in 0.11 M Na$^+$ (ref. 27)

reflects a slower dissociation at a given temperature [Figure 5(b)]. However,
for most sequences and independently of the temperature, potassium also
increases the association rate constant by a factor of 20–50 [Figure 5(a)].
The ratio $k_{on}^{K^+}/k_{on}^{Na^+}$ varies between 1 (for T$_2$AG$_3$) and 400 (TG$_5$T at 4°C).
In other words, potassium affects k_{on} *and* k_{off}. This difference in k_{on} between sodium and potassium is more or less conserved over a 2–37°C -
temperature range, suggesting that the activation energies of association are
similar.

Various concentrations of sodium and potassium have been tested in the
50–400 mM range. Increasing ionic strength plays little, if any role in the

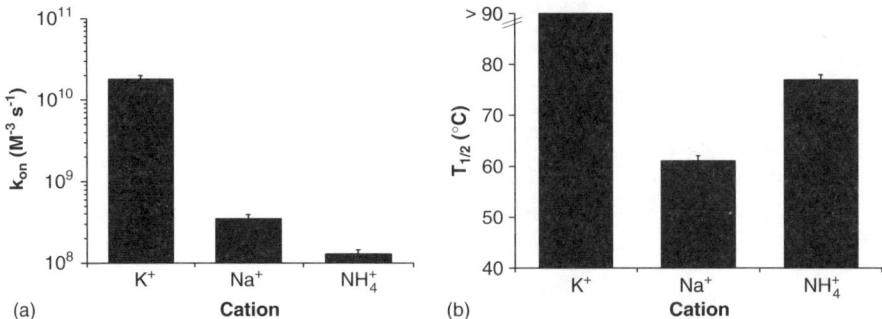

Figure 5 *Impact of ion nature on association rate and thermal stability. (a) Association rate constants (k_{on}) for the TG_4T tetramolecular quadruplex in the presence of 0.11 M Na^+ K^+ or NH_4^+ (log scale). (b) Apparent melting temperature ($T_{1/2}$, in °C) for the TG_4T tetramolecular quadruplex in the presence of 0.11 M Na^+, K^+ or NH_4^+. As shown in this figure, although ammonium leads to slower quadruplex formation than sodium, the NH_4^+ tetramolecular quadruplex has a higher thermal stability than the same quadruplex in Na^+*

thermal dissociation of the TG_4T tetramolecular quadruplex ($\Delta T_{1/2} < 2$°C; unpublished observations). However, varying the sodium concentration has a dramatic effect on the association process. This illustrates the fact that the stability of parallel quadruplexes is indeed dependent on the cation concentration, and that this effect is mainly reflected in k_{on}, as for many other nucleic acid structures such as duplexes or triplexes. A 10-fold increase in Na^+ concentration leads to a 1–2×10^3 increase in the association rate constant, in agreement with the involvement of several (≈ 3) Na^+ or K^+ ions in the association process.

The association rates are also strongly dependent on sequence length: the longer the G-tract, the faster the association.[27] This effect is surprisingly large: each extra guanine leads to an approximately 10-fold larger association rate constant [Table 1(a)]. As expected, a stabilizing effect may also be seen on dissociation: the longer the G-tract, the higher the $T_{1/2}$. Unfortunately, this effect is harder to quantify as many quadruplexes do not dissociate even at 90°C. Extra non G-bases at the 5′ or 3′ end play a detrimental role in the association of DNA and RNA [Table 1(b)]. The addition of one thymine/uracil to a DNA or RNA oligonucleotide decreases the association rate constant by a factor of roughly 10. This effect is cumulative; the slowest forming quadruplexes involve strands containing four thymines. The effect on k_{off} is more variable. As shown in Table 1(b), the presence of extra 5' or 3' thymines is often stabilizing (compare TG_4T and $TTTG_4T$, $T_{1/2} = +12.5$°C in Na^+). However, the parallel tetraplex formed by single copies of the sequences dT_nG_4 is most stable for $n=1$ and least stable for $n=8$, the longest tract that has been studied.[141] In other words, the presence of extra thymines is always detrimental to the association rate constant but often beneficial to the thermal stability of the quadruplex.

Table 1 *Kinetic parameters for tetramolecular quadruplexes*

(a) Effect of G tract length

Sequence (5'->3')	G-tract length	$k_{on}{}^a$	$T_{1/2}{}^b$ (Na⁺)	$T_{1/2}{}^c$ (K⁺)	Reference(s)
TGGGT	3	5.2×10^7	16	48	29,191,205
TGGGGT	4	3.8×10^8	54.5	–	12,33,38,51,69,147,149,153
TGGGGGT	5	6.1×10^9	–	–	
TGGGGGGT	6	1.4×10^{11}	–	–	

(b) Effect of terminal bases

Sequence (5'->3')	G-tract length	$k_{on}{}^a$	$T_{1/2}{}^b$ (Na⁺)	$T_{1/2}{}^c$ (K⁺)	Reference(s)
TGGGG	4	2.8×10^9	47	–	155
AGGGG	4	3.9×10^{10}	54	–	
AGGGGT	4	1.3×10^{10}	59.5	–	
TGGGGT	4	3.8×10^8	54.5	–	12,33,38,51,69,147,149,153
TTGGGGT	4	4.3×10^7	nd	nd	12,37,100
TTGGGGTT	4	1.5×10^6	71	–	12,125,130
TTTGGGGT	4	1.3×10^6	67	nd	

(c) Effect of single guanine substitution

Sequence (5'->3')	$k_{on}{}^a$ Na⁺	K⁺	NH₄⁺	$T_{1/2}{}^b$ Na⁺
TGGGGGT	9.8×10^9	2.3×10^{12}	1.1×10^9	>90
TIGGGGT	2.4×10^8	2.6×10^{10}	9.4×10^7	70
T7GGGGT	6.1×10^7	2.1×10^9	1.7×10^7	64
T8GGGGT	1.6×10^9	4.5×10^{10}	4.4×10^9	67

^{a} Association rate constant at 4°C (3°C in panel C), pH 7, with 0.11 M Na⁺. k_{on} is given in M^{-3} s^{-1}±30% and determined as in Wyatt *et al.*[130]
^{b} Non-equilibrium melting temperature of the preformed quadruplex, expressed in °C, in 0.11 M Na⁺ and determined with a temperature gradient of 0.18°C min^{-1}.
^{c} Non-equilibrium melting temperature of the preformed quadruplex, expressed in °C, in 0.11 M K⁺ and determined with a temperature gradient of 0.18°C min^{-1}.Note: '-' indicates no melting of the quadruplex, even at the highest temperature recorded (94°C); 'nd' indicates not done.

2.9   Higher Order Structures

"Simple" quadruplexes accommodate one, two or four independent strands. It is however possible to form more complex structures based on G-quartets that involve more than four strands.

2.9.1   Quadruplex Dimers

The simplest quadruplex-based higher order structure is a double-quadruplex, in which two tetramolecular quadruplexes associate to form an octaplex. One may distinguish two types of dimers (a) those based on simple stacking of terminal quartets and (b) those involving interlocked base quartets.

(a) Dimerization of a tetramolecular quadruplex occurs for oligomers that have a single multi-guanine motif at their 3' or 5' end, with a guanine as the terminal base. Thus, the oligomer T_8G_3T forms a unique quadruplex in the presence of Na^+, K^+ or Rb^+; however, its isomeric counterpart T_9G_3 in K^+ or Rb^+ generates an additional ladder of products of substantially lower gel mobility.[157] This observation is probably valid for most oligonucleotides starting or ending with a guanine stretch.[157,158] The dynamics and thermodynamics of dimerization of parallel G-quadruplexed DNA was recently studied by Kato et al.[159] Dimerization occurs with T_2AG_n ($n = 3$–5) but not with T_2AG_nT. Saturation transfer NMR experiments give a value of $6 \times 10^{-1}\,s^{-1}$ for the rate of conversion of the dimer of the T_2AG_3 tetrameric quadruplex to the quadruplex "monomer". Shift changes are consistent with dimeric assembly through interaction between the 3' terminal quartets of the two tetrameric quadruplexes by end-to-end stacking. The slow inter-exchange between monomers and dimers of the G-quadruplexes formed by the T_2AG_n sequences allows the quantitative characterization of the energetics of the process. The equilibrium constant (K_{eq}) of quadruplex dimerization may be obtained from the NMR signal intensities of the two forms, and the temperature dependence of the K_{eq} value allows the determination of thermodynamic parameters associated with the process. The enthalpy change (ΔH) clearly indicates that the dimerization of the G-quadruplexes is largely enthalpic-driven in origin. The ΔH value ($-8.9\,kcal\,mol^{-1}$) obtained for the dimerization of the G-quadruplexes is close to one-half of the value estimated for the formation of a single G-tetrad, reflecting the significant contribution of G-quartet stacking interactions to the stability of the G-quadruplexes. Negative charge–charge repulsions disrupt the dimerization of G-quadruplexes when a phosphate group is added directly to the 5'-terminal G of the G_5CTA oligomer,[160] demonstrating the importance of electrostatic interactions on the formation of these higher order structures. Finally, the addition of a G4-ligand may promote the formation of such complexes in which two quadruplexes are orientated in a tail-to-tail manner with the ligand sandwiched between terminal G4 planes.[161]

(b) A tighter dimer occurs when two quadruplexes interact after slippage. In this case, dimerization not only involves stacking of terminal quartets, but also sharing of H-bonds between guanines belonging to strands originally involved in different quadruplexes. The simplest example of such a dimeric form is represented by the G_3T tetranucleotide sequence, which has been shown to self-assemble into an interlocked quadruplex dimer.[158] Two "slipped" quadruplexes dimerize via free G-bases at the 5' ends of different strands by forming an extra G-tetrad. Octamerization of the G_3T oligonucleotide allows the formation of five G-quartets and two GTGT mixed tetrads. Formation of this octamer requires high strand concentrations ($>100\,\mu M$) and depends on the experimental protocol chosen for self-assembly. Once formed, this complex is much more thermally stable than the tetramolecular quadruplex, as shown by an apparent increase in melting temperature by 40°C.[158] The driving force for oligomerization is probably the hydrophobic stabilization, as a dimerized quadruplex has less aromatic quartet faces exposed to the solvent than two monomeric quadruplexes. This type of oligomerization may well be biologically relevant, as an interlocked dimeric, parallel-stranded DNA quadruplex has recently been described as a potent inhibitor of HIV-1 integrase.[162]

2.9.2 G-Wires

A number of guanine-rich oligonucleotides may fold into supermolecular assemblies called G-wires[18] or frayed-wires.[163] $G_4T_2G_4$,[18] $A_{15}G_{15}$,[163] T_9G_3,[157] $C_4T_4G_4T_2G_4$[164] or $G_{11}T$[165] spontaneously assemble into large superstructures that resolve as discrete bands on native and denaturing electrophoresis gels. G-wires formed by $G_4T_2G_4$ are up to 1 μm long, with a diameter (2.5 nm) consistent with the diameter of the G-quadruplex. Formation of G-wires is dependent on the presence of Na^+, K^+ and/or Mg^{2+} and, once formed, G-wires are resistant to denaturation. These complexes form within minutes[163] or hours even at moderate concentrations. G-wire formation is also temperature-dependent, and usually is faster at high temperatures. For the $G_4T_2G_4$ sequence incubated in sodium, G-wire formation is more pronounced at 37°C than at 23 or 3°C.[18] Self-association of the $A_{15}G_{15}$ oligonucleotide proceeds faster at 81°C than at 64°C.[166] In the latter case, a "stem" is formed through interactions between the guanine residues of the associated oligonucleotides, whereas the adenine "arms" remain single-stranded and may be hybridized to a T_{10} or T_{15} oligomer without disrupting the quadruplex stem.[163,167]

The type and concentration of the cation present in solution determines which conformation the oligonucleotide will adopt.[163] Assembly of $G_4T_2G_4$ G-wires occurs more efficiently in Na^+, yet G-wires formed in potassium are more stable.[18] The relative inefficiency of potassium to promote G-wires is probably the result of its higher stabilizing effect on G′2 DNA, thus preventing its conversion into a wire. For $C_4T_4G_4T_2G_4$, the data suggest that these higher order species arise from successive additions of parent oligomer to an initially formed quadruplex. Since the self-assembly requires the presence of 100 mM K^+ plus 20 mM Mg^{2+} (it is not observed with K^+ or Mg^{2+} alone), these cations behave in a synergistic manner in the formation and/or stability of the supermolecular self-assemblies.[164] Several reports indicate the crucial role played by divalent cations such as magnesium[166–168] or calcium[169] on G-wire formation. Ca^{2+} induces a structural transition of the $G_4T_4G_4$ oligonucleotide from the antiparallel to the parallel G-quadruplex, and finally G-wire formation.[169] The kinetic parameters also indicate that $G_4T_4G_4$ undergoes this transition through multiple steps involving Ca^{2+} binding, isomerization and oligomerization of $G_4T_4G_4$. Note that the presence of a 5′ phosphate group inhibits G-wire formation, as shown for the $G_4T_2G_4$ sequence.[18]

For $(CGG)_4$, molar concentrations of potassium induce aggregates formation, with kinetic profiles that resemble those of autocatalytic reacting systems. The kinetics of this transformation is extremely slow at pH 8 but greatly facilitated in acidic conditions (pH 5.4), allowing the onset of aggregation at 20°C within 1 day. An interesting (but speculative) new model for multimerization called "G-lego" has recently been proposed for $G_{11}T$.[165] This model requires the rearrangement and sharing of hydrogen bonds to form a new quartet between two interacting quartets.

Once formed, all these complexes are extremely resistant. G-wires prepared in the presence of potassium resist to standard denaturation conditions (40–50% formamide/7–8 M Urea, 100°C).[18,163] However, $A_{15}G_{15}$ frayed wires may be disrupted, even at low temperature, by the addition of a fully complementary sequence $C_{15}T_{15}$.[166] The reaction proceeds rapidly at temperatures below the melting point of the Watson–Crick A_{15}–T_{15} duplex, indicating that, once bound to an A_{15} arm of the wire, the $C_{15}T_{15}$ oligonucleotide rapidly invades the guanine stem.[166] This result is rather surprising, given the high stability of G-quartets and the difficulty to hybridize a complementary strand to a quadruplex-folded strand, even if a single-stranded overhang is present.[64] These observations, coupled with the dication dependency of many wires formation and accessibility of their guanine residues to methylation[170] (in contrary to what is observed in a classical quadruplex) indicate that G-wires do not simply behave as large quadruplexes, but have unique properties.

2.10   Effect of Chemical Modifications

Quadruplex formation is not restricted to DNA. A number of different oligomers with an altered sugar phosphate backbone are able to form G-quadruplexes.[171]

2.10.1   RNA and 2′-Sugar Modified Analogues

Poly rG can form four-stranded helices[172] and RNAs containing short runs of Gs can also tetramerize. Four-stranded RNA guanine quadruplexes are suspected to be involved in crucial biological functions for cellular and retroviral RNA, such as translational repression, splicing and mRNA turnover.[173–181] An RNA tetraplex containing G-quartet and U-quartet structures can be very stable.[182] A 19-base oligoribonucleotide r-$GC_2GAUG_2UAGUGUG_4U$, forms a K^+-stabilized tetrameric aggregate.[183] In some cases, stability of the RNA quadruplex may be significantly higher than the stability of the corresponding DNA.[21,27] This stabilization results in part from a stabilizing role of the T → U base substitution: a DNA oligonucleotide with thymines replaced by deoxyuracyls forms a more stable quadruplex.[21] Uracil and thymine tetrads show significantly different characteristics which may contribute to the differences between DNA and RNA.[184] However, the sugar itself plays a crucial role on the stability and structure of a quadruplex. One of the most striking examples of the difference between DNA and RNA was provided by Uesugi and co-workers.[185,186] They demonstrated that the solution structure of r-$G_2AG_2U_4G_2AG_2$ is completely different from its corresponding DNA $G_2AG_2T_4G_2AG_2$. The RNA sample forms an intrastrand parallel quadruplex with a G-tetrad and a hexad, whereas the DNA sample forms a classical dimeric antiparallel quadruplex. A concentration of 0.1 mM K^+ is required for half-folding of the RNA sample, as compared to 60 mM for the DNA oligonucleotide. In this case, the melting temperature of r-$G_2AG_2U_4G_2AG_2$ is approximately 30°C higher that the T_m of its corresponding DNA.[185]

The same 2′-sugar modifications seem to have different structural effects depending on the molecularity of the G-quadruplex. In fact, substitution of the ribose 2′-H with a methoxy group can yield, not only to a destabilization of short intramolecular G-quadruplex structures, but also to a complete change of their conformation, as shown by absorbance differential spectra.[21,152] However, the same modification in the tetramolecular complexes (TG_4T and r-UG_4U) leads to a strong stabilization of the structure.[27] Short RNA, DNA or 2′-Omethyl strands with a single guanine block form well-defined, tetramolecular G-quadruplexes, which exhibit a fourth-order dependence of the association rate, a negative energy of activation for association and a strongly positive energy of activation for dissociation.[27] Interestingly, the RNA strand forms a much more stable complex than its DNA equivalent. This extreme stability results from a much faster (10^3-fold) association and a slower (10^{-3}-fold) dissociation. Likewise, the ribonucleotide analogues (2′-OH substitution) show quite different thermal stabilities in the intra- and tetramolecular compounds, giving rise to hysteresis in the former and to incomplete denaturation in the latter (r-UG_4U).[21]

These results may have notable implications for a better understanding of the role of the 2′-position of the sugar moiety. From these findings one can presume the participation of the 2′-position of the sugar moiety in crucial interactions, which may be modulated by the possibility to form hydrogen bonds. This difference might be explained by the greater difficulty in adopting a *syn* conformation for the ribonucleoside residues, making the formation of parallel quadruplexes favourable as compared to antiparallel quadruplexes. However, other molecularity-dependent structural features seem to be involved in the stabilization of G-quadruplexes and additional studies would be necessary to investigate the role of this factor.

2.10.2   Further Backbone Modifications

2.10.2.1   Methylphosphonate Analogues

For all the G-quadruplexes analysed, loss of the negative charge at the level of the phosphate backbone, as in the methylphosphonate analogues, leads to a strong destabilization of the G-quadruplex structure, independently from its molecularity.[21] This result is in agreement with the evidence accumulated from several crystal structures, in which the negatively charged oxygen atoms of the backbone are found to be involved in a complex pattern of water-bridges with the sugar groups and the edges of the guanine units. These extensive hydrogen-bonding interactions give rise to a well-ordered distribution of water molecules along the G-quadruplex grooves, also called hydration spines, which seem to be relevant for the stabilization of the structure. However, this does not necessarily imply that non-negatively charged oligomers cannot form G-quadruplexes, as recently demonstrated for PNAs (see below).

2.10.2.2 Phosphorothioate Analogues

Not only the charge but also the ionic radius of the atom of the oligonucleotide backbone may have an impact in the formation and stabilization of G-quadruplex structures. Substitution of the oxygen atoms of the phosphate backbone with sulfur atoms appears to affect the stability of the G-quadruplex structure in a molecularity-dependent way. For both bimolecular and tetramolecular phosphorothioate oligonucleotides, the larger size of the S^- in comparison to the O^- destabilizes the G-quadruplex.[21] However, the different spatial arrangement of the intramolecular complexes seems to be less affected by the presence of sulfur ions, leading to quasi equivalent melting temperatures with respect to those of the unmodified species.[21,171]

2.10.2.3 LNA and PNA

The solution structure of a locked nucleic acid (LNA) quadruplex, formed by the oligomer d(TGGGT), containing only conformationally restricted LNA residues has been reported.[187] The molecule adopts a parallel stranded conformation with a 4-fold rotational symmetry, right-handed helicity and three well-defined G-tetrads. The thermal stability of this parallel LNA quadruplex has been found to be comparable with that of r-UG_3U but much higher that the corresponding DNA quadruplex structure formed by TG_3T.[187] In contrast, in the case of a thrombin aptamer sequence in which the guanine residues have been replaced by LNAs, the thermal stability of the modified quadruplex is much lower that that of the unmodified DNA sequence.[188] The presence of a methylene bridge confers an RNA-like C3'-*endo* conformation to the sugar and reduces its conformational flexibility. One may then propose that this conformation is favourable towards tetramolecular parallel quadruplexes, but unfavourable towards antiparallel quadruplexes which, involve guanines in the *syn* conformation. According to this hypothesis, it has been recently demonstrated that a single LNA residue can induce a change in the thermodynamically preferred structure of an intramolecular G-quadruplex (formed by Oxy-3.5 = $(G_4T_4)_3G_4$) from an antiparallel to a parallel structure.[189]

PNAs are oligonucleotide mimics in which the entire sugar-phosphate backbone has been replaced with a pseudopeptide, resulting in a neutral oligomer. Quadruplex formation with PNAs was first investigated for PNA. DNA hybrids, Datta and Armitage[190] designed a 2 PNA. 2 DNA G-quadruplex bearing the *Oxytricha nova* telomeric sequence $G_4T_4G_4$. The resulting quadruplex shows a preference for Na^+ over Li^+. FRET experiments revealed that (i) the two PNA strands are parallel to each other, (ii) the two DNA strands are parallel to each other, and (iii) the 5'-termini of the DNA strands align with the N-termini of the PNA strands. It is also possible to form a "pure" PNA quadruplex.[191,192] The ion dependency of these structures is analogous to that reported for DNA quadruplexes. It is also possible to design a chimeric strand bearing DNA and PNA residues such as TGGGt (lower case letter refers to PNAs). The thermal stability of the chimera is quite similar to that observed for the DNA only sequence TG_3T.[193] However, TGGgt and tgGGT do not form well defined structures: this is attributed to the flexibility of the PNA backbone which probably has an unfavourable contribution to the energetics of quadruplex formation. In a recent article, Petraccone *et al.* analysed the folding of tGGGGT and TGGGGt.[31] These chimera are thermally more stable (by 4–7°C) than the corresponding TG_4T DNA strand.

2.10.3 Base Substitutions

The driving force for G-quadruplex formation is clearly the canonical G-quartet, with four guanines. However, DNA quadruplex formation is not restricted to G-repeat sequences, with their characteristic stacked uniform G·G·G·G tetrad architectures. Rather, the quadruplex fold has a more versatile and robust architecture, accessible to a range of mixed sequences, with the potential to form A·A·A·A, G·C·G·C and A·T·A·T tetrads or even hexads, heptads or octads.[194–198] Many articles analysed these "non-G quartets", often in the context of parallel tetramolecular quadruplexes. NMR studies show that the thymine in the centre of the TG_2TG_2C four-stranded quadruplex forms a thymine quartet[199] and the cytosine in the TG_3CGT quadruplex forms a cytosine quartet.[200] Adenine quartets,[201] U-quartets[182] and bulges may also be accommodated in RNA quadruplexes,[202] expanding the structural repertoire of G-quadruplexes.[203] Stable mixed G·C·G·C tetrads can also be present.[204] A base hexad involving two adenines and four guanines may be formed.[186]

From the examples shown above and from many others, it is clear that quadruplexes may incorporate unusual quartets. However, their contributions to the kinetics and energetics of the quadruplex are often poorly understood and structural methods only give clues on the effect of these modifications. Unlike G-quartets, only one hydrogen bond is observed between most base quartets. This explains in part the lower stability of these quartets as compared to the canonical G-quartet. Comparison of the thermal stability of a quadruplex incorporating a modified tetrad with a canonical G-quadruplex is a convenient method to estimate the impact of such tetrad on the quadruplex structure. However, the increase or decrease in apparent melting temperature only partially reflects the energetics of the modified tetrad. As mentioned above, melting of tetramolecular quadruplex is very often associated with a hysteresis, and the melting profile reflects the kinetic stability rather than the thermodynamic stability of the structure. In other words, variations in apparent T_m rather reflect effects on dissociation (k_{off}) as a function of temperature than on the equilibrium constant. For example, $T^{Br}GGGT$ and $TG^{Br}GGT$ form a more stable quadruplex than the unmodified sequence, whereas $TGG^{Br}GT$ is much less stable than the natural counterpart.[205]

In order to fully understand the contribution of these modified bases, one also has to determine their effects on k_{on}. In order to obtain quantitative information, one can perform isothermal association experiments on these modified quadruplexes.[156] We are currently analysing the kinetics of simple variants of the TG_4T and TG_5T quadruplexes in which a single guanine has been replaced by a different base. We have demonstrated huge differences (up to 10^5-fold) in the association constants of these quadruplexes depending on their primary sequence [Table 1(c)]. For example, a single guanine leads to an approximately 1000-fold slower association in K^+, whereas 8-oxo guanine and inosine have a more limited effect. Most modifications indeed lead to a strong decrease in the association rate. This indicates that the sole observation of the variation of the apparent T_m underestimates the penalty imposed by a modified quartet: in general, the incorporation of a single modified quartet not only leads to a decrease in thermal stability, but also to a decrease in the association rate (Gros *et al.*, in preparation). Stefl *et al.*[206] performed MD simulations of DNA quadruplex molecules containing modified bases. The incorporation of 6-thioguanine and 6-thiopurine sharply destabilizes four-stranded G-DNA structures, whereas inosine has a limited effect. In summary, non-classical quartets are rarely favourable to the energetics of the quadruplexes: they are tolerated at best!

2.11   Effect of Quadruplex Ligands

Ligands that selectively bind to G-quadruplex structures may modulate telomerase activity, alter telomere structure or repress the transcription of key oncogenes (for a review, see this book's chapters by M. Searle and J.F. Riou). The way ligands interact with quadruplexes also give clues on the dynamics of the G-quartets: with the possible exception of a porphyrin derivative,[207] true intercalation (between two quartets) is considered unfavourable or impossible. This indicates that, contrary to DNA base pairs, transient unstacking/opening of two G-quartets is an extremely rare or short event, incompatible with the incorporation of a planar chromophore. End stacking, which does not require the separation of two quartets nor the release of cation is therefore the most frequently observed mode of interaction.[161,208]

An important point concerning small quadruplex ligands would be to demonstrate that these molecules may not only bind to preformed quadruplexes, but also induce the formation of the multistranded structures starting from single strands. Kinetic analyses are well suited to answer to this question, as these modes of interactions should be reflected by differences in dissociation and association rates of the quadruplex, respectively.

2.11.1   Small Ligands and Bimolecular Quadruplexes

A perylene derivative called PIPER was the first example of a small ligand behaving as a driver in the assembly of quadruplex structures.[144] Gel-shift experiments demonstrate that PIPER can dramatically accelerate the association of a DNA oligomer containing two tandem repeats of the human telomeric sequence ($TTAG_3$) into di- and tetrameric G-quadruplexes. In doing so, PIPER alters the oligomer dimerization kinetics from second to first order and accelerates the assembly of varied dimeric G-quadruplexes by an estimated 100-fold. Note, however, that the proposed pathway for bimolecular quadruplex formation in the presence of ligands is unlikely to be applicable for the DNA-only situation, as it seems to involve tetramolecular intermediates, which form extremely slowly at the strand concentration used.[144] Ethidium derivatives,[209] as well as a number of different molecules[210] also behave as drivers for the assembly of nucleic acids secondary structures. Among G-quadruplex ligands, telomestatin is extremely potent, as it selectively facilitates the formation of the human telomeric G-quadruplex, in the nM concentration range.[211]

2.11.2 Small Ligands and Intramolecular Quadruplexes

Many ligands interact with intramolecular quadruplexes, and several laboratories studied in detail their interaction with the human telomeric quadruplex (for a review see ref. 210). Insights into the impact of such ligands on quadruplex kinetics is scarce. It is not yet clear whether these ligands can accelerate the folding of intramolecular quadruplexes, but some of them hamper duplex formation between a quadruplex-forming oligonucleotide and its complementary strand, probably as a result of a slower quadruplex unfolding process.[210] Several ligands are also able to induce a conformational change of the quadruplex. For example, a tetracationic porphyrin and a core-modified expanded porphyrin analogue called Se2SAP may convert the parallel *c-myc* G-quadruplex into a different G-quadruplex structure.[124,212] Se2SAP is also able to convert the preformed basket human telomeric G-quadruplex to a hybrid quadruplex structure, that has strong parallel and antiparallel characteristics.[92]

2.11.3 Small Ligands and Tetramolecular Quadruplexes

In the case of tetramolecular complexes, the uncoupling between the association and dissociation processes allows easy determination of k_{on} and k_{off}. We are currently taking advantage of this property to study the interaction of small ligands or proteins with quadruplexes: are these molecules able to accelerate the kinetics of association of monomers giving the tetrameric form, or do they only bind to the structured quadruplex, lowering its dissociation rate constant? In the example provided in Figure 6(a), 360A, a potent quadruplex ligand which preferentially binds to chromosomal extremities in living cells,[213,214] is able to promote the formation of a TG_3T tetramolecular quadruplex. Isothermal absorbance measurements confirm this accelerated quadruplex formation [Figure 6(b)] (De Cian *et al.*, in preparation). However, both methods require relatively high strand and ligand concentrations (10 μM or more). Other techniques will be required to confirm these findings in the submicromlar range.

2.11.4 Effects of Proteins

Proteins may also modulate the formation of G-quadruplexes. The beta-subunit of the *Oxytricha* telomere-binding protein acts as a molecular chaperone to

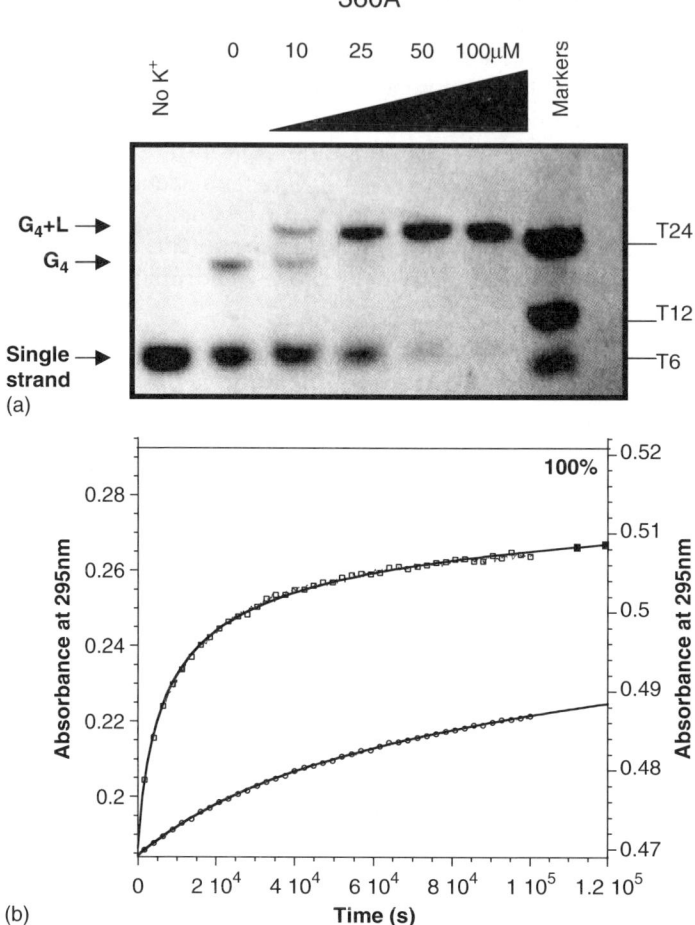

Figure 6 *A small G-quadruplex ligand accelerates quadruplex formation. (a) Gel electrophoresis experiment. 100 µM of TG₃T was incubated at 4°C for 4.5 h in a pH 7.2 10 mM lithium cacodylate buffer supplemented with 110 mM KCl, in the presence of various concentrations (0–100 µM) of a specific quadruplex ligand, 360A. Samples were then loaded on a 20% acrylamide TBE 1X gel supplemented with 20 mM KCl. Nucleic acids were revealed by UV-shadow using a 254 nm UV lamp. A retarded band reveals the formation of a quadruplex; the presence of the quadruplex ligand induces a further retardation. (b) Isothermal (4°C) absorbance measurements at 295 nm of 31 µM TG₄T in the absence (circle, left Y-scale) or presence (squares, right Y-square) of 10 µM 360A in a pH 7.2 10 mM lithium cacodylate buffer supplemented with 110 mM NaCl*

promote the formation of dimers and specific higher order complexes of telomeric DNA stabilized by G-quartets. These reactions occur under physiological conditions *in vitro*,[215,216] and cooperative action of the alpha and beta subunits has recently been demonstrated *in vivo*.[14] The HOPI gene product, which plays a crucial role in the formation of synaptonemal complex in *Saccharomyces cerevisiae*, binds robustly to G4 DNA and facilitates its formation.[217] Topoisomerase I is also able to promote the formation of four-stranded intermolecular DNA structures when added to single-stranded DNA containing a stretch of at least five guanines.[126]

In a reciprocal fashion, some proteins may disrupt quadruplexes, such as a resolvase from placental tissue,[218] SV40 large T-antigen[219] or helicases from the RecQ family.[220–224] These helicases are proposed to function in dissociating alternative DNA structures during recombination and/or replication at telomeric ends. Less clear is the role of proteins such as POT1 or UP1, which favour the unfolded *vs.* the folded state of an intramolecular quadruplex:[225,226] do these proteins simply displace the equilibrium towards the single-strand by binding preferentially to it, or are they able to actively disrupt the quadruplex structure? A comparison of the kinetics of protein binding with that of quadruplex unfolding should answer this question.

2.12 Implications for the Folding Pathway

The thermodynamic, kinetic and structural data presented above give some indications about the folding pathway required to form G-quadruplexes. In terms of folding intermediates, the best-studied systems concern tetramolecular quadruplexes. Several experimental and theoretical studies provide clues on the possible ways to a fully formed quadruplex.

2.12.1 Tetramolecular Quadruplexes

We[27] and others[31,130] have demonstrated that association of short strands containing a single guanine-repeat seems to obey a fourth-order kinetics model. Third or fourth-order reactions are not common in biochemistry, and the practical consequences of this reaction order are important. A fourth-order reaction does not imply that an elementary kinetic step involves a four-body collision. Such mechanism is extremely unlikely and other processes could lead to this fourth order.[130] The structure of these elusive intermediates remains unknown: Stefl *et al.*[151] have recently demonstrated that a Hoogsteen G-G duplex is an improbable intermediate. Its identification will be experimentally difficult, as numerical simulations indicate that it may not be present at detectable levels.[130]

Concerning the activation energy of association, Pörschke and Eigen[227] obtained negative "apparent" activation energies for duplex formation and interpreted their results within the now so-called nucleation-zipping model. Our data are consistent with the observed fourth-order dependence of the association rate and with the negative energy of activation for association, which implies a rapid pre-equilibrium step.[130] The negative values we have obtained for E_{on} are in favour of the nucleation-zipping model. Our finding that k_{on} strongly decreases upon reducing ionic strength whereas k_{off} is independent of NaCl concentration is also in good agreement with what is observed in the duplex case. Other nucleic acid structures such as duplexes,[227,228] triplexes[145] and i-DNA[146] also exhibit a negative E_{on}. The observation that ionic strength and nature of the monocation plays an important role in the value of k_{on} indicates that several (possibly three) ions are involved in the rate-limiting step and participate in the early stages of the quadruplex stem assembly:[151] k_{on} strongly decreases upon reducing the NaCl concentration but k_{off} is independent of ionic strength. Ions must be present at an early stage of quadruplex formation and cannot be considered as late-arrival guests once the 4-strands are gathered!

In most (but not all cases, see TG_n and G_nT), there is no clear evidence for complicated folding pathways proceeding through stable intermediate states. Nevertheless, the order of the reaction n, when left unknown in equation (1) is often experimentally found between 3.4 and 4 (*i.e.* lower than 4; data not shown); this could be the signature of small amounts of stable intermediates. Folding of chimera containing a few PNA residues reveals intermediate species, as shown by single value deconvolution (SVD) analysis.[31] However, in most cases, the concentration of these intermediates is probably very low (which is fortunate for the validity of the two state model assumed for van't Hoff analysis of melting curves). This assumption has been experimentally tested in a limited number of cases. Using SVD analysis of CD spectra, Petraccone *et al.* found that only two major species (*i.e.*, single-strand and quadruplex) are present during the thermal denaturation and the isothermal association of TG_4T strands.

One way to identify these elusive intermediates is mass spectrometry. Electrospray allows the isolation of species of various molecular weights, (Figure 7) (Rosu *et al.*, in preparation). These dimeric and trimeric species might be intermediates leading to the formation of the tetrameric structure. However, we cannot exclude yet that they represent "dead ends" in the pathway towards quadruplexes: their progressive disappearance at long incubation times do not unambiguously demonstrate they are true intermediates.

2.12.2 Are G-G Hairpins Plausible Intermediates for Bimolecular Quadruplexes?

More complicated is the case of tetramolecular structures formed by longer sequences such as $T_4G_4T_4G_4$. In this case, the presence of two guanine runs allows the formation of a bimolecular complex. As soon as an oligonucleotide is long enough to fold back on itself, one can imagine that the formation of intramolecular G-G base pairs may act as an intermediate for quartet formation. Raman spectroscopy suggests that the $T_4G_4T_4G_4$ sequence adopts a highly compact conformation, consistent with a hairpin secondary structure[229] with G.G basepairs of the Hoogsteen type. This telomeric hairpin secondary structure – which remains controversial – could be used as the recognition motif for quadruplex formation.[230]

2.12.3 Molecular Modelling of Intermediate States

Using MD simulations with explicit inclusion of counterions and solvent Stefl *et al.* investigated the stability of supramolecular intermediates (including

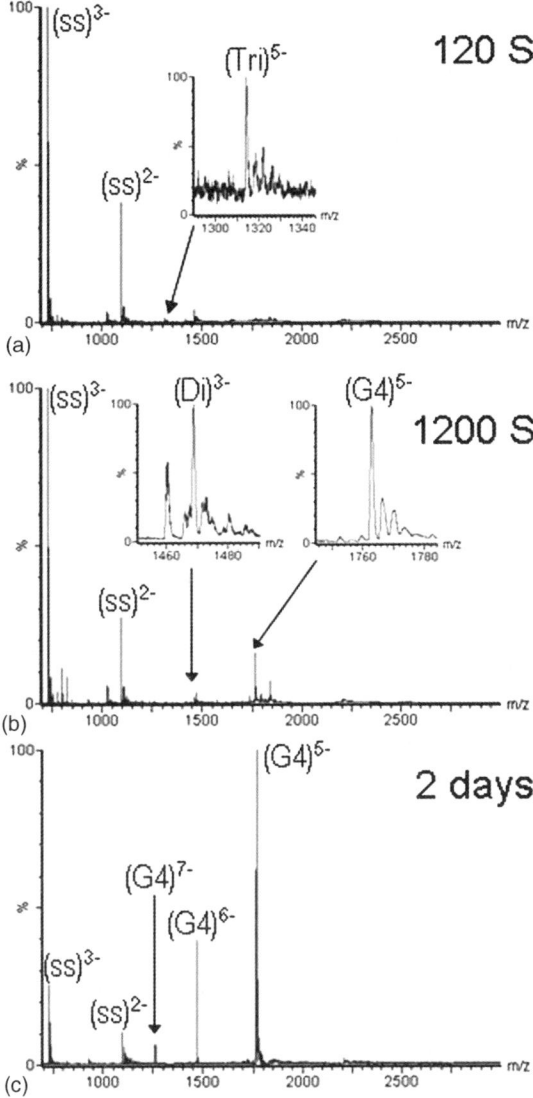

Figure 7 *ESI-MS studies of tetramolecular quadruplexes. The TG$_5$T oligonucleotide was incubated in 0.15 M pH 7.0 ammonium acetate buffer for various amounts of time, then analysed by electrospray mass spectrometry (negative mode). From the initial peak corresponding to the TG$_5$T single-strand (SS), one can see the gradual conversion to a quadruplex(G4), as well as the presence of dimeric (Di) and trimeric (Tri) species*

two-, three and four-stranded assemblies with out-of-register basepairing between guanines) possibly involved in the formation of a parallel stranded G-DNA stem consisting of four strands of G$_4$.[151] The simulations suggest that "cross-like" two-stranded assemblies may serve as nucleation centres in the initial formation of parallel stranded G-DNA quadruplexes, proceeding through a series of rearrangements involving trapping of cations, association of additional strands and progressive slippage of strands towards the full stem. The approach applied here also demonstrated that some putative intermediates such as G-G duplexes or triplexes are inherently unstable and therefore unlikely to be intermediates in the folding pathway. This makes the putative G.G hairpin presented in the previous paragraph an improbable intermediate.

2.13   Conclusion

Despite major differences between the various classes of quadruplexes, some general conclusions may be drawn. For these oligomers, it is clear that:

(i) Most quadruplexes are stable under physiological conditions.

(ii) These structures are strongly enthalpy-driven: as a result, the equilibrium constant is strongly temperature-dependent. The favourable enthalpy is always compensated by negative entropy of formation.

(iii) Except for intramolecular quadruplexes involving only two quartets and tetramolecular quadruplexes with three quartets, the lifetime of the quadruplexes is relatively long in the 0–40°C temperature range. k_{off} values at 37°C range between 10^{-2} and 10^{-6} s^{-1}.

(iv) All quadruplexes have strongly positive activation energy for dissociation E_{off} (*i.e.*, k_{off} increases with temperature, whereas the lifetime of the complexes decreases with temperature).

(v) Quadruplexes probably have negative activation energy for association E_{on}. Contrary to the previous point, this is a matter of debate, as others found quasi-nul or even positive values for this parameter.

(vi) At a given temperature, k_{off} is smaller and k_{on} is greater in potassium compared to sodium (both effects lead to a higher equilibrium affinity constant in K^+).

(vii) A decrease in salt concentration has little effect on G4 dissociation but strongly decreases the rate constant of association hence, lowering the association equilibrium constant.

We have summarized various time constants collected during this study (Figure 8) and the impact of a number of parameters on the kinetics of association and dissociation of these complexes (Table 2). This might be helpful for a variety of applications. Finally, one should emphasize the extraordinary stability of RNA quadruplexes, coupled with their relatively fast kinetics of association.

As a final word, although our understanding of the thermodynamic, kinetic and structural properties of quadruplexes has improved, considerable work lies ahead to fully understand the rules governing their formation and to identify the intermediates state.

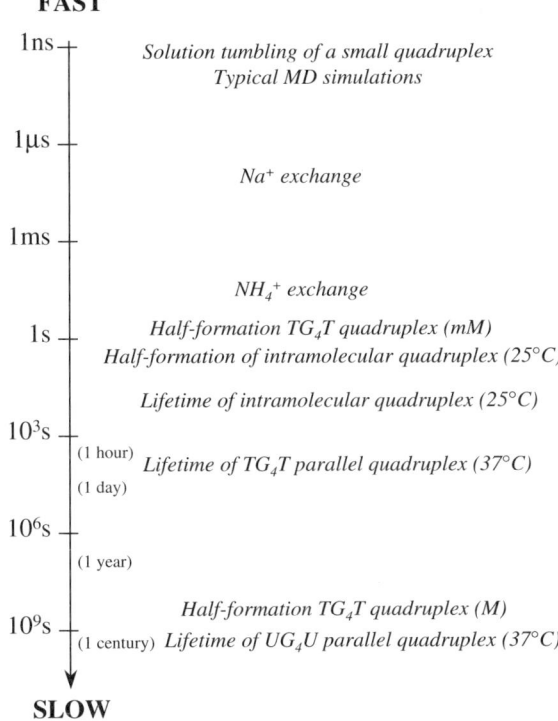

FAST

1ns — Solution tumbling of a small quadruplex
Typical MD simulations

1μs —

Na+ exchange

1ms —

NH₄+ exchange

1s — Half-formation TG₄T quadruplex (mM)
Half-formation of intramolecular quadruplex (25°C)

Lifetime of intramolecular quadruplex (25°C)

10³s —
(1 hour) Lifetime of TG₄T parallel quadruplex (37°C)
(1 day)

10⁶s —
(1 year)

Half-formation TG₄T quadruplex (M)
10⁹s —
(1 century) Lifetime of UG₄U parallel quadruplex (37°C)

SLOW

Figure 8   *Timescales for quadruplexes. We summarize in this graph (that spans 18 orders of magnitudes) the different time constants found for various processes involving quadruplexes (ion exchange (μs to ms); quadruplex lifetimes). The values provided here are only examples; actual durations may vary significantly, depending on experimental conditions or sequence. For example, depending on sequence and 5' or 3' modification, Merkina & Fox calculated half-lives for tetramolecular quadruplexes that varies between ≈ 10 min and ≈ 1 month![128]*

Table 2   *Summary of the effects of various parameters on G-quadruplexes*

Parameter	Association k_{on}	Dissociation k_{off}	Equilibrium $K_a = k_{on}/k_{off}$
Increased temperature	−	++	−−
Increased ionic strength	+	0	+
Mg²⁺ addition	+	+	≈0
Spermine/Spermidine add.	+	+	≈0
Na⁺ → K⁺	++	−−	++
DNA → RNA	++	−−	++
DNA → PS[a]	−	+	−
Longer G-stretch	++	−−	++
Longer non-G overhang	−	+/−[b]	Varies[b]

Note: 0 indicates the no effect; +/++ increase or strong increase (> 30-fold), respectively of the association or dissociation rate constant or of the equilibrium constant; −/−− decrease (or strong decrease) of the parameter.
[a] From Wyatt *et al.* and B. Saccà *et al.*
[b] Generally, dissociation is slower, but sequence-dependent effects may be observed, depending on length, base composition and side (5' or 3') of the extra thymines.

Acknowledgments

We thank M. Rougée, A. Ghelab, P. Arimondo, T. Garestier (MNHN, Paris, France) J.F. Riou (Université de Reims, France), P. Mailliet and E. Mandine (Sanofi-Aventis, Vitry, France) for helpful discussions and M. Hoarau and D. Labit for skilled technical assistance. S.A. is the recipient of a "Fondation Jérôme Lejeune" doctoral fellowship. This work was supported by ARC (# 3365 to J.L.M). and E.U. FP6 "MolCancerMed" (LSHC-CT-2004-502943) grants.

References

1. R.K. Ralph, W.J. Connors and H.G. Khorana, *J. Am. Chem. Soc.*, 1962, **84**, 2265–2266.
2. M. Gellert, M.N. Lipsett and D.R. Davies, *Proc. Natl. Acad. Sci. USA*, 1962, **48**, 2013–2018.
3. J.R. Fresco and J. Massoulié, *J. Am. Chem. Soc.*, 1963, **85**, 1352–1353.
4. W. Guschlbauer, J.F. Chantot and D. Thiele, *J. Biomol. Struct. Dyn.*, 1990, **8**, 491–511.
5. J.R. Williamson, *Annu. Rev. Biophys. Biomol. Struc.*, 1994, **23**, 703–730.
6. E. Henderson, C.C. Hardin, S.K. Walk, I. Tinoco Jr. and E.H. Blackburn, *Cell*, 1987, **51**, 899–908.
7. Y. Oka and C.A. Thomas Jr., *Nucleic Acids Res.*, 1987, **15**, 8877–8898.
8. W.I. Sundquist and A. Klug, *Nature*, 1989, **342**, 825–829.
9. J.R. Williamson, M.K. Raghuraman and T.R. Cech, *Cell*, 1989, **59**, 871–880.
10. A.M. Zahler, J.R. Williamson, T.R. Cech and D.M. Prescott, *Nature*, 1991, **350**, 718–720.
11. A. Siddiqui-Jain, C.L. Grand, D.J. Bearss and L.H. Hurley, *Proc. Natl. Acad. Sci. USA*, 2002, **99**, 11593–11598.
12. C. Schaffitzel, I. Berger, J. Postberg, J. Hanes, H.J. Lipps and A. Plückthun, *Proc. Natl. Acad. Sci. USA*, 2001, **98**, 8572–8577.
13. M.L. Duquette, P. Handa, J.A. Vincent, A.F. Taylor and N. Maizels, *Genes Dev.*, 2004, **18**, 1618–1629.
14. K. Paeschke, T. Simonsson, J. Postberg, D. Rhodes and H. Lipps, *Nat. Struct. Mol. Biol.*, 2005, **12**, 847–854.
15. J.T. Davis, *Angew. Chem. Int. Ed.*, 2004, **43**, 668–698.
16. E.A. Venczel and D. Sen, *J. Mol. Biol.*, 1996, **257**, 219–224.
17. R.P. Fahlman and D. Sen, *J. Mol. Biol.*, 1998, **280**, 237–244.
18. T.C. Marsh and E. Henderson, *Biochemistry*, 1994, **33**, 10718–10724.
19. K. Poon and R.B. Macgregor, *Biophys. Chem.*, 2000, **84**, 205–216.
20. J.L. Mergny, A.T. Phan and L. Lacroix, *FEBS Lett.*, 1998, **435**, 74–78.
21. B. Saccà, L. Lacroix and J.L. Mergny, *Nucleic Acids Res.*, 2005, **33**, 1182–1192.
22. M. Vorlickova, J. Chladkova, I. Kejnovska, M. Fialova and J. Kypr, *Nucleic Acids Res.*, 2005, **33**, 5851–5860.
23. W. Li, P. Wu, T. Ohmichi and N. Sugimoto, *FEBS Lett.*, 2002, **526**, 77–81.

24. M. Lu, Q. Guo and N.R. Kallenbach, *Biochemistry*, 1992, **31**, 2455–2459.
25. R. Jin, K.J. Breslauer, R.A. Jones and B.L. Gaffney, *Science*, 1990, **250**, 543–546.
26. M. Lu, Q. Guo and N.R. Kallenbach, *Biochemistry*, 1993, **32**, 598–601.
27. J.L. Mergny, A. de Cian, A. Ghelab, B. Saccà and L. Lacroix, *Nucleic Acids Res.*, 2005, **33**, 81–94.
28. L. Petraccone, E. Erra, I. Duro, V. Esposito, A. Randazzo, L. Mayol, C.A. Matia, G. Barone and C. Giancola, *Nucleos. Nucleot. Nucleic Acids*, 2005, **24**, 757–760.
29. R.Z. Jin, B.L. Gaffney, C. Wang, R.A. Jones and K.J. Breslauer, *Proc. Natl. Acad. Sci. USA*, 1992, **89**, 8832–8836.
30. D.S. Pilch, G.E. Plum and K.J. Breslauer, *Curr. Opin. Struct. Biol.*, 1995, **5**, 334–342.
31. L. Petraccone, B. Pagano, V. Esposito, A. Randazzo, G. Piccialli, G. Barone, C.A. Mattia and C. Giancola, *J. Am. Chem. Soc.*, 2005, **127**, 16215–16223.
32. S. Haider, G.N. Parkinson and S. Neidle, *J. Mol. Biol.*, 2002, **320**, 189–200.
33. K. Phillips, Z. Dauter, A.I.H. Murchie, D.M.J. Lilley and B. Luisi, *J. Mol. Biol.*, 1997, **273**, 171–182.
34. D. Miyoshi, A. Nakao and N. Sugimoto, *Biochemistry*, 2002, **41**, 15017–15024.
35. D. Miyoshi, A. Nakao and N. Sugimoto, *Nucleos. Nucleot. Nucleic Acids*, 2003, **22**, 1591–1594.
36. Y. Wang and D.J. Patel, *Structure*, 1993, **1**, 263–282.
37. Y. Wang and D.J. Patel, *J. Mol. Biol.*, 1993, **234**, 1171–1183.
38. F. Aboul-ela, A.I.H. Murchie, D.G. Norman and D.M.J. Lilley, *J. Mol. Biol.*, 1994, **243**, 458–471.
39. G.D. Strahan, M.A. Keniry and R.H. Shafer, *Biophys. J.*, 1998, **75**, 968–981.
40. N. Spackova, I. Berger and J. Sponer, *J. Am. Chem. Soc.*, 1999, **121**, 5519–5534.
41. P. Schultze, F.W. Smith and J. Feigon, *Structure*, 1994, **2**, 221–233.
42. M.A. Keniry, G.D. Strahan, E.A. Owen and R.H. Shafer, *Eur. J. Biochem.*, 1995, **233**, 631–643.
43. M.A. Keniry, E.A. Owen and R.H. Shafer, *Nucleic Acids Res.*, 1997, **25**, 4389–4392.
44. A. Risitano and K.R. Fox, *Biochemistry*, 2003, **42**, 6507–6513.
45. A. Risitano and K.R. Fox, *Org. Biomol. Chem.*, 2003, **1**, 1852–1855.
46. P. Hazel, J. Huppert, S. Balasubramanian and S. Neidle, *J. Am. Chem. Soc.*, 2004, **126**, 16405–16415.
47. A. Risitano and K.R. Fox, *Nucleic Acids Res.*, 2004, **32**, 2598–2606.
48. A. Rich, *Gene*, 1993, **135**, 99–109.
49. M. Mills, L. Lacroix, P. Arimondo, J.L. Leroy, J.C. François, H.H. Klump and J.L. Mergny, *Curr. Med. Chem. Anti-Cancer Agents*, 2002, **2**, 627–644.
50. A.T. Phan and J.L. Mergny, *Nucleic Acids Res.*, 2002, **30**, 4618–4625.

51. L. Petraccone, E. Erra, V. Esposito, A. Randazzo, L. Mayol, L. Nasti, G. Barone and C. Giancola, *Biochemistry*, 2004, **43**, 4877–4884.
52. J.L. Mergny and L. Lacroix, *Oligonucleotides*, 2003, **13**, 515–537.
53. E.A. Venczel and D. Sen, *Biochemistry*, 1993, **32**, 6220–6228.
54. D. Sen and W. Gilbert, *Nature*, 1990, **344**, 410–414.
55. D. Miyoshi, H. Karimata and N. Sugimoto, *Angew. Chem. Int. Ed.*, 2005, **44**, 3740–3744.
56. J. Li, J.J. Correia, L. Wang, J.O. Trent and J.B. Chaires, *Nucleic Acids Res.*, 2005, **33**, 4649–4659.
57. G. Oliviero, N. Borbone, A. Galeone, M. Varra, G. Piccialli and L. Mayol, *Tetrahedron Lett.*, 2004, **45**, 4869–4872.
58. G. Oliviero, J. Amato, N. Borbone, A. Galeone, M. Varra, G. Piccialli and L. Mayol, *Nucleos. Nucleot. Nucleic Acids*, 2005, **24**, 739–741.
59. C.C. Hardin, T. Watson, M. Corregan and C. Bailey, *Biochemistry*, 1992, **31**, 833–841.
60. T. Miura and G.J. Thomas, *Biochemistry*, 1994, **33**, 7848–7856.
61. B. Datta and B.A. Armitage, *J. Am. Chem. Soc.*, 2001, **123**, 9612–9619.
62. H. Deng and W.H. Braunlin, *Biopolymers*, 1995, **35**, 677–681.
63. M. Salazar, B.D. Thompson, S.M. Kerwin and L.H. Hurley, *Biochemistry*, 1996, **35**, 16110–16115.
64. P. Alberti and J.L. Mergny, *Proc. Natl. Acad. Sci. USA*, 2003, **100**, 1569–1573.
65. P. Alberti and J.L. Mergny, *Cell. Mol. Biol.*, 2004, **50**, 241–253.
66. D. Miyoshi, S. Matsumura, S. Nakano and N. Sugimoto, *J. Am. Chem. Soc.*, 2004, **126**, 165–169.
67. A. Rangan, O.Y. Fedoroff and L.H. Hurley, *J. Biol. Chem.*, 2001, **276**, 4640–4646.
68. D. Sun, K. Guo, J.J. Rusche and L.H. Hurley, *Nucleic Acids Res.*, 2005, **33**, 6070–6080.
69. G. Laughlan, A.I.H. Murchie, D.G. Norman, M.H. Moore, P.C.E. Moody, D.M.J. Lilley and B. Luisi, *Science*, 1994, **265**, 520–524.
70. Q. Xu, H. Deng and W.H. Braunlin, *Biochemistry*, 1993, **32**, 13130–13137.
71. A. Wlodarczyk, P. Grzybowski, A. Patkowski and A. Dobek, *J. Phys. Chem. B*, 2005, **109**, 3594–3605.
72. I. Smirnov and R.H. Shafer, *J. Mol. Biol.*, 2000, **296**, 1–5.
73. J.A. Mondragon-Sanchez, J. Liquier, R.H. Shafer and E. Taillandier, *J. Biomol. Struct. Dyn.*, 2004, **22**, 365–373.
74. N. Nagesh and D. Chatterji, *J. Biochem. Biophys. Methods*, 1995, **30**, 1–8.
75. F. Rosu, V. Gabelica, C. Houssier, P. Colson and E. De Pauw, *Rapid Commun. Mass Spectrom.*, 2002, **16**, 1729–1736.
76. F. Rosu, V. Gabelica, K. Shin-ya and E. DePauw, *Chem. Commun.*, 2003, **34**, 2702–2703.
77. A. Wong and G. Wu, *J. Am. Chem. Soc.*, 2003, **125**, 13895–13905.
78. M.A. Keniry, *FEBS Lett.*, 2003, **542**, 153–158.
79. S.W. Blume, V. Guarcello, W. Zacharias and D.M. Miller, *Nucleic Acids Res.*, 1997, **25**, 617–625.

80. C.C. Hardin, A.G. Perry and K. White, *Biopolymers*, 2000, **56**, 147–194.
81. D. Miyoshi, A. Nakao, T. Toda and N. Sugimoto, *FEBS Lett.*, 2001, **496**, 128–133.
82. T. Simonsson, *Biol. Chem.*, 2001, **382**, 621–628.
83. N.V. Hud, F.W. Smith, F.A.L. Anet and J. Feigon, *Biochemistry*, 1996, **35**, 15383–15390.
84. J. Gu and J. Leczczynski, *J. Phys. Chem. A*, 2002, **106**, 529–532.
85. S. Takenaka, H. Ueyama, T. Nojima and M. Takagi, *Anal. Bioanal. Chem.*, 2003, **375**, 1006–1010.
86. N.V. Hud, P. Schultze, V. Sklenar and J. Feigon, *J. Mol. Biol.*, 1999, **285**, 233–243.
87. F. Federiconi, P. Ausili, G. Fragneto, C. Ferrero and P. Mariani, *J. Phys. Chem. B*, 2005, **109**, 11037–11045.
88. M. Borzo, C. Detellier, P. Laszlo and A. Paris, *J. Am. Chem. Soc.*, 1980, **102**, 1124–1134.
89. J. Gu, J. Leczczynski and M. Bansal, *Chem. Phys. Lett.*, 1999, **311**, 209–214.
90. J.L. Sessler, M. Sathiosatham, K. Doerr, V. Lynch and K.A. Abboud, *Angew. Chem. Int. Ed.*, 2000, **39**, 1300–1303.
91. M.Y. Kim, M. Gleason Guzman, E. Izbicka, D. Nishioka and L.H. Hurley, *Cancer Res.*, 2003, **63**, 3247–3256.
92. E.M. Rezler, J. Seenisamy, S. Bashyam, M.Y. Kim, E. White, W.D. Wilson and L.H. Hurley, *J. Am. Chem. Soc.*, 2005, **127**, 9439–9447.
93. E. Baldrich and C.K. OSullivan, *Anal Biochem.*, 2005, **341**, 194–197.
94. H. Deng and W.H. Braunlin, *J. Mol. Biol.*, 1996, **255**, 476–483.
95. A. Delville, C. Detellier and P. Laszlo, *J. Magn. Reson.*, 1979, **34**, 301–315.
96. T. van Mourik and A.J. Dingley, *Chem. Eur. J.*, 2005, **11**, 6064–6079.
97. S. Chowdhury and M. Bansal, *J. Biomol. Struct. Dyn.*, 2001, **18**, 647–669.
98. P. Sket, M. Crnugelj, W. Kozminski and J. Plavec, *Org. Biomol. Chem.*, 2004, **2**, 1970–1973.
99. T. Miura and G.J. Thomas, *Biochemistry*, 1995, **34**, 9645–9654.
100. Y. Wang and D.J. Patel, *Biochemistry*, 1992, **31**, 8112–8119.
101. F.W. Smith and J. Feigon, *Nature*, 1992, **356**, 164–168.
102. G. Gupta, A.E. Garcia, Q. Guo, M. Lu and N.R. Kallenbach, *Biochemistry*, 1993, **32**, 7098–7103.
103. X. Mao and W.H. Gmeiner, *Biophys. Chem.*, 2005, **113**, 155–160.
104. A.J. Dingley, R.D. Peterson, S. Grzesiek and J. Feigon, *J. Am. Chem. Soc.*, 2005, **127**, 14466–14472.
105. V. Gabelica, F. Rosu, M. Witt, G. Baykut and E. De Pauw, *Rapid Com. Mass Spec.*, 2005, **19**, 201–208.
106. C.H. Kang, X. Zhang, R. Ratliff, R. Moyzis and A. Rich, *Nature*, 1992, **356**, 126–131.
107. F.W. Smith and J. Feigon, *Biochemistry*, 1993, **32**, 8682–8692.
108. R.F. Macaya, P. Schultze, F.W. Smith, J.A. Roe and J. Feigon, *Proc. Natl. Acad. Sci. USA*, 1993, **90**, 3745–3749.

109. P. Schultze, R.F. Macaya and J. Feigon, *J. Mol. Biol.*, 1994, **235**, 1532–1547.

110. J.A. Kelly, J. Feigon and T.O. Yeates, *J. Mol. Biol.*, 1996, **256**, 417–422.

111. G.N. Parkinson, M.P.H. Lee and S. Neidle, *Nature*, 2002, **417**, 876–880.

112. S. Neidle and G.N. Parkinson, *Curr. Opin. Struct. Biol.*, 2003, **13**, 275–283.

113. S. Redon, S. Bombard, M.A. Elizondo-Riojas and J.C. Chottard, *Nucleic Acids Res.*, 2003, **31**, 1605–1613.

114. I. Ourliac- Garnier, M.A. Elizondo-Riojas, S. Redon, N.P. Farrell and S. Bombard, *Biochemistry*, 2005, **44**, 10620–10634.

115. A. Risitano and K.R. Fox, *Bioorg. Med. Chem Lett.*, 2005, **15**, 2047–2050.

116. I.N. Rujan, J.C. Meleney and P.H. Bolton, *Nucleic Acids Res.*, 2005, **33**, 2022–2031.

117. J.Y. Qi and R.H. Shafer, *Nucleic Acids Res.*, 2005, **33**, 3185–3192.

118. A.T. Phan and D.J. Patel, *J. Am. Chem. Soc.*, 2003, **125**, 15021–15027.

119. J. Ren, X. Qu, J.O. Trent and J.B. Chaires, *Nucleic Acids Res.*, 2002, **30**, 2307–2315.

120. A.T. Phan, Y.S. Modi and D.J. Patel, *J. Mol. Biol.*, 2004, **338**, 93–102.

121. P. Balagurumoorthy, S.K. Brahmachari, D. Mohanty, M. Bansal and V. Sasisekharan, *Nucleic Acids Res.*, 1992, **20**, 4061–4067.

122. F.M. Chen, *Biochemistry*, 1992, **31**, 3769–3776.

123. T. Miura, J.M. Benevides and G.J. Thomas, *J. Mol. Biol.*, 1995, **248**, 233–238.

124. J. Seenisamy, E.M. Rezler, T.J. Powell, D. Tye, V. Gokhale, C.S. Joshi, A. Siddiqui- Jain and L.H. Hurley, *J. Am. Chem. Soc.*, 2004, **126**, 8702–8709.

125. T. Shida, K. Yokoyama, S. Tamai and J. Sekiguchi, *Chem. Pharm. Bull. (Tokyo)*, 1991, **39**, 2207–2211.

126. P. Arimondo, J.F. Riou, J.L. Mergny, J. Tazi, J.S. Sun, T. Garestier and C. Hélène, *Nucleic Acids Res.*, 2000, **28**, 4832–4838.

127. S. Lyonnais, C. Hounsou, M.P. Teulade-F ichou, J. Jeusset, E. LeCam and G. Mirambeau, *Nucleic Acids Res.*, 2002, **30**, 5276–5283.

128. E.E. Merkina and K.R. Fox, *Biophys. J.*, 2005, **89**, 365–373.

129. C.C. Hardin, M. Corregan, B.A. Brown and L.N. Frederick, *Biochemistry*, 1993, **32**, 5870–5880.

130. J.R. Wyatt, P.W. Davis and S.M. Freier, *Biochemistry*, 1996, **35**, 8002–8008.

131. R. Darby, M. Sollogoub, C. McKeen, L. Brown, A. Risitano, N. Brown, C. Barton, T. Brown and K. Fox, *Nucleic Acids Res.*, 2002, **30**, e39.

132. Y. Zhao, Z.Y. Kan, Z.X. Zeng, Y.H. Hao, H. Chen and Z. Tan, *J. Am. Chem. Soc.*, 2004, **126**, 13255–13264.

133. K. Halder and S. Chowdhury, *Nucleic Acids Res.*, 2005, **33**, 4466–4474.

134. K. Halder, V. Mathur, D. Chugh, A. Verma and S. Chowdhury, *Biochem. Biophys. Res. Commun.*, 2005, **327**, 49–56.

135. A. Ambrus, D. Chen, J.X. Dai, R.A. Jones and D.Z. Yang, *Biochemistry*, 2005, **44**, 2048–2058.

136. M.K. Raghuraman and T.R. Cech, *Nucleic Acids Res.*, 1990, **18**, 4543–4552.
137. W. Li, D. Miyoshi, S. Nakano and N. Sugimoto, *Biochemistry*, 2003, **42**, 11736–11744.
138. J.J. Green, L.M. Ying, D. Klenerman and S. Balasubramanian, *J. Am. Chem. Soc.*, 2003, **125**, 3763–3767.
139. L.M. Ying, J.J. Green, H.T. Li, D. Klenerman and S. Balasubramanian, *Proc. Natl. Acad. Sci. USA*, 2003, **100**, 14629–14634.
140. A.T. Phan, Y.S. Modi and D.J. Patel, *J. Am. Chem. Soc.*, 2004, **126**, 8710–8716.
141. Q. Guo, M. Lu and N.R. Kallenbach, *Biochemistry*, 1993, **32**, 3596–3603.
142. F.W. Smith, F.W. Lau and J. Feigon, *Proc. Natl. Acad. Sci. USA*, 1994, **91**, 10546–10550.
143. P. Sket, M. Crnugelj and J. Plavec, *Bioorgan. Med. Chem.*, 2004, **12**, 5735–5744.
144. H.Y. Han, C.L. Cliff and L.H. Hurley, *Biochemistry*, 1999, **38**, 6981–6986.
145. M. Rougée, B. Faucon, J.L. Mergny, F. Barcelo, C. Giovannangeli, T. Garestier and C. Hélène, *Biochemistry*, 1992, **31**, 9269–9278.
146. J.L. Mergny and L. Lacroix, *Nucleic Acids Res.*, 1998, **26**, 4797–4803.
147. V. Dapic, V. Abdomerovic, R. Marrington, J. Peberdy, A. Rodger, J.O. Trent and P.J. Bates, *Nucleic Acids Res.*, 2003, **31**, 2097–2107.
148. M. Cevec and J. Plavec, *Biochemistry*, 2005, **44**, 15238–15246.
149. F. Aboul-ela, A.I.H. Murchie and D.M.J. Lilley, *Nature*, 1992, **360**, 280–282.
150. J.P. Deng, Y. Xiong and M. Sundaralingam, *Proc. Natl. Acad. Sci. USA*, 2001, **98**, 13665–13670.
151. R. Stefl, T.E. Cheatham, N. Spackova, E. Fadrna, I. Berger, J. Koca and J. Sponer, *Biophys. J.*, 2003, **85**, 1787–1804.
152. J.L. Mergny, J. Li, L. Lacroix, S. Amrane and J.B. Chaires, *Nucleic Acids Res.*, 2005, **33**, e138.
153. C. Caceres, G. Wright, C. Gouyette, G. Parkinson and J.A. Subirana, *Nucleic Acids Res.*, 2004, **32**, 1097–1102.
154. P. Wallimann, R.J. Kennedy, J.S. Miller, W. Shalango and D.S. Kenp, *J. Am. Chem. Soc.*, 2003, **125**, 1203–1220.
155. D.V. Lieberman and C.C. Hardin, *Biochim. Biophysica. Acta*, 2004, **1679**, 59–64.
156. J. Gros, M. Webba da Silva, A. De Cian, S. Amrane, F. Rosu, A. Bourdoncle, B. Saccà, P. Alberti, L. Lacroix and J.L. Mergny, *Nucleic Acids Symp. Series*, 2005, **49**, 61–62.
157. D. Sen and W. Gilbert, *Biochemistry*, 1992, **31**, 65–70.
158. Y. Krishnan-Ghosh, D. Liu and S. Balasubramanian, *J. Am. Chem. Soc.*, 2004, **126**, 11009–11016.
159. Y. Kato, T. Ohyama, H. Mita and Y. Yamamoto, *J. Am. Chem. Soc.*, 2005, **127**, 9980–9981.
160. M.K. Uddin, Y. Kato, Y. Takagi, T. Mikuma and K. Taira, *Nucleic Acids Res.*, 2004, **32**, 4618–4629.
161. O.Y. Fedoroff, M. Salazar, H. Han, V.V. Chemeris, S.M. Kerwin and L.H. Hurley, *Biochemistry*, 1998, **37**, 12367–12374.

162. A.T. Phan, V. Kuryavyi, J.B. Ma, A. Faure, M.L. Andreola and D.J. Patel, *Proc. Natl. Acad. Sci. USA*, 2005, **102**, 634–639.

163. E. Protozanova and R.B. Macgregor, *Biochemistry*, 1996, **35**, 16638–16645.

164. T.Y. Dai, S.P. Marotta and R.D. Sheardy, *Biochemistry*, 1995, **34**, 3655–3662.

165. M. Biyani and K. Nishigaki, *Gene.*, 2005, **364**, 130–138.

166. E. Protozanova and R.B. Mac Gregor, *Biophys. Chem.*, 2000, **84**, 137–147.

167. E. Protozanova and R.B. Mac Gregor, *Biophys. J.*, 1998, **75**, 982–989.

168. S.P. Marotta, P.A. Tamburri and R.D. Sheardy, *Biochemistry*, 1996, **35**, 10484–10492.

169. D. Miyoshi, A. Nakao and N. Sugimoto, *Nucleic Acids Res.*, 2003, **31**, 1156–1163.

170. K. Poon and R.B. Macgregor, *Biopolymers*, 1998, **45**, 427–434.

171. V. Dapic, P.J. Bates, J.O. Trent, A. Rodger, S.D. Thomas and D.M. Miller, *Biochemistry*, 2002, **41**, 3676–3685.

172. S. Arnott, R. Chandrasekaran and M. Martilla, *Biochem. J.*, 1974, **141**, 537–543.

173. W.I. Sundquist and S. Heaphy, *Proc. Natl. Acad. Sci. USA*, 1993, **90**, 3393–3397.

174. J. Christiansen, M. Kofod and F.C. Nielsen, *Nucleic Acids Res.*, 1994, **22**, 5709–5716.

175. J.C. Darnell, K.B. Jensen, P. Jin, V. Brown, S.T. Warren and R.B. Darnell, *Cell*, 2001, **107**, 489–499.

176. C. Schaeffer, B. Bardoni, J.L. Mandel, B. Ehresmann, C. Ehresmann and H. Moine, *Embo J.*, 2001, **20**, 4803–4813.

177. H. Moine and J.L. Mandel, *Science*, 2001, **294**, 2487–2488.

178. S. Bonnal, C. Schaeffer, L. Creancier, S. Clamens, H. Moine, A.C. Prats and S. Vagner, *J. Biol. Chem.*, 2003, **278**, 39330–39336.

179. D. Gomez, T. Lemarteleur, L. Lacroix, P. Mailliet, J.L. Mergny and J.F. Riou, *Nucleic Acids Res.*, 2004, **32**, 371–379.

180. B.C. Horsburgh, H. Kollmus, H. Hauser and D.M. Coen, *Cell*, 1996, **86**, 949–959.

181. V. Brown, P. Jin, S. Ceman, J.C. Darnell, W.T. O'Donnell, S.A. Tenenbaum, X. Jin, Y. Feng, K.D. Wilkinson, J.D. Keene, R.B. Darnell and S.T. Warren, *Cell*, 2001, **107**, 477–487.

182. C.J. Cheong and P.B. Moore, *Biochemistry*, 1992, **31**, 8406–8414.

183. J. Kim, C. Cheong and P.B. Moore, *Nature*, 1991, **351**, 331–332.

184. M. Meyer and J. Suhnel, *J. Biomol. Struct. Dyn.*, 2003, **20**, 507–517.

185. H. Liu, A. Kugimiya, T. Sakurai, M. Katahira and S. Uesugi, *Nucleos. Nucleot. Nucleic Acids*, 2002, **21**, 785–801.

186. H. Liu, A. Matsugami, M. Katahira and S. Uesugi, *J. Mol. Biol.*, 2002, **322**, 955–970.

187. A. Randazzo, V. Esposito, O. Ohlenschlager, R. Ramachandran and L. Mayol, *Nucleic Acids Res.*, 2004, **32**, 3083–3092.

188. A. Randazzo, V. Esposito, O. Ohlenschlager, R. Ramachandran, A. Virgilio and L. Mayol, *Nucleos. Nucleot. Nucleic Acids*, 2005, **24**, 780–795.
189. P.K. Dominick and M.B. Jarstfer, *J. Am. Chem. Soc.*, 2004, **126**, 5050–5051.
190. B. Datta, C. Schmitt and B.A. Armitage, *J. Am. Chem. Soc.*, 2003, **125**, 4111–4118.
191. Y. Krishnan-Ghosh, E. Stephens and S. Balasubramanian, *J. Am. Chem. Soc.*, 2004, **126**, 5944–5945.
192. B. Datta, M.E. Bier, S. Roy and B.A. Armitage, *J. Am. Chem. Soc.*, 2005, **127**, 4199–4207.
193. V. Esposito, A. Galeone, L. Mayol, A. Messere, G. Piccialli and A. Randazzo, *Nucleos. Nucleot. Nucleic Acids*, 2003, **22**, 1681–1684.
194. N. Zhang, A. Gorin, A. Majumdar, A. Kettani, N. Chernichenko, E. Skripkin and D.J. Patel, *J. Mol. Biol.*, 2001, **312**, 1073–1088.
195. N. Zhang, A. Gorin, A. Majumdar, A. Kettani, N. Chernichenko, E. Skripkin and D.J. Patel, *J. Mol. Biol.*, 2001, **311**, 1063–1079.
196. E. Gavathiotis and M.S. Searle, *Org. Biomol. Chem.*, 2003, **1**, 1650–1656.
197. M. Webba da Silva, *Biochemistry*, 2003, **42**, 14356–14365.
198. M. Webba da Silva, *Biochemistry*, 2005, **44**, 3754–3764.
199. P.K. Patel and R.V. Hosur, *Nucl. Acid Res.*, 1999, **27**, 2457–2464.
200. P.K. Patel, N.S. Bhavesh and R.V. Hosur, *Biochem. Biophys. Res. Commun.*, 2000, **270**, 967–971.
201. B.C. Pan, Y. Xiong, K. Shi, J.P. Deng and M. Sundaralingam, *Structure*, 2003, **11**, 815–823.
202. B.C. Pan, Y. Xiong, K. Shi and M. Sundaralingam, *Structure*, 2003, **11**, 1423–1430.
203. J.A. Berglund, *Structure*, 2003, **11**, 1315–1316.
204. A. Kettani, S. Bouaziz, A. Gorin, H. Zhao, R.A. Jones and D.J. Patel, *J. Mol. Biol.*, 1998, **282**, 619–636.
205. V. Esposito, A. Randazzo, G. Piccialli, L. Petraccone, C. Giancola and L. Mayol, *Org. Biomol. Chem.*, 2004, **2**, 313–318.
206. R. Stefl, N. Spackova, I. Berger, J. Koca and J. Sponer, *Biophys. J.*, 2001, **80**, 455–468.
207. I. Haq, J.O. Trent, B.Z. Chowdhry and T.C. Jenkins, *J. Am. Chem. Soc.*, 1999, **121**, 1768–1779.
208. G.R. Clark, P.D. Pytel, C.J. Squire and S. Neidle, *J. Am. Chem. Soc.*, 2003, **125**, 4066–4067.
209. F. Koeppel, J.F. Riou, A. Laoui, P. Mailliet, P.B. Arimondo, D. Labit, O. Petigenet, C. Hélène and J.L. Mergny, *Nucleic Acids Res.*, 2001, **29**, 1087–1096.
210. P. Alberti, M. Hoarau, L. Guittat, M. Takasugi, P.B. Arimondo, L. Lacroix, M. Mills, M.P. Teulade Fichou, J.P. Vigneron, J.M. Lehn, P. Mailliet and J.L. Mergny, in *Small Molecule DNA and RNA Binders: From Synthesis to Nucleic Acid Complexes*, C. Bailly, M. Demeunynck and D. Wilson (eds), Wiley, Weinheim, 315–336.
211. M.Y. Kim, H. Vankayalapati, K. Shin-ya, K. Wierzba and L.H. Hurley, *J. Am. Chem. Soc.*, 2002, **124**, 2098–2099.

212. J. Seenisamy, S. Bashyam, V. Gokhale, H. Vankayalapati, D. Sun, A. SiddiquiJain, N. Streiner, K. Shinya, E. White, W.D. Wilson and L.H. Hurley, *J. Am. Chem. Soc.*, 2005, **127**, 2944–2959.

213. C. Granotier, G. Pennarun, L. Riou, F. Hoffschir, L.R. Gauthier, A. DeCian, D. Gomez, E. Mandine, J.F. Riou, J.L. Mergny, P. Mailliet, B. Dutrillaux and F.D. Boussin, *Nucleic Acids Res.*, 2005, **33**, 4182–4190.

214. G. Pennarun, C. Granotier, L.R. Gauthier, D. Gomez, F. Hoffschir, E. Mandine, J.F. Riou, J.L. Mergny, P. Mailliet and F.D. Boussin, *Oncogene*, 2005, **24**, 2917–2928.

215. G.W. Fang and T.R. Cech, *Cell*, 1993, **74**, 875–885.

216. G.W. Fang and T.R. Cech, *Biochemistry*, 1993, **32**, 11646–11657.

217. K. Muniyappa, S. Anuradha and B. Byers, *Mol. Cell. Biol.*, 2000, **20**, 1361–1369.

218. C. Harrington, Y. Lan and S.A. Akman, *J. Biol. Chem.*, 1997, **272**, 24631–24636.

219. N. Baran, L. Pucshansky, Y. Marco, S. Benjamin and H. Manor, *Nucleic Acids Res.*, 1997, **25**, 297–303.

220. H. Sun, J.K. Karow, I.D. Hickson and N. Maizels, *J. Biol. Chem.*, 1998, **273**, 27587–27592.

221. H. Sun, R.J. Bennett and N. Maizels, *Nucleic Acids Res.*, 1999, **27**, 1978–1984.

222. P. Mohaghegh, J.K. Karow, R.M. Brosh Jr., V.A. Bohr Jr. and I.D. Hickson, *Nucleic Acids Res.*, 2001, **29**, 2843–2849.

223. X. Wu and N. Maizels, *Nucleic Acids Res.*, 2001, **29**, 1765–1771.

224. M.D. Huber, D.C. Lee and N. Maizels, *Nucleic Acids Res.*, 2002, **30**, 3954–3961.

225. A.J. Zaug, E.R. Podell and T.R. Cech, *Proc. Natl Acad. Sci. USA*, 2005, **102**, 10864–10869.

226. H. Fukuda, M. Katahira, N. Tsuchiya, Y. Enokizono, T. Sugimura, M. Nagao and H. Nakagama, *Proc. Natl. Acad. Sci. USA*, 2002, **99**, 12685–12690.

227. D. Pörschke and M. Eigen, *J. Mol. Biol.*, 1971, **62**, 361–381.

228. M.E. Craig, D.M. Crothers and P. Doty, *J. Mol. Biol.*, 1971, **62**, 383–401.

229. L. Laporte and G.J. Thomas, *J. Mol. Biol.*, 1998, **281**, 261–270.

230. L. Laporte and G.J. Thomas, *Biochemistry*, 1998, **37**, 1327–1335.

CHAPTER 3

Structural Diversity of G-Quadruplex Scaffolds

ANH TUÂN PHAN, VITALY KURYAVYI,
KIM NGOC LUU AND DINSHAW J. PATEL

Structural Biology Program, Memorial Sloan-Kettering Cancer Center, New York NY 10021, USA

3.1 Introduction

Guanine-rich DNA and RNA sequences can adopt a noncanonical structure called the G-quadruplex,[1–3] which is a four-stranded structure built from the stacking of several $G \cdot G \cdot G \cdot G$ tetrads and stabilized by cations, such as K^+ and Na^+ (Chapter 1). Evidence is mounting that this type of nucleic acid structure exists in nature and is biologically relevant[4–14] (see Chapters 6–8). The G-tetrad (or G-quartet) is a cyclic hydrogen-bonded square planar alignment of four guanines [Figure 1(a)]. The G-tetrad was identified[15] in 1962 and structural studies on G-quadruplexes were first reported[16–22] in the early 1990s.

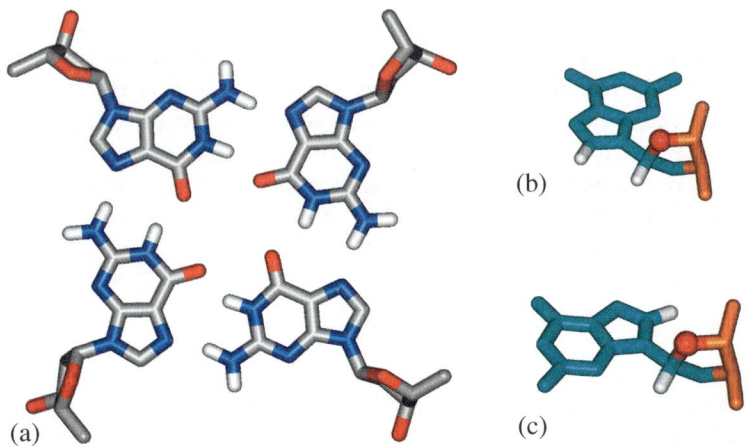

Figure 1 *(a) G-tetrad alignment: (b, c) Guanine in (b) syn and (c) anti glycosidic conformations*

G-quadruplex structures are polymorphic regarding the G-tetrad core and the loops. Recent G-quadruplex structures have been reported to contain a variety of new structural elements, as opposed to "simpler" structures that were reported or proposed a decade ago.[16–22] This Chapter first describes and classifies the core and loop elements in G-quadruplexes, then examines several examples of recently reported structures.

3.2   Core and Loops in G-Quadruplexes

3.2.1   G-Tetrad Core

A G-tetrad core involves the stacking of several tetrads. The backbone strands (or columns) that support a tetrad can be of different orientations.[1–3] There are four possibilities (i) four strands are oriented in the same direction; (ii) three strands are oriented in one direction and the fourth in the opposite direction; (iii) two neighboring strands are oriented in one direction and the two remaining in the opposite direction (as a result, each strand has both parallel and antiparallel adjacent strands); and (iv) two strands across one diagonal are oriented in one direction and the two remaining strands across the other diagonal are oriented in the opposite direction (as a result, each strand has two antiparallel adjacent strands). The glycosidic conformations of guanines within a G-tetrad can be either *syn* [Figure 1(b)] or *anti* [Figure 1(c)]. The relative orientations of strands are geometrically related with the glycosidic conformations of guanines. The four possibilities of strand orientations just described are associated with the following possibilities of G-tetrad alignments, respectively: (i) *anti · anti · anti · anti* or *syn · syn · syn · syn*; (ii) *syn · anti · anti · anti* or *anti · syn · syn · syn*; (iii) *syn · syn · anti · anti*; and (iv) *syn · anti · syn · anti*.

Besides the differences in strand orientations and *syn/anti* conformations of guanines, G-tetrad cores can be polymorphic with respect to the continuity of the strands. In this regard, there are three major classes of G-tetrad core: (i) continuous; (ii) bulged; and (iii) interrupted. The "normal" continuous G-tetrad core was observed in most of the reported G-quadruplex structures. Interrupted G-tetrad cores, lacking one or two connecting backbone strands between adjacent stacked G-tetrads, were observed recently, for example, in interlocked G-quadruplexes.[23,24] A bulged G-tetrad core with flipped-out U residues was observed in the crystal structure of a G-quadruplex formed by a RNA/DNA chimera.[25] Other types of tetrad cores include the intercalation of tetrads[26] and mixed tetrads.[27–30]

3.2.2   Loops

Loops in G-quadruplexes are linkers connecting G-stretches that support the G-tetrad core. The loops can be classified into four major families (i) edgewise loops connecting two adjacent antiparallel strands [Figure 2(a)]; (ii) diagonal loops connecting two opposing antiparallel strands [Figure 2(b)]; (iii) double-chain-reversal loops connecting adjacent parallel strands [Figure 2(c)]; and (iv)

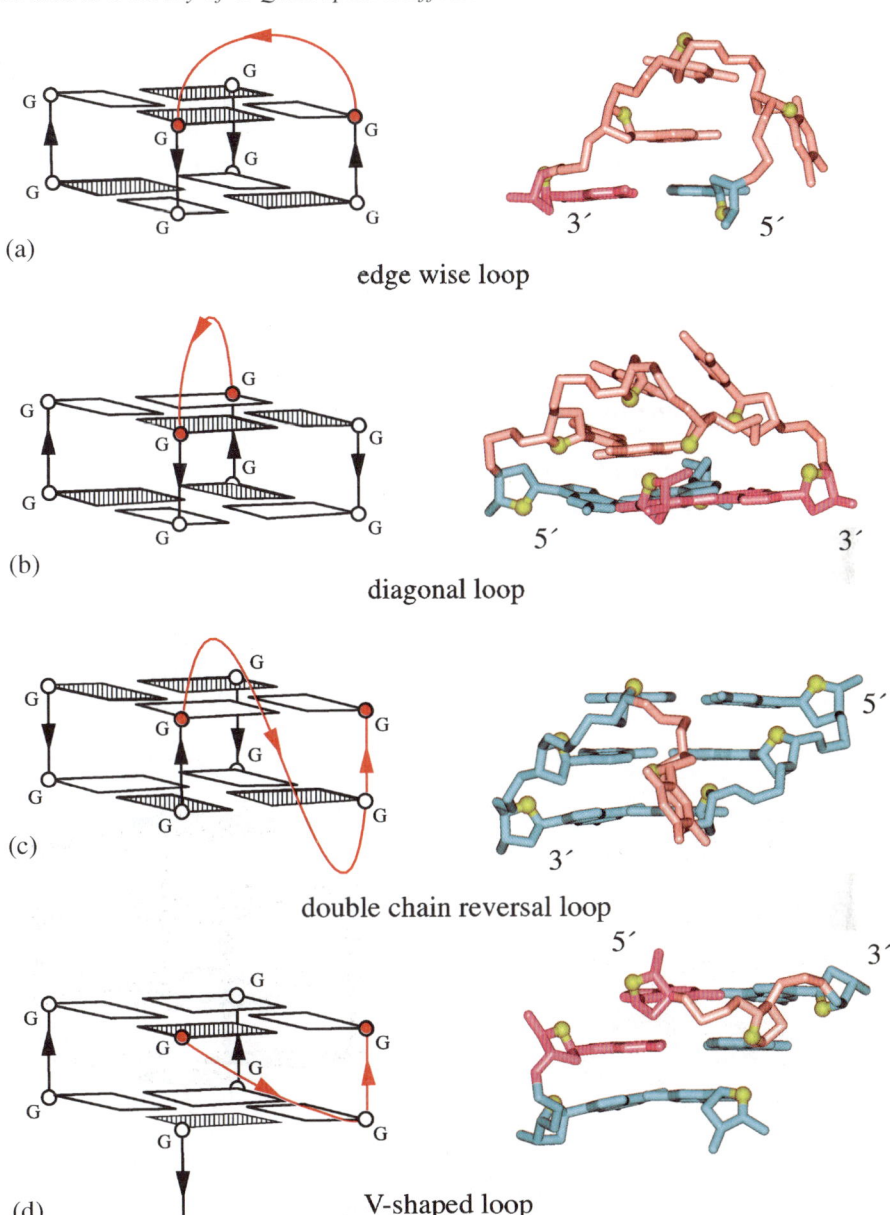

Figure 2 *Schematic and representative structures of different loops in G-quadruplexes: (a) edgewise loop, (b) diagonal loop, (c) double-chain-reversal loop, and (d) V-shaped loop*

V-shaped loops connecting two corners of a G-tetrad core in which one supporting column is lacking [Figure 2(d)]. The types of loops depend strongly on the size and sequence of the linkers. Diagonal loops contain three or more residues.[18,31–35] Besides one example of a quadruplex with mixed tetrads,[30] in which edgewise loops contain one residue, most other edgewise loops were observed to contain at least two residues.[27–29,31,32,36–40] Double-chain-reversal loops were observed to bridge two or three G-tetrad layers and contain from one to six residues;[23,24,34,37,40–46] shorter loops are more stable than longer loops.[44,45] The adenine in single-residue double-chain-reversal loops that bridge two G-tetrad layers can form hydrogen bonds with one edge of the G-tetrad, and as a result, pentad,[23,24] hexad,[41] and heptad[42] base alignments are formed. Residues in the loops can form base pairing alignments that further stabilize the G-quadruplex structures. There are examples of triad-containing edgewise[47] and diagonal loops.[34,46,48]

3.3   G-Quadruplexes with Different Structural Elements

3.3.1   Triad-Containing Edgewise Loops

Triads are hydrogen-bonded planar alignments of three bases, of which two are consecutive in the sequence.[49] The d(GGGTTCAGG) sequence forms a dimeric G-quadruplex with edgewise loops.[47] The particularity of this structure is the formation of a G · (C–A) triad in the GTTCA edgewise loop (Figure 3). In this structure, two G-tetrads are sandwiched between two G · (C–A) triads, while the latter cap the two ends of the G-tetrad core. In the G · (C–A) triad, the guanine is in the central position, forming hydrogen-bonds with the cytosine

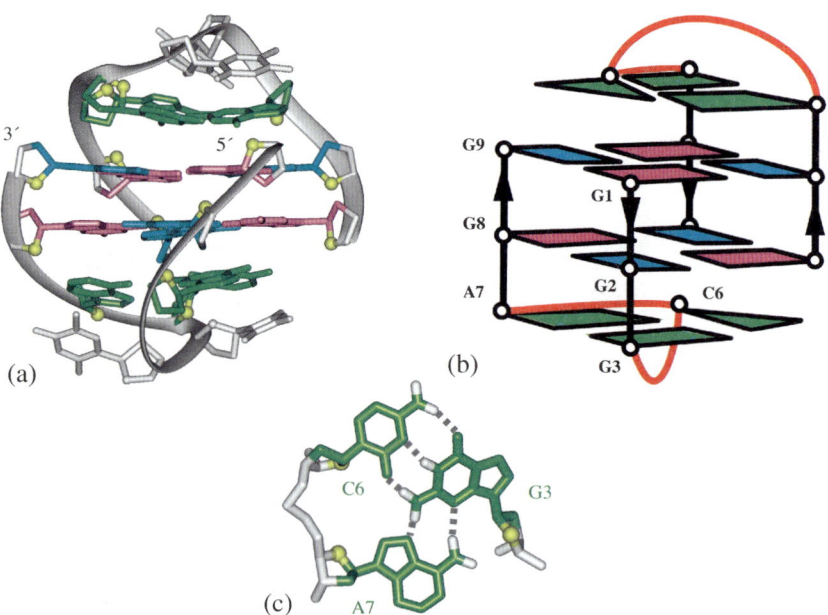

Figure 3   *Structure of the d(GGGTTCAGG) quadruplex with triad-containing edgewise loops.[47] (a) Ribbon and (b) schematic view of the structure. (c) Alignment of G · (C-A) triad. Loops are colored red; anti and syn guanines are colored cyan and magenta, respectively*

and the adenine within a Watson–Crick G·C and a sheared G·A pairs, respectively [Figure 3(c)].

3.3.2 Triad-Containing Diagonal Loops

The d($A_2G_2T_4A_2G_2$) sequence forms a diamond-shaped dimeric G-quadruplex[48] with diagonal loops (Figure 4). Formation of a T·(A-A) triad [Figure 4(c)] was observed in the A_2T_4 diagonal loops. This triad is stabilized by stacking on the G-tetrad core.

The d($G_2T_4G_2CAG_2GT_4G_2T$) sequence forms an intramolecular G-quadruplex[34] defined by one double-chain-reversal and two diagonal loops (Figure 5). The two-residue (CA) double-chain-reversal loop bridges two G-tetrad layers (Figure 5a and b). A G·(T–T) triad [Figure 5(c)] was observed in the GT_4 diagonal loop [Figure 5(b)].

Formation of a G·(A–G) triad in the context of the GAAG diagonal loop[46] is discussed in paragraph 3.3.9 below.

3.3.3 V-Shaped Loop

The d($G_3AG_2T_3G_3AT$) forms a dimeric G-quadruplex (Figure 6) with several new structural elements.[23] There is a V-shaped loop connecting two corners

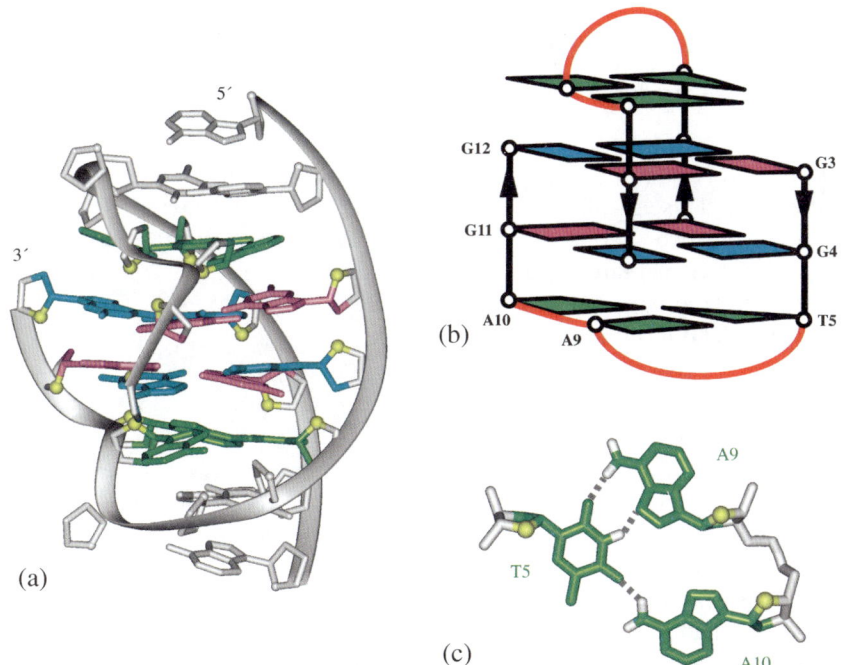

Figure 4 *Structure of the d($A_2G_2T_4A_2G_2$) diamond-shaped dimeric G-quadruplex with triad-containing diagonal loops.[48] (a) Ribbon and (b) schematic view of the structure. (c) Alignment of T·(A-A) triad*

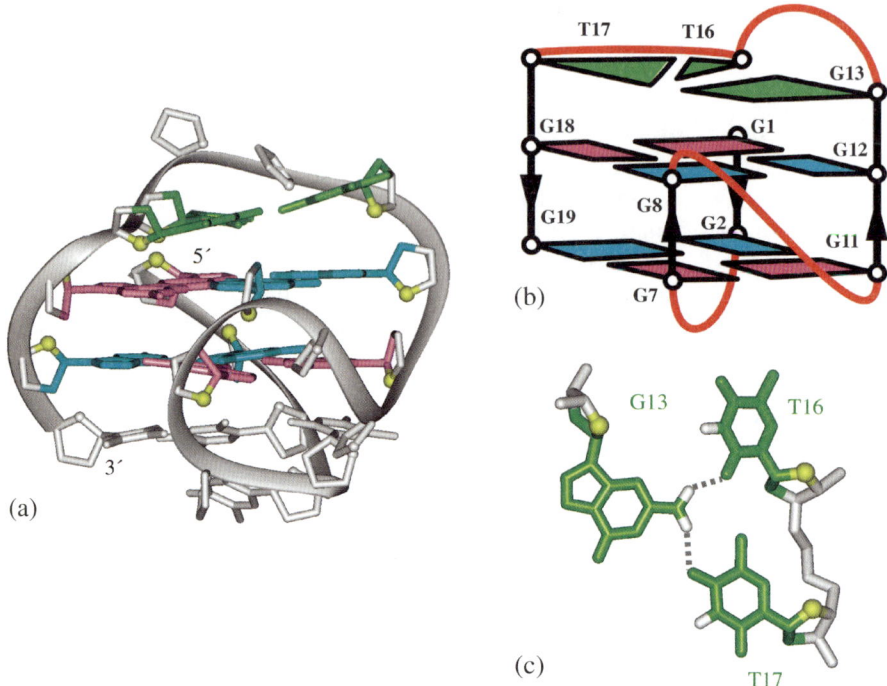

Figure 5 *Structure of the intramolecular G-quadruplex with a triad-containing diagonal loop formed by the d($G_2T_4G_2CAG_2GT_4G_2T$) sequence.[34] (a) Ribbon and (b) schematic view of the structure. (c) Alignment of $G \cdot (T\text{-}T)$ triad*

(G10 and G11) of the G-tetrad core where one supporting column is missing [Figure 6(b)]. This established an interrupted core together with the interlocking between two monomer subunits. The structure also contains a $A \cdot (G \cdot G \cdot G \cdot G)$ pentad [Figure 6(c)]. Similar feature of a V-shaped loop was observed in the structure of the dimeric G-quadruplex formed by the d(GGGTTTGGGG) sequence.[50]

3.3.4 Interlocked G-Quadruplexes

The *93del* d(GGGGTGGGAGGAGGGT) oligonucleotide, an inhibitor of HIV-1 integrase, was shown to form in K^+ solution a very stable two-fold symmetric dimeric quadruplex[24] (Figure 7). Each monomer subunit contains two $G \cdot G \cdot G \cdot G$ tetrads and one $A \cdot (G \cdot G \cdot G \cdot G)$ pentad, where all the G-stretches are parallel and linked by three single-nucleotide double-chain-reversal loops. Dimer formation is achieved through mutual pairing of G1 of one monomer, with G2, G6, and G13 of the other monomer, to complete $G \cdot G \cdot G \cdot G$ tetrad formation. The compact interlocking of symmetry-related subunits through $G2 \cdot G6 \cdot G1 \cdot G13$ tetrad formation across the dimeric interface (Figure 7) constitutes an interesting feature of the *93del* architecture and

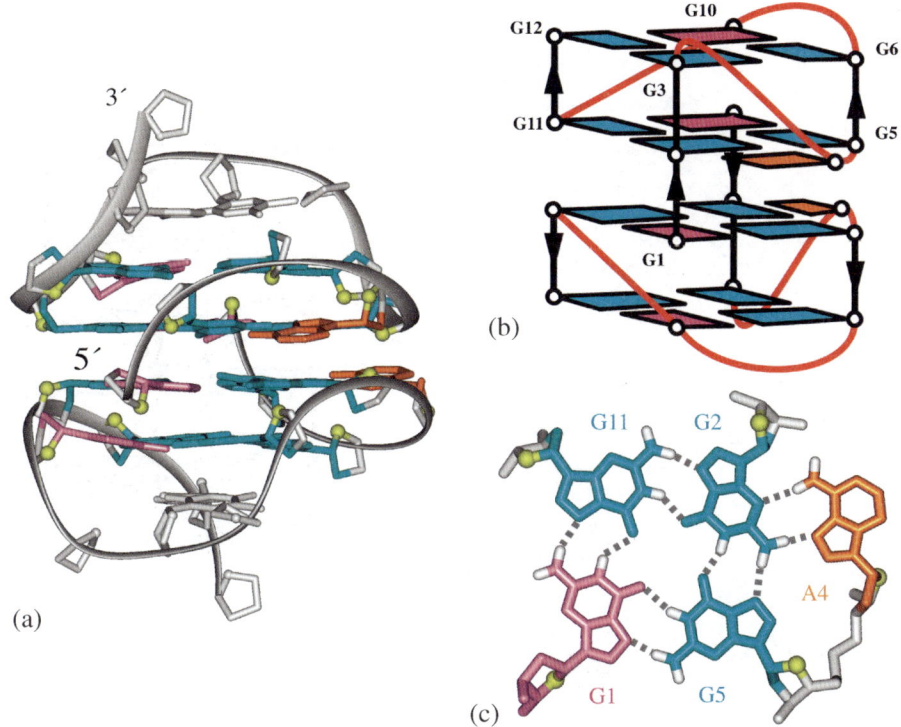

Figure 6 *Structure of a G-quadruplex with a V-shaped loop.[23] (a) Ribbon and (b) schematic view of the structure. (c) Alignment of A · (G · G · G · G) pentad*

highlights a new principle for robust dimeric quadruplex folding. The core this G-quadruplex structure constitutes an interrupted G-tetrad core [Figure 7(b)].

3.3.5 Pentads, Hexads, Heptads, and Octads

The adenine in a single-residue double-chain-reversal loop that bridges two G-tetrad layers can form hydrogen bonds with one edge of the G-tetrad. This hydrogen-bond alignment was first observed in the A · (G · G · G · G) · A hexad (Figure 8) formed by the d(GGAGGAG) sequence.[41] Depending on the number of single-adenine double-chain-reversal loops in the quadruplex structures, pentad,[23,24] hexad,[51] and heptad[42] base alignments were observed later in other contexts. G–U octad[52,53] and A–U octad[54] base alignments were observed in the crystal structures of G-quadruplexes formed by short RNA/DNA chimera sequences.

3.3.6 Mixed Tetrads

The d(GAGCAGGT) sequence forms a head-to-head dimeric quadruplex involving a G · G · G · G tetrad and a G · C · G · C [Figure 9(c)] and an

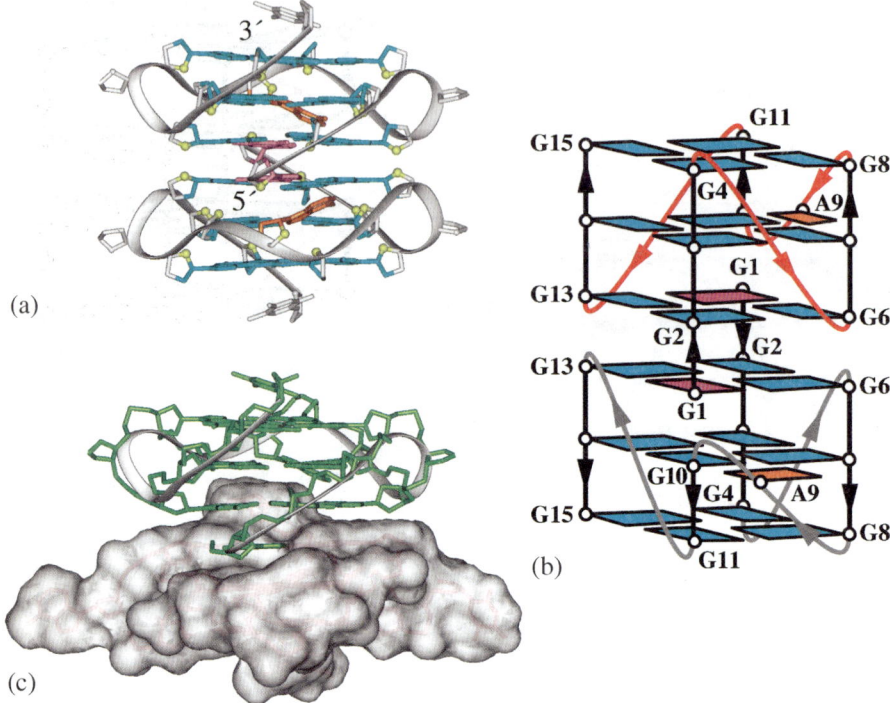

Figure 7 *Structure of the 93del dimeric quadruplex.[24] (a) Ribbon and (b) schematic view of the structure. (c) Structure of the 93del dimeric quadruplex with one subunit shown in surface view, thereby emphasizing the recognition feature between the two subunits*

A·T·A·T [Figure 9(d)] mixed tetrad aligned through their major groove edges.[30] Both Watson–Crick G·C and A·T pairings occur between two monomer subunits. The G·G·G·G tetrad is sandwiched between the two mixed tetrads (Figure 9a and b). It appears that the central G-tetrad helps to stabilize and template formation of the G·C·G·C and A·T·A·T mixed tetrads by stacking interaction. Other types of homo-base tetrads, such as A·A·A·A and U·U·U·U tetrads,[25,52–55] have also been observed and are stabilized by stacking on G-tetrads. It should also be noted that G·C·G·C and G·C·A·T mixed tetrads can be formed through minor groove alignments.[61,62]

3.3.7 Four-Repeat Human Telomeric Sequences

The four-repeat human telomeric d[AGGG(TTAGGG)₃] sequence was shown to form an intramolecular G-quadruplex in Na⁺ solution,[31] in which guanines around each tetrad are *syn·syn·anti·anti*, loops are successively edgewise–diagonal–edgewise and each G-tract has both a parallel and an antiparallel adjacent strands [Figure 10(a–c)]. By contrast, the crystal structure of the same

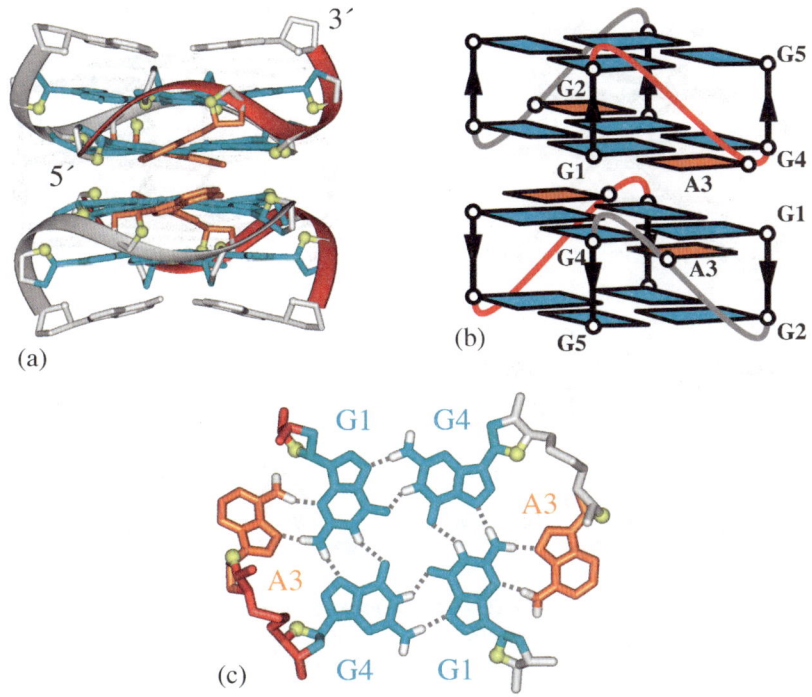

Figure 8 *Structure of a G-quadruplex containing a A·(G·G·G·G)·A hexad.[51] (a) Ribbon and (b) schematic view of the structure. (c) Alignment of A·(G·G·G·G)·A hexad*

sequence in the presence of K[+] revealed a completely different intramolecular G-quadruplex,[43] where all strands are parallel, guanines are *anti* and loops are double-chain-reversal [Figure 10(d–f)] – see Chapter 1. This, the so-called, "propeller-type" structure could readily facilitate higher order telomere folding and unfolding.[43] Recently, it has been shown[56,59,60] that the four-repeat human telomeric sequence forms in K[+] solution an intramolecular G-quadruplex, which is distinctly different from previously reported structures in Na[+] solution and in a K[+]-containing crystal. This G-quadruplex contains the (3+1) G-tetrad core (see below) and involves: one *anti·syn·syn·syn* and two *syn·anti·anti·anti* G-tetrads; one double-chain-reversal and two edgewise loops; three G-tracts oriented in one direction and the fourth in the opposite direction (Figure 11c).

3.3.8 (3+1) G-Tetrad Core Topology

The four-repeat *Tetrahymena* telomeric d(T$_2$G$_4$)$_4$ sequence was shown to form an intramolecular G-quadruplex (Figure 11a) with one double-chain-reversal and two edgewise loops.[40] This is the first observation of a double-chain-reversal loop: sequence T$_2$ bridges three G-tetrad layers. Furthermore this

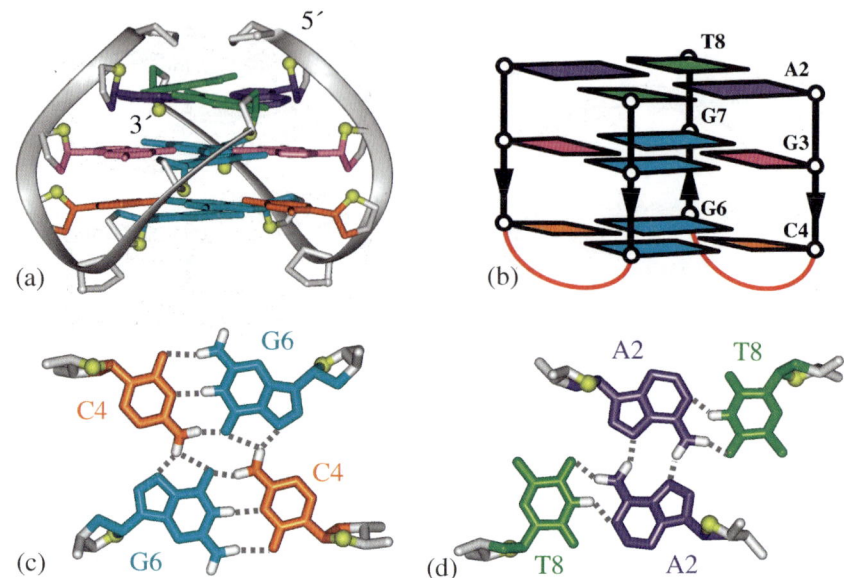

(a)   (b)

(c)   (d)

Figure 9   *Structure of the d(GAGCAGGT) dimeric quadruplex containing $G \cdot C \cdot G \cdot C$ and $A \cdot T \cdot A \cdot T$ mixed tetrads.*[30] *(a) Ribbon and (b) schematic view of the structure. (c, d) Alignment of (c) $G \cdot C \cdot G \cdot C$ and (d) $A \cdot T \cdot A \cdot T$ mixed tetrads*

structure contains a (3+1) G-tetrad core, in which three strands are oriented in one direction and the fourth in the opposite direction and G-tetrads are *anti·syn·syn·syn* or *syn·anti·anti·anti*. More than a decade later, the (3+1) G-tetrad core was observed in a dimeric G-quadruplex formed by the three-repeat human telomeric sequence in Na⁺ solution,[57] and in an intramolecular G-quadruplex formed by the four-repeat human telomeric sequence in K⁺ solution.[56] However, there are differences in the order of G-tracts in these three structures: when their G-tetrad cores are aligned, the 5′-end in each structure starts from a distinct corner of the core (Figure 11). Recently, a variant sequence of the four G-tract human *bcl-2* promoter[58] was shown to adopt the G-quadruplex fold with the same (3+1) core topology as the G-quadruplex fold formed by the four-repeat *Tetrahymena* telomeric sequence. Thus, the (3+1) core topology of the G-quadruplex, initially identified in 1994, and thought to be an anomaly at that time, now appears to be a robust G-quadruplex scaffold, adopted both by telomeric and oncogenic promoter sequences.

3.3.9 Fold-Back G-Quadruplexes

A five-guanine-tract sequence from the *c-myc* promoter was found to form a distinct novel parallel-stranded fold-back G-quadruplex topology [Figure 12(b)] in K⁺ solution.[46] The structure revealed several new motifs not observed previously. In particular, a guanine (G24) of the 3′-end is plugged back into the

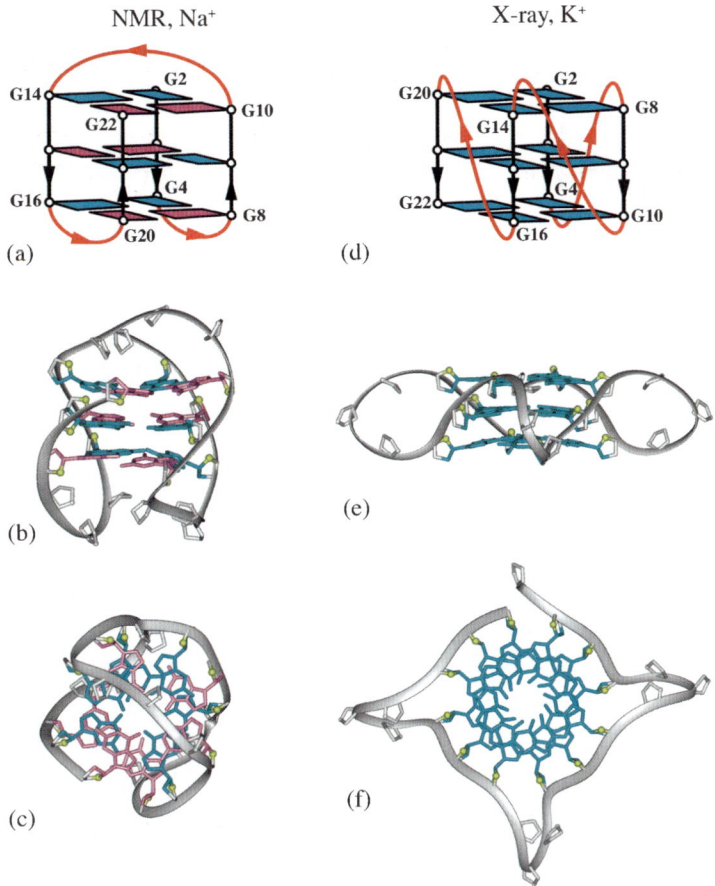

Figure 10 *Structure of intramolecular G-quadruplexes formed by the human telomeric sequence: (a–c) in Na$^+$ solution;[31] (d–f) in a K$^+$-containing crystal.[43] (a, d) Schematic, (b, e) side, and (c, f) top views*

G-tetrad core by participating in G-tetrad formation and displacing another guanine (G10) of a continuous G-tract into a loop. This configuration is provided by formation of a diagonal loop [Figure 12(c)], which contains a G·(A–G) triad [Figure 12(d)] stacking on and capping the G-tetrad core. In this topology there is an interruption in the G-tetrad core between G24 and G9 (Figure 12a and b).

3.3.10 Bulged and Interrupted G-Tetrad Core

Bulges of uracil residues were observed in the crystal structure of the tetrameric parallel quadruplex formed by the RNA/DNA chimera r(U)(^{Br}dG)r(UGGU) sequence.[25] U3 residues flip out between stacked G2(*syn*)·G2(*syn*)·G2 (*syn*)·G2(*syn*) and G4(*anti*)·G4(*anti*)·G4(*anti*)·G4(*anti*) [Figure 13(b)]. The bulged U3 residues adopt *syn* glycosidic conformation and their O2 and N3

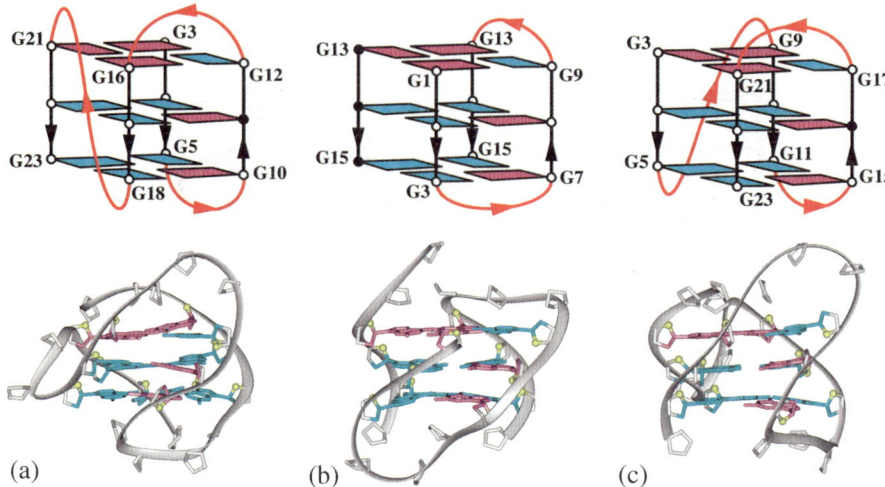

Figure 11   *Structures of the (3+1) core-containing G-quadruplexes formed by (a) the four-repeat Tetrahymena telomeric d[(TTGGGG)₄] sequence in Na⁺ solution[40], (b) the three-repeat human telomeric d[GGG(TTAGGG)₂T] sequence in Na⁺ solution,[57] and (c) the four-repeat human telomeric d[TAG-GG(TTAGGG)₃] sequence in K⁺ solution[56]*

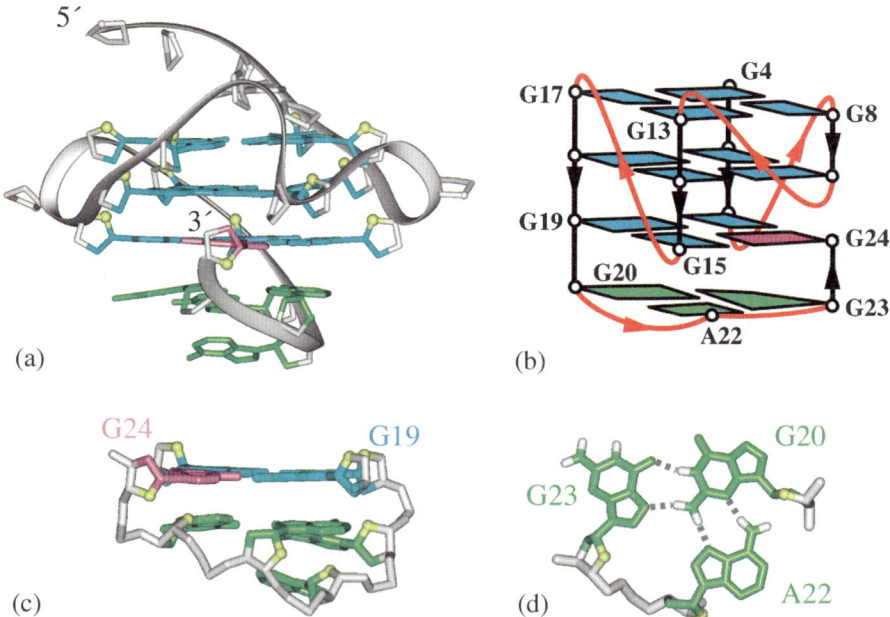

Figure 12   *Fold-back G-quadruplex structure formed by a five-G-tract sequence of the c-myc promoter.[46] (a) Ribbon and (b) schematic view of the structure. (c) Triad-containing diagonal loop. (d) Alignment of G · (A-G) triad*

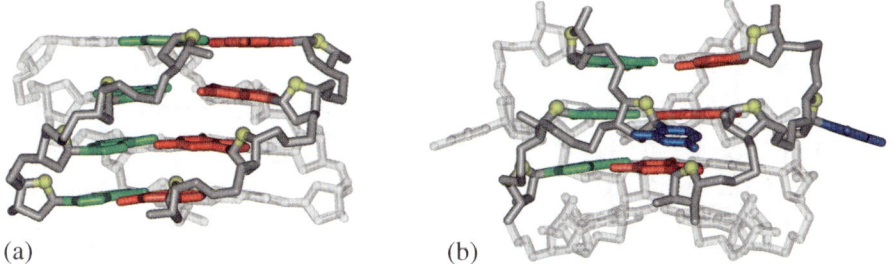

Figure 13 *Comparison between (a) continuous and (b) bulged G-tetrad[25] cores*

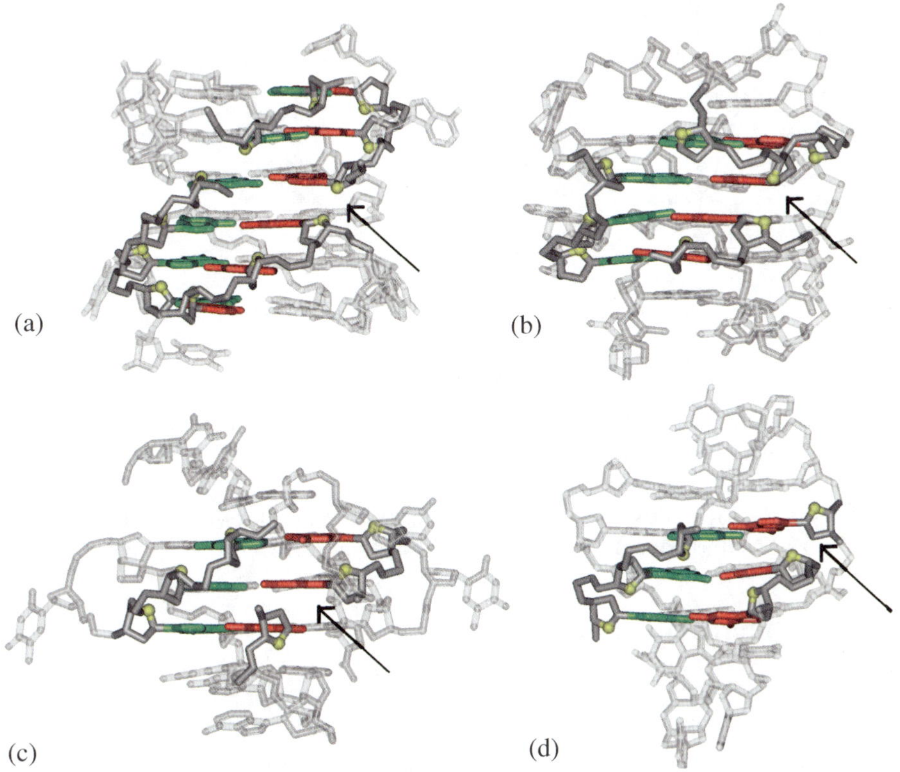

(a)

(b)

(c)

(d)

Figure 14 *Examples of interrupted G-tetrad cores in G-quadruplex structures from (a) ref. 24, (b) ref. 23, (c) ref. 46, and (d) ref. 50. The break points are highlighted by arrows. Relative sugar orientations across the break are (a, b) 5'-to-5', (c) 3'-to-3', and (d) 5'-to-3'*

atoms face outwards, serving as potential effective recognition and interaction sites. The bulge formation widens the groove width, but does not make any bends or kinks in the quadruplex structure.[25] The comparison between this bulged G-tetrad core and a continuous G-tetrad core is shown in Figure 13.

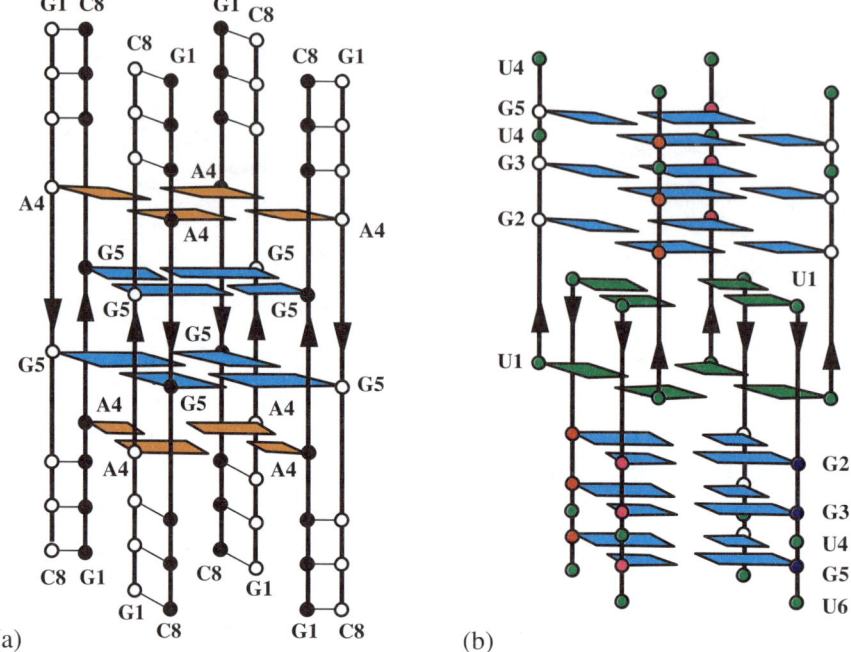

Figure 15 *Intercalation of (a) central G · G · G · G tetrads[26] and (b) terminal U · U · U · U tetrads[55]*

The structure of the bulged G-tetrad core can be considered as an intermediate between the continuous and the interrupted cores mentioned above. Figure 14 shows different examples of interrupted G-tetrad cores. The relative sugar orientations between two residues across the break point can be (i) 5′-to-5′ (Figure 14a and b); (ii) 3′-to-3′ [Figure 14(c)]; and (iii) 5′-to-3′ [Figure 14(d)].

3.3.11   Intercalation of Tetrads

Intercalation of G-tetrads was observed in a crystal structure of d(GCGA-GAGC) at low K^+ concentration,[26] termed "i-motif of G-quartets", in which four base-intercalated duplexes are assembled to form an octaplex with eight G5 forming a stacked double G-tetrads in the central part of the structure [Figure 15(a)]. This intercalation feature was also observed for terminal U · U · U · U tetrads,[55] providing a way for interlocking two G-quadruplex subunits [Figure 15(b)].

3.4   Conclusion

G-quadruplex architectures can contain a large variety of G-tetrad core and loop elements. This structural polymorphism depends strongly on sequences and experimental conditions, such as the nature of cations. Some general rules for G-quadruplex folding have emerged, such as the robustness of the single-residue double-chain-reversal loop in K^+ solution. Systematic studies of different structural elements are needed to fully understand the folding rules of G-quadruplexes and to be able to predict a G-quadruplex topology from sequence. Recognition of different structural motifs in G-quadruplexes by small molecules and proteins is also a subject of future research.

References

1. D.J. Patel, S. Bouaziz, A. Kettani and Y. Wang, Structures of guanine-rich, and cytosine-rich quadruplexes formed *in vitro* by telomeric centromeric and triplet repeat disease DNA sequences, , in *Oxford Handbook of Nucleic Acid Structures*, S. Neidle (ed), Oxford University Press, Oxford, 1999, 389–453.
2. T. Simonsson, G-quadruplex DNA structures – variations on a theme, *Biol. Chem*, 2001, **382**, 621–628.
3. J.T. Davis, G-quartets 40 years later: From 5′-GMP to molecular biology and supramolecular chemistry, *Angew. Chem. Int. Ed. Engl.*, 2004, **43**, 668–698.
4. C. Schaffitzel, I. Berger, J. Postberg, J. Hanes, H.J. Lipps and A. Pluckt-hun, *In vitro* generated antibodies specific for telomeric guanine-quadru-plex DNA react with *Stylonychia lemnae* macronuclei, *Proc. Natl. Acad. Sci. USA*, 2001, **98**, 8572–8577.
5. K. Paeschke, T. Simonsson, J. Postberg, D. Rhodes and H.J. Lipps, Telomere end-binding proteins control the formation of G-quadruplex DNA structures *in vivo*, *Nat. Struct. Mol. Biol.*, 2005, **12**, 847–854.
6. M.L. Duquette, P. Handa, J.A. Vincent, A.F. Taylor and N. Maizels, Intracellular transcription of G-rich DNAs induces formation of G-loops, novel structures containing G4 DNA, *Genes Dev.*, 2004, **18**, 1618–1629.
7. J.D. Wen, C.W. Gray and D.M. Gray, SELEX selection of high-affinity oligonucleotides for bacteriophage Ff gene 5 protein, *Biochemistry*, 2001, **40**, 9300–9310.
8. K. Muniyappa, S. Anuradha and B. Byers, Yeast meiosis-specific protein Hop1 binds to G4 DNA and promotes its formation, *Mol. Cell Biol.*, 2000, **20**, 1361–1369.
9. P. Mohaghegh, J.K. Karow, R.M. Brosh Jr., V.A. Bohr and I.D. Hickson, The Bloom's and Werner's syndrome proteins are DNA structure-specific helicases, *Nucleic Acids Res.*, 2001, **29**, 2843–2849.
10. H. Sun, A. Yabuki and N. Maizels, A human nuclease specific for G4 DNA, *Proc. Natl. Acad. Sci. USA*, 2001, **98**, 12444–12449.
11. A. Siddiqui-Jain, C.L. Grand, D.J. Bearss and L.H. Hurley, Direct evi-dence for a G-quadruolex in a promoter region and its targeting with a small molecule to repress c-MYC transcription, *Proc. Natl. Acad. Sci. USA*, 2002, **99**, 11593–11598.
12. A.M. Zahler, J.R. Williamson, T.R. Cech and D.M. Prescott, Inhibition of telomerase by G-quartet DNA structures, *Nature*, 1991, **350**, 718–720.

13. A.J. Zaug, E.R. Podell and T.R. Cech, Human POT1 disrupts telomeric G-quadruplexes allowing telomerase extension *in vitro*, *Proc. Natl. Acad. Sci. USA*, 2005, **102**, 10864–10869.

14. L. Oganesian, I.K. Moon, T.M. Bryan and M.B. Jarstfer, Extension of G-quadruplex DNA by ciliate telomerase, *EMBO J.*, 2006, **25**, 1148–1159.

15. M. Gellert, M.N. Lipsett and D.R. Davies, Helix formation by guanylic acid, *Proc. Natl. Acad. Sci. USA*, 1962, **48**, 2013–2018.

16. J.R. Williamson, M.K. Raghuraman and T.R. Cech, Monovalent cation-induced structure of telomeric DNA: The G-quartet model, *Cell*, 1989, **59**, 871–880.

17. W. Guschlbauer, J.F. Chantot and D. Thiele, Four-stranded nucleic acid structures 25 years later: From guanosine gels to telomer DNA, *J. Biomol. Struct. Dyn.*, 1990, **8**, 491–511.

18. F.W. Smith and J. Feigon, Quadruplex structure of *Oxytricha* telomeric DNA oligonucleotides, *Nature*, 1992, **356**, 164–168.

19. C. Kang, X. Zhang, R. Ratliff, R. Moyzis and A. Rich, Crystal structure of four-stranded *Oxytricha* telomeric DNA, *Nature*, 1992, **356**, 126–131.

20. Y. Wang and D.J. Patel, Guanine residues in $d(T_2AG_3)$ and $d(T_2G_4)$ form parallel-stranded potassium cation stabilized G-quadruplexes with anti glycosidic torsion angles in solution, *Biochemistry*, 1992, **31**, 8112–8119.

21. F. Aboul-ela, A.I. Murchie and D.M. Lilley, NMR study of parallel-stranded tetraplex formation by the hexadeoxynucleotide $d(TG_4T)$, *Nature*, 1992, **360**, 280–282.

22. G. Laughlan, A.I. Murchie, D.G. Norman, M.H. Moore, P.C. Moody, D.M. Lilley and B. Luisi, The high-resolution crystal structure of a parallel-stranded guanine tetraplex, *Science*, 1994, **265**, 520–524.

23. N. Zhang, A. Gorin, A. Majumdar, A. Kettani, N. Chernichenko, E. Skripkin and D.J. Patel, V-shaped scaffold: A new architectural motif identified in an A.(G.G.G.G) pentad-containing dimeric DNA quadruplex involving stacked G(*anti*).G(*anti*).G(*anti*).G(*syn*) tetrads, *J. Mol. Biol.*, 2001, **311**, 1063–1079.

24. A.T. Phan, V. Kuryavyi, J.B. Ma, A. Faure, M.L. Andreola and D.J. Patel, An interlocked dimeric parallel-stranded DNA quadruplex: A potent inhibitor of HIV-1 integrase, *Proc. Natl. Acad. Sci. USA*, 2005, **102**, 634–639.

25. B. Pan, Y. Xiong, K. Shi and M. Sundaralingam, Crystal structure of a bulged RNA tetraplex at 1.1 Å resolution: Implications for a novel binding site in RNA tetraplex, *Structure*, 2003, **11**, 1423–1430.

26. J. Kondo, W. Adachi, S. Umeda, T. Sunami and A. Takenaka, Crystal structures of a DNA octaplex with I-motif of G-quartets and its splitting into two quadruplexes suggest a folding mechanism of eight tandem repeats, *Nucleic Acids Res.*, 2004, **32**, 2541–2549.

27. A. Kettani, R.A. Kumar and D.J. Patel, Solution structure of a DNA quadruplex containing the fragile X syndrome triplet repeat, *J. Mol. Biol.*, 1995, **254**, 638–656.

28. A. Kettani, S. Bouaziz, A. Gorin, H. Zhao, R.A. Jones and D.J. Patel, Solution structure of a Na cation stabilized DNA quadruplex containing G.G.G.G and G.C.G.C tetrads formed by G–G–G–C repeats observed in adeno-associated viral DNA, *J. Mol. Biol.*, 1998, **282**, 619–636.

29. S. Bouaziz, A. Kettani and D.J. Patel, A K cation-induced conformational switch within a loop spanning segment of a DNA quadruplex containing G–G–G–C repeats, *J. Mol. Biol.*, 1998, **282**, 637–652.

30. N. Zhang, A. Gorin, A. Majumdar, A. Kettani, N. Chernichenko, E. Skripkin and D.J. Patel, Dimeric DNA quadruplex containing major groove-aligned A.T.A.T and G.C.G.C tetrads stabilized by inter-subunit Watson–Crick A.T and G.C pairs, *J. Mol. Biol.*, 2001, **312**, 1073–1088.

31. Y. Wang and D.J. Patel, Solution structure of the human telomeric repeat d[AG$_3$(T$_2$AG$_3$)$_3$] G-tetraplex, *Structure*, 1993, **1**, 263–282.

32. Y. Wang and D.J. Patel, Solution structure of the *Oxytricha* telomeric repeat d[G$_4$(T$_4$G$_4$)$_3$] G-tetraplex, *J. Mol. Biol.*, 1995, **251**, 76–94.

33. S. Haider, G.N. Parkinson and S. Neidle, Crystal structure of the potassium form of an *Oxytricha nova* G-quadruplex, *J. Mol. Biol.*, 2002, **320**, 189–200.

34. V. Kuryavyi, A. Majumdar, A. Shallop, N. Chernichenko, E. Skripkin, R. Jones and D.J. Patel, A double chain reversal loop and two diagonal loops define the architecture of a unimolecular DNA quadruplex containing a pair of stacked G(*syn*).G(*syn*).G(*anti*).G(*anti*) tetrads flanked by a G.(T-T) triad and a T.T.T triple, *J. Mol. Biol.*, 2001, **310**, 181–194.

35. M. Crnugelj, N.V. Hud and J. Plavec, The solution structure of d(G$_4$T$_4$G$_3$)$_2$: A bimolecular G-quadruplex with a novel fold, *J. Mol. Biol.*, 2002, **320**, 911–924.

36. P. Schultze, R.F. Macaya and J. Feigon, Three-dimensional solution structure of the thrombin-binding DNA aptamer d(GGTTGGTGTGG TTGG), *J. Mol. Biol.*, 1994, **235**, 1532–1547.

37. A.T. Phan and D.J. Patel, Two-repeat human telomeric d(TAG-GGTTAGGGT) sequence forms interconverting parallel and antiparallel G-quadruplexes in solution: Distinct topologies thermodynamic, properties, and folding/unfolding kinetics, *J. Am. Chem. Soc.*, 2003, **125**, 15021–15027.

38. A.T. Phan, Y.S. Modi and D.J. Patel, Two-repeat *Tetrahymena* telomeric d(TGGGGTTGGGGT) sequence interconverts between asymmetric dimeric G-quadruplexes in solution, *J. Mol. Biol.*, 2004, **338**, 93–102.

39. P. Hazel, G.N. Parkinson and S. Neidle, Topology variation and loop structural homology in crystal and simulated structures of a bimolecular DNA quadruplex, *J. Am. Chem. Soc.*, 2006, **128**, 5480–5487.

40. Y. Wang and D.J. Patel, Solution structure of the *Tetrahymena* telomeric repeat d(T$_2$G$_4$)$_4$ G-tetraplex, *Structure*, 1994, **2**, 1141–1156.

41. A. Kettani, A. Gorin, A. Majumdar, T. Hermann, E. Skripkin, H. Zhao, R. Jones and D.J. Patel, A dimeric DNA interface stabilized by stacked A.(G.G.G.G).A hexads and coordinated monovalent cations, *J. Mol. Biol.*, 2000, **297**, 627–644.

42. A. Matsugami, K. Ouhashi, M. Kanagawa, H. Liu, S. Kanagawa, S. Uesugi and M. Katahira, An intramolecular quadruplex of (GGA)$_4$ triplet repeat DNA with a G:G:G:G: tetrad and a G(:A)G(:A)G(:A)G: heptad, and its dimeric interaction, *J. Mol. Biol.*, 2001, **313**, 255–269.

43. G.N. Parkinson, M.P. Lee and S. Neidle, Crystal structure of parallel quadruplexes from human telomeric DNA, *Nature*, 2002, **417**, 876–880.

44. A.T. Phan, Y.S. Modi and D.J. Patel, Propeller-type parallel-stranded G-quadruplexes in the human c-myc promoter, *J. Am. Chem. Soc.*, **126**, 8710–8716.

45. P. Hazel, J. Huppert, S. Balasubramanian and S. Neidle, Loop-length-dependent folding of G-quadruplexes, *J. Am. Chem. Soc.*, 2004, **126**, 16405–16415.

46. A.T. Phan, V. Kuryavyi, H.Y. Gaw and D.J. Patel, Small-molecule interaction with a five-guanine-tract G-quadruplex structure from the human MYC promoter, *Nat. Chem. Biol.*, 2005, **1**, 167–173.

47. A. Kettani, G. Basu, A. Gorin, A. Majumdar, E. Skripkin and D.J. Patel, A two-stranded template-based approach to G.(C–A) triad formation: Designing novel structural elements into an existing DNA framework, *J. Mol. Biol.*, 2000, **301**, 129–146.

48. V. Kuryavyi, A. Kettani, W. Wang, R. Jones and D.J. Patel, A diamond-shaped zipper-like DNA architecture containing triads sandwiched between mismatches and tetrads, *J. Mol. Biol.*, 2000, **295**, 455–469.

49. V.V. Kuryavyi and T.M. Jovin, Triad-DNA: A model for trinucleotide repeats, *Nat. Genet.*, 1995, **9**, 339–341.

50. M. Crnugelj, P. Sket and J. Plavec, Small change in a G-rich sequence, a dramatic change in topology: New dimeric G-quadruplex folding motif with unique loop orientations, *J. Am. Chem. Soc.*, 2003, **125**, 7866–7871.

51. A. Majumdar, A. Kettani, E. Skripkin and D.J. Patel, Pulse sequences for detection of NH$_2\cdots$N hydrogen bonds in sheared G.A mismatches via remote, non-exchangeable protons, *J. Biomol. NMR*, 2001, **19**, 103–113.

52. J. Deng, Y. Xiong and M. Sundaralingam, X-ray analysis of an RNA tetraplex (UGGGGU)$_4$ with divalent Sr^{2+} ions at subatomic resolution (0.61 Å), *Proc. Natl. Acad. Sci. USA*, 2001, **98**, 13665–13670.

53. B. Pan, Y. Xiong, K. Shi, J. Deng and M. Sundaralingam, Crystal structure of an RNA purine-rich tetraplex containing adenine tetrads: Implications for specific binding in RNA tetraplexes, *Structure*, 2003, **11**, 815–823.

54. B. Pan, Y. Xiong, K. Shi and M. Sundaralingam, An eight-stranded helical fragment in RNA crystal structure: Implications for tetraplex interaction, *Structure*, 2003, **11**, 825–831.

55. B. Pan, K. Shi and M. Sundaralingam, Base-tetrad swapping results in dimerization of RNA quadruplexes: Implications for formation of the i-motif RNA octaplex, *Proc. Natl. Acad. Sci. USA*, 2006, **103**, 3130–3134.

56. K.N. Luu, A.T. Phan, V. Kuryavyi, L. Lacroix and D.J. Patel, Structure of the human telomere in K$^+$ solution: An intramolecular (3+1) G-quadruplex scaffold, *J. Am. Chem. Soc.*, 2006, in press.

57. N. Zhang, A.T. Phan and D.J. Patel, (3+1) Assembly of three human telomeric repeats into an asymmetric dimeric G-quadruplex, *J. Am. Chem. Soc.*, 2005, **127**, 17277–17285.

58. J. Dai, T.S. Dexheimer, D. Chen, M. Carver, A. Ambrus, R.A. Jones and D. Yang, An intramolecular G-quadruplex structure with mixed parallel/antiparallel G-strands formed in the human BCL-2 promoter region in solution, *J. Am. Chem. Soc.*, 2006, **128**, 1096–1098.

59. Y. Xu, Y. Noguchi and H. Sugiyama, The new models of the human telomere d[AGGG(TTAGGG)$_3$] in K$^+$ solution, *Bioorg. Med. Chem.*, 2006, in press.

60. A. Ambrus, D. Chen, J. Dai, T. Bialis, R.A. Jones and D. Yang, Human telometric sequence forms a hybrid-type intramolecular G-quadruplex structure with mixed parallel/antiparallel strands in potassium solution, *Nucl. Acids Res.*, 2006, **34**, 2723–2735.

61. N. Escaja, J.L. Gelpi, M. Orozco, M. Rico, E. Pedroso and C. Gonzalez, Four-stranded DNA structure stabilized by a novel G:C:A:T tetrad, *J. Am. Chem. Soc.*, 2003, **125**, 5654–5662.

62. N. Escaja, I. Gomez-Pinto, J. Viladoms, M. Rico, E. Pedroso and C. Gonzalez, Induced-fit recognition of DNA by small circular oligonucleotides, *Chemistry*, 2006, **12**, 4035–4042.

CHAPTER 4

The Role of Cations in Determining Quadruplex Structure and Stability

NICHOLAS V. HUD[a] AND JANEZ PLAVEC[b]

[a] School of Chemistry and Biochemistry, Georgia Institute of Technology, Atlanta GA, USA 30332
[b] Slovenian NMR Center, National Institute of Chemistry, Slovenia SI-1001 Ljubljana

4.1 Introduction

G-quadruplex formation absolutely requires the participation of cations.[1–9] Thus, no in-depth treatise on quadruplex nucleic acids would be complete without ample discussion of G-quadruplex–cation interactions. The basic building block of a G-quadruplex is the $G \cdot G \cdot G \cdot G$ quartet, which is composed of four hydrogen-bonded guanine nucleotides in a horizontal planar arrangement (Figure 1). The coordination of cations by the closely spaced

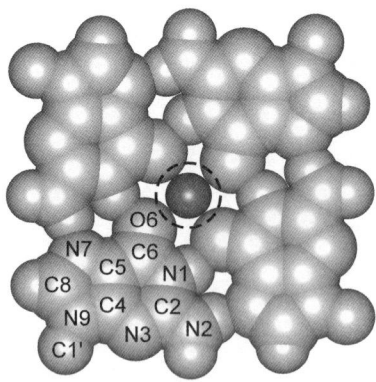

Figure 1 *The G-quartet. A Na$^+$ is shown coordinated in the plane of the quartet by the O6 oxygen atoms. The dashed circle represents the ionic radius of a K$^+$, which is too large for in-plane coordination*

carbonyl oxygen atoms of a G-quartet was postulated long before the first high-resolution structure of a G-quadruplex was determined. Unlike other known DNA structures, G-quartets interact directly with dehydrated cations *via* inner sphere coordination.[10] It is therefore not surprising that the formation, stability, and structural details of G-quadruplexes are dependent on cation species and cation concentration. DNA and RNA G-quadruplexes can be formed with different numbers of oligonucleotide strands (*i.e.*, 1, 2, or 4) and with different strand orientations (*i.e.*, parallel or antiparallel), but in all cases G-quadruplex structures are stabilized by direct cation coordination, and the particular fold adopted by a G-rich oligonucleotide sequence can be cation-dependent.[4,9,11]

Owing to their physiological importance, K^+ and Na^+ ions are the most extensively characterized cations with respect to their ability to stabilize G-quadruplex structures. However, a number of other cations are also known to promote G-quadruplex formation. The list of cations thus far includes the monovalent cations Rb^+, Cs^+,[12,13] NH_4^+,[14,15] and Tl^+,[16] as well as the divalent cations Sr^{2+},[13,17] Ba^{2+},[18] and Pb^{2+}.[14,19] In contrast, Ca^{2+} and Mg^{2+} ions apparently do not promote the formation of G-quadruplex structures in the absence of other cations. In general, low divalent cation concentrations initially stabilize G-quadruplexes, while increasing concentrations eventually become destabilizing.

High-resolution X-ray crystallographic and NMR structures have now provided valuable insights into the effects of cation size and charge on G-quadruplex structure, and on the location of ions within G-quadruplexes.[10,16,20-26] Solution-state NMR studies have demonstrated that cations undergo dynamic exchange between coordination sites in the interior of a G-quadruplex and bulk solution.[27,28] A combination of techniques has also provided insights into the thermodynamics of the cation selectivity exhibited by G-quadruplexes. The results of these investigations, and other studies regarding the nature of G-quadruplex–cation interactions, are the topic of this chapter.

4.2 Crystallographic Analysis of Cation Coordination within G-Quadruplexes

The strong correlation observed between the melting temperature of guanosine gels and the ionic radii of cations was an early indication of site-specific ion binding by G-quartets.[29] Numerous subsequent studies of G-quadruplexes have confirmed the inseparability of G-quartet formation and cation coordination. The strong interaction between cations and G-quartets originates from electrostatic interactions involving the free electron pairs of the guanine O6 oxygen atoms. However, cation coordination is not restricted to a particular geometry within a G-quadruplex. A series of stacked G-quartets produces a regular geometry, and potential cation coordination sites, with four O6 atoms within the plane of a G-quartet, or with eight O6 atoms between two stacked G-quartets. Ions such as K^+ and NH_4^+ (ionic radii 1.33 Å and 1.43 Å, respectively) are too large to coordinate in the plane of a G-quartet, whereas Na^+ (ionic radius 0.95 Å) is small enough to be coordinated within the plane of a G-quartet (Figure 1). Thus, multiple cation coordination geometries are possible and observed. Electrostatic interactions between guanine O6 atoms and cations contribute significantly to the overall stability of a G-quadruplex. On the other hand, electrostatic repulsions between cations within a G-quadruplex are substantial. Thus, the exact locations and coordination geometries of cations within a given G-quadruplex are the result of a balance between attractive interactions with carbonyl oxygen atoms and mutual cation repulsion.

The determination of G-quadruplex crystal structures has provided a detailed view of cation coordination that is unmatched by any other DNA structure. A prime example is the X-ray crystal structure of the DNA hexamer d(TG$_4$T), which as been refined to a resolution of 0.95 Å (PDB ID 352D). This crystal structure confirmed that d(TG$_4$T) forms a tetramolecular parallel-stranded G-quadruplex in the presence of Na$^+$ ions.[10] The asymmetric unit of the d(TG$_4$T) crystal consists of four G-quadruplexes, with two quadruplexes being coaxially stacked in a 5′ to 5′ orientation such that a continuous stack of eight G-quartets is formed (Figure 2).[10] Seven Na$^+$ ions are coordinated along the axial channel formed by the O6 atoms of eight G-quartets (Figure 2). Close inspection of the crystal structure reveals that the central Na$^+$ in this channel is positioned equidistant between the planes of two stacked G-quartets (Figure 2). The coordination geometry of this particular Na$^+$ is bipyrimidal, with eight equidistant guanine O6 atoms. Na$^+$ ions located away from this central position are displaced incrementally further from this perfect symmetrical bipyrimidal coordination. At the ends of this channel, terminal Na$^+$ ions are in fact located in the planes of the end G-quartets, coordinated by the carbonyl groups of only four guanine residues. Displacement of the Na$^+$ ions from the central eight-coordinate geometry to the in-plane four-coordinate geometry is believed to arise from electrostatic repulsion between adjacent Na$^+$ ions. The mutual repulsion of Na$^+$ ions along the central axis of the G-quadruplex has also been hypothesized to be the origin of an out of plane bending observed for the terminal G-quartets of the eight stacked quartets.[10]

The high-resolution crystal structure of d(TG$_4$T) also revealed the location of cations and water molecules located around the outside of the G-quadruplex. Crystals of d(TG$_4$T) were formed from a solution that contained both NaCl and CaCl$_2$. Similar to many other nucleic acid crystals, divalent cations, in this case Ca^{2+} ions, were found at crystal lattice contact points between nucleic acid assemblies (*i.e.*, bridging between backbone phosphates of two G-quadruplexes). Single water molecules were found to provide a fifth oxygen atom for coordination of the outermost Na$^+$ ions that are coordinated in the planes of the terminal G-quartets. Additionally, water molecules were found along the phosphate backbone, in the four grooves, and as part of hydration shell of Ca^{2+} ions. There was no evidence that Ca^{2+} ions substituted for Na$^+$ ions at any sites within the G-quadruplex.

d(TG$_4$T) has also been found to produce two distinct crystal forms when crystallized from solutions containing both Na$^+$ and Tl$^+$ ions.[30] These crystals did not provide as high-resolution data (*i.e.*, 2.2 Å and 2.5 Å), but the very different electron densities of Tl$^+$ and Na$^+$ ions allowed these two cations to be

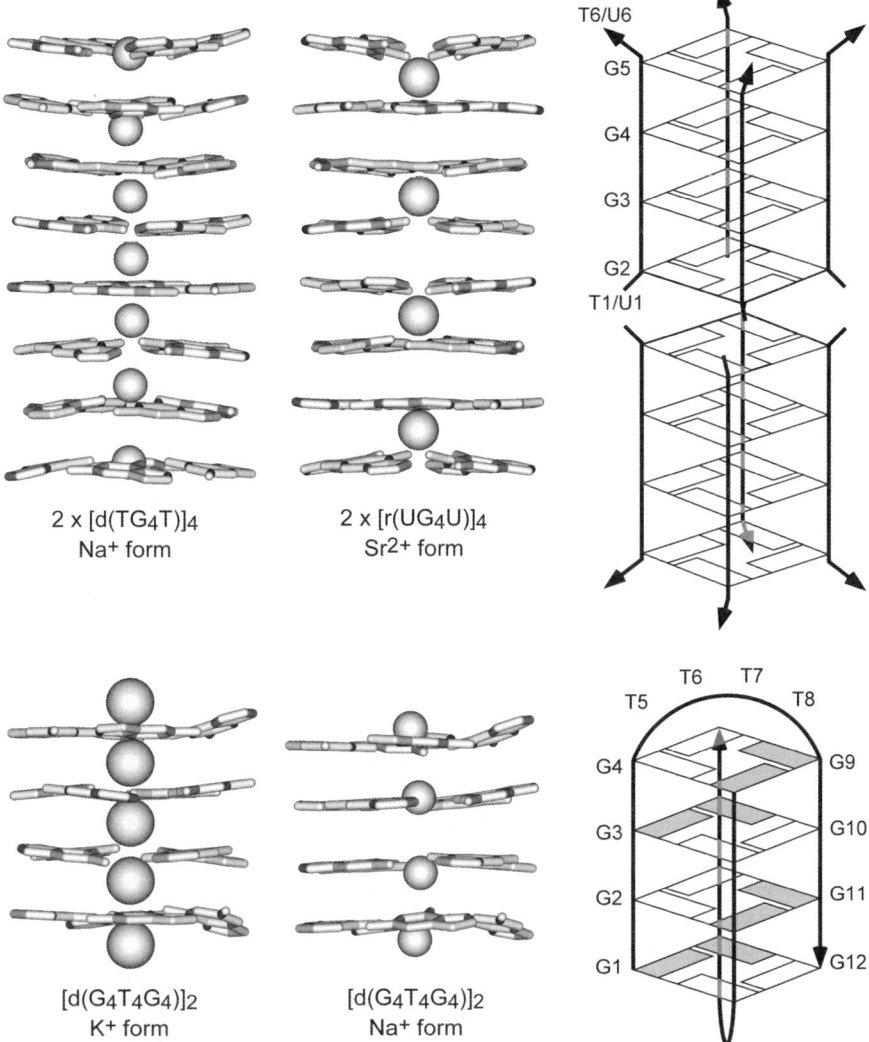

Figure 2 *Cation localization within G-quadruplexes as revealed by X-ray crystal structures. For clarity, only the guanine bases (stick representation) and cations (ionic radii) are shown for the crystal structures. The diagrams to the right of the crystal structures indicate DNA/RNA strand orientation and assembly for the crystal structures shown. White rectangles represent guanine residues with* anti *glycosidic bond angles and gray rectangles represent guanine residues with* syn *glycosidic bond angles. The structures shown were selected to illustrate that* Na^+ *can vary in coordination geometry from being within the plane of a G-quartet to being equidistant between two adjacent G-quartets;* K^+ *ions are exclusively coordinated between two adjacent G-quartets; and* Sr^{2+} *ions, which are also coordinated between two adjacent G-quartets, only occupy every other potential coordination site. Crystal structures shown are* Na^+ *form of* $d(TG_4T)$: *PDB ID 352D;* Sr^{2+} *form of* $r(UG_4U)$: *PDB ID 1J8G;* K^+ *form of* $d(G_4T_4G_4)$: *PDB ID 1JRN;* Na^+ *form of* $d(G_4T_4G_4)$: *PDB ID 1JB7*

distinguished in the G-quadruplex structure. One crystal form contained two G-quadruplexes in the asymmetric unit (PDB ID 1S45), whereas the other form contained three G-quadruplexes (PDB ID 1S47). In all cases, G-quadruplexes are stacked in a head-to-head fashion with a 5′ to 5′ orientation, as found in the Na^+ crystal structure.[10] The G-quadruplex structure observed in both crystal forms is also very similar to the tetramolecular parallel-stranded structure described above. The main difference lies in the position of metal ions within the G-quadruplex, and the low occupancy of inner-G-quadruplex cation coordination sites by Tl^+ ions (*i.e.*, occupancy levels between 0.15 and 0.70). The Tl^+ ions coordinated within the G-quadruplex are positioned between G-quartet planes.[30] This coordination geometry would be expected for Tl^+ ions, as the ionic radius of Tl^+ (1.44 Å) is even slightly larger than that of K^+, and certainly too large to be coordinated within the plane of a G-quartet. Low Tl^+ ion occupancy has been attributed to the higher concentration of Na^+ ions in the crystallization solution.

The bimolecular quadruplex formed by the DNA sequence $d(G_4T_4G_4)$ in the presence of K^+ ions has been crystallized in two different space groups, and their structures have been refined to 2.0 Å and 1.49 Å resolution.[22] One of these crystal structures (PDB ID 1JRN) has two G-quadruplexes per asymmetric unit, whereas the other crystal structure (PDB ID 1JPQ) has only one G-quadruplex per asymmetric unit. All three G-quadruplex structures exhibit the same bimolecular diagonally looped folding topology (Figure 2). This fold is identical to the structure determined for the same DNA sequence 10 years earlier by NMR in the presence of both Na^+ and K^+ ions.[31–33] There are four G-quartets in the bimolecular quadruplex formed by $d(G_4T_4G_4)$. Unlike the tetramolecular structure formed by $d(TG_4T)$, in which all guanine resides have *anti* glycosidic torsion angles, the glycosidic torsion angles for guanine bases in $[d(G_4T_4G_4)]_2$ are alternating *syn–anti* along each strand and *syn–syn–anti–anti* within each G-quartet (Figure 2). The central two G-quartets in each G-quadruplex are approximately coplanar. One guanine base in each terminal G-quartet is slightly more tilted than the others and stacks effectively with the adjacent 3′-thymine base. The average rise between adjacent G-quartets is consistently 3.3 Å.[22] A linear row of five equidistant K^+ ions lies along the helical axis within the central core of all three G-quadruplex crystal structures (Figure 2). All K^+ ions have full occupancy. The average K^+–K^+ distance is 3.38 Å. All three G-quadruplexes have three octahedrally coordinated K^+ ions that are located equidistant between the planes of two adjacent G-quartets, with each K^+ and eight O6 atoms forming a square antiprismatic arrangement. The outer K^+ ions are located within the loops, where they also achieve octahedral coordination with the outer G-quartet O6 carbonyl oxygen atoms and the O2 atoms from the adjacent thymine bases, together with two oxygen atoms provided by water molecules. These K^+ ions have only slightly increased mobilities (*i.e.*, thermal factors) compared to the three central K^+ ions.[22]

The X-ray crystal structure at 1.86 Å resolution of the *Oxytricha nova* telomere end binding protein (OnTEBP) in complex with DNA and in the presence of NaCl (PDB ID 1JB7) has also shown that $d(G_4T_4G_4)$ adopts a bimolecular G-quadruplex structure[21] that is identical to the structure formed in the absence of the protein.[22] Again, four guanine bases form four G-quartets, and four thymine residues form loops that span across the diagonal of the terminal G-quartets. The structural features of the G-quadruplex in the protein:DNA complex are also very similar to those of the NMR structure,[31,32] including the stacking of bases between G-quartets and the alternating *syn–anti* orientations across glycosidic bonds. The high-resolution X-ray diffraction data also reveal the positions of Na^+ ions in the center of the G-quadruplex.[21] The two central Na^+ ions are nearly coplanar within the central G-quartets and are in distorted, octahedral coordination environments (Figure 2). The outer two Na^+ ions are positioned above and below the planes of the outer G-quartets toward the T_4 loops, and are coordinated by two O2 atoms of bases T5 and T7 in addition to the carbonyl oxygen atoms of the terminal G-quartet. Na^+-thymine O2 coordination is consistent with the earlier proposal that changes observed by NMR in T nucleotide positions, when Na^+ ions are exchanged for K^+ or NH_4^+ ions, are driven by the loss of thymine O2 coordination, which is feasible for Na^+ but not for the larger K^+ or NH_4^+ ions (discussed in Section 4.6).[33] Electron density maps also revealed potential cation binding sites between the outside G-quartets and loops that could not be unequivocally attributed to water molecules or Na^+ ions.[21] Partial occupancy of these sites suggested that Na^+ ions exchange between the binding sites inside the G-quadruplex and bulk solution (a phenomenon that is discussed in Section 4.3).

A crystal structure has also been determined, at 2.1 Å resolution, for the monomolecular G-quadruplex formed by d[AGGG(TTAGGG)$_3$] in the presence of K^+ ions (PDB ID 1KF1).[23] Similar to the crystal structures described above, K^+ ions were found in this structure to be positioned equidistant between stacked G-quartets. The distances between K^+ ions and each of the eight carbonyl oxygen atoms in a bipyramidal antiprismatic arrangement is 2.7 Å.[23] The G-quadruplex formed by d[AGGG(TTAGGG)$_3$] is of particular interest because it represents four units of the human telomere repeat sequence. Details of the crystal structure formed by this sequence in the presence of K^+ ions and in the presence of Na^+ ions are discussed below in Section 4.6. The crystal structure of the bimolecular G-quadruplex adopted by the dodecamer sequence d(TAGGGTTAGGGT), also derived from the human telomere repeat sequence, has been determined and similarly contains K^+ ions coordinated between adjacent G-quartets (PDB ID 1K8P).[23]

It has been appreciated for sometime that Sr^{2+} ions can promote G-quadruplex formation.[17] The fact that Sr^{2+} can stabilize G-quadruplexes to a similar degree as K^+ seemed somewhat enigmatic. The ionic radius of Sr^{2+} (1.13 Å) is in between that of Na^+ and K^+; however, the energy from electrostatic repulsions between Sr^{2+} ions within a G-quadruplex would be four times that of monovalent cations, if Sr^{2+} ions were coordinated with a similar spacing along the central axis of a G-quadruplex. The crystal structure of the RNA G-quadruplex formed by the sequence r(UGGGGU) in the presence of Sr^{2+} ions (PDB ID 1J8G), which was determined at ultra-high resolution (0.61 Å), seems to have provided an answer to this apparent enigma.[25] Four strands of r(UGGGGU) form a parallel-stranded quadruplex, such as that described above for the analogous DNA sequence d(TGGGGT), but in the case of r(UGGGGU), all four strands in the asymmetric unit are symmetry-related molecules. The tetramolecular G-quadruplexes of r(UGGGGU) stack with one another in opposite polarity (head-to-head or tail-to-tail) to form a pseudo-continuous column. Sr^{2+} ions are coordinated between *every other* pair of stacked G-quartets, without any other cations coordinated at the intervening positions (Figure 2). The Sr^{2+} ions observed are associated with eight carbonyl oxygen atoms of adjacent G-quartets in a bipyramidal-antiprism geometry.[25] Thus, it appears that the greater electrostatic repulsion between Sr^{2+} ions forces these ions to leave vacant cation coordination sites within G-quadruplexes.

In general, X-ray crystal structures support the conclusion that cations with ionic radii larger than Na^+, such as K^+ and Tl^+, are coordinated exclusively between adjacent G-quartets because they are too large to be coordinated within the plane of a G-quartet. In contrast, Na^+ ions can be coordinated within a G-quartet with three distinct ligand geometries: bipyrimidal coordination sites directly between two G-quartet layers, octahedral coordination sites where the ion is coplanar with the G bases, and intermediate less symmetric positions. Inter-Na^+ distances in the parallel-stranded tetramolecular d(TGGGGT) structure and the bimolecular d($G_4T_4G_4$) structure described above are greater than the average distance between stacked G-quartets. Na^+ ions, being less constrained by steric clashes than K^+ ions, can therefore occupy a range of positions and in doing so reduce electrostatic repulsions between adjacent ions. Studies on the relationship between G-quadruplex stability and cation species have primarily considered ionic radius as the most important factor. However, the fact that Na^+ and Ca^{2+} have almost the same ionic radii (0.95 Å and 0.99 Å, respectively) illustrates that there are other factors, including the energy of dehydration and/or coordination number, which must be considered. These factors are discussed below.

4.3   Localization of Cation Coordination Sites in G-Quadruplexes in the Solution State

The first experimental evidence to demonstrate the direct coordination of dehydrated cations between G-quartets in the solution state was provided by the observation of broadening and upfield shift of the ^{23}Na resonance in solutions of 5′-GMP.[34,35] With respect to folded G-quadruplex structures, one of the first studies to determine the number of cations bound within a G-quadruplex in the solution state was performed using the DNA sequence d($G_3T_4G_3$).[36] In the presence of either NaCl or KCl, d($G_3T_4G_3$) forms a bimolecular fold-back structure containing three stacked G-quartets.[37,38] This topology of the d($G_3T_4G_3$) quadruplex is similar to that shown in Figure 1 for the closely related DNA sequence d($G_4T_4G_4$), except with one less G-quartet. ^1H NMR spectroscopy was used to follow the competition between Na^+ and

K^+ ions for coordination sites within the G-quadruplex.[36] Changes in ^1H NMR spectra during these titration experiments indicated a gradual transition of $[d(G_3T_4G_3)]_2$ from the G-quadruplex structure observed in the presence of only NaCl to the G-quadruplex structure observed in presence of only KCl. Although the Na^+ and K^+ forms of $[d(G_3T_4G_3)]_2$ are the same molecular folds, there are relatively small structural differences between the two forms that are manifested by changes in ^1H chemical shifts. No separate or additional ^1H resonances were observed in samples of $[d(G_3T_4G_3)]_2$ that contained various mixtures of NaCl and KCl, which demonstrated that the rate of exchange of Na^+ and K^+ from this quadruplex at 25°C is fast on the NMR time scale (*ca.* < 10 ms). ^1H chemical shift changes observed over the course of KCl titration into a sample of $[d(G_3T_4G_3)]_2$ initially in the Na^+ form were fit perfectly by a model in which two Na^+ ions are replaced by two K^+ ions within the G-quadruplex. Furthermore, quantitative analysis of chemical shift changes during the same titration experiments gave important insights regarding the thermodynamic basis for the preferential coordination of K^+ over Na^+ by G-quadruplexes (see Section 4.5).

Along with the alkali metal ions, the NH_4^+ ion had been shown to stabilize G-quartets to an extent that is similar to that observed for Na^+.[1,14] Hud *et al.*[27] used this fact to develop $^{15}NH_4^+$ as a solution-state probe of monovalent cation coordination sites.[39] These authors used ^1H–^1H and ^1H–^{15}N NMR cross-relaxation experiments, as well as ^{15}N-filtered ^1H experiments, to characterize NH_4^+ ion localization within the bimolecular G-quadruplex $[d(G_4T_4G_4)]_2$. Detection and analysis of dipolar interactions between the protons of the bound $^{15}NH_4^+$ ions and G-imino protons revealed that $[d(G_4T_4G_4)]_2$ coordinates three $^{15}NH_4^+$ ions within its symmetric architecture, one in each of two symmetry-related sites (termed the "outer sites") and one on the axis of symmetry of the bimolecular G-quadruplex structure (termed the "inner site") (Figure 3). The observed NOE cross-peak intensities also revealed that $^{15}NH_4^+$ ions in a G-quadruplex are positioned equidistant between stacked G-quartets.[27,40]

NMR spectroscopy has also been used to monitor the competition between Na^+ and NH_4^+ ions for coordination within G-quadruplexes at specific sites. For the G-quadruplex formed by $d(G_4T_4G_4)$ in the presence of NH_4Cl, Na^+ ions will replace NH_4^+ ions at all three of the above-mentioned outer and inner binding sites (Figure 3) as the NaCl concentration is increased.[28] Moreover, Na^+ ions preferentially replace NH_4^+ ions coordinated at the inner site of $[d(G_4T_4G_4)]_2$ quadruplexes before NH_4^+ ions at the outer binding sites are replaced.[28] The 11-mer $d(G_3T_4G_4)$, a single nucleotide deletion of $d(G_4T_4G_4)$, has been shown by NMR spectroscopy to fold into an unusual asymmetric, bimolecular structure in the presence of NH_4^+, K^+, and Na^+ ions. The structure of $[d(G_3T_4G_4)]_2$ contains three G-quartets, similar to the structure of $[d(G_3T_4G_3)]_2$, but with one diagonal and one lateral-type loop.[41,42] Multinuclear NMR studies have demonstrated that the three G-quartets of $[d(G_3T_4G_4)]_2$ create two cation coordination sites.[40] Titration of KCl into a solution of $[d(G_3T_4G_4)]_2$ folded in the presence of $^{15}NH_4^+$ ions revealed a

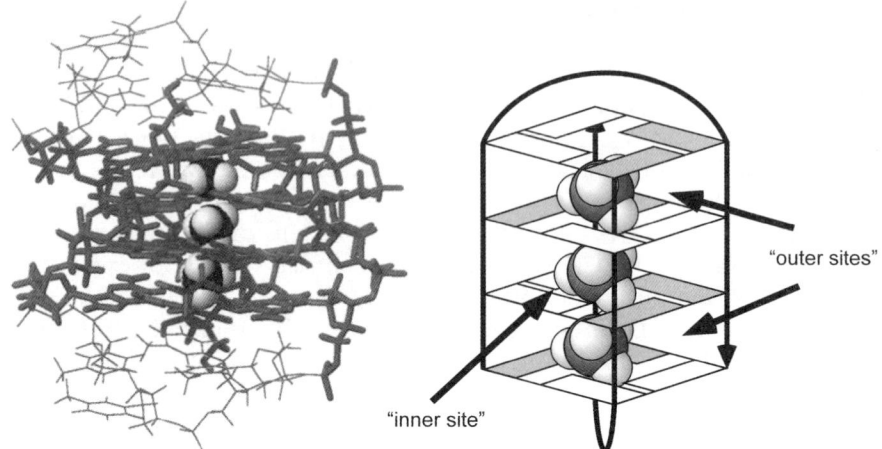

Figure 3 *Left: Calculated structure of [d(G₄T₄G₄)]₂ with internal ¹⁵NH₄⁺ ions based on NMR-derived distance restraints.²⁷ For clarity, the dT residues of the loops are drawn with thinner lines than the dG residues. Right: Topology diagram of [d(G₄T₄G₄)]₂ indicating the two distinct coordination sites, the symmetry-related outer sites and the unique inner site that lies on the axis of rotational symmetry*

mixed mono-K^+/mono-$^{15}NH_4^+$ form that represents an intermediate in the conversion of the di-$^{15}NH_4^+$ form into the di-K^+ form. $^{15}NH_4^+$ ions were similarly found to replace Na^+ ions inside the quadruplex. The preference for $^{15}NH_4^+$ over Na^+ ions for the two internal coordination sites is considerably less than the preference for K^+ over $^{15}NH_4^+$ ions. The two coordination sites within the G-quadruplex differ to such a degree that $^{15}NH_4^+$ ions bound to the site that is closer to the lateral-type loop are always replaced first during titration by KCl. That is, the second binding site is not occupied by a K^+ ion until a K^+ ion already resides at the first binding site. Quantitative analysis of the relative concentrations of three di-cation forms (*i.e.*, 2 $^{15}NH_4^+$, $^{15}NH_4^+$/ K^+, and 2 K^+) at equilibrium, which are in slow exchange on the NMR time scale, showed that differences in standard Gibbs free energy between the di-$^{15}NH_4^+$ form and the $^{15}NH_4^+$/K^+ form is 5.7 kcal mol^{-1}, and between the $^{15}NH_4^+$/K^+ form and the di-K^+ form is 4.3 kcal mol^{-1}.[40]

In agreement with X-ray crystallographic and NMR spectroscopy studies, molecular dynamics calculations on the bimolecular G-quadruplex [d(G₃T₄G₃)]₂ have demonstrated the localization of K^+ ions between the neighboring G-quartets and in octahedral coordination by guanine O6 oxygens.[43] Two K^+ ions were found to be slightly offset from a central location between pairs of adjacent G-quartet planes because of repulsive forces between the ion pairs.[43] We note that solid-state NMR spectroscopy has also been used to locate cations in G-quadruplexes.[44–47] A recent ^{23}Na NMR study has confirmed that three Na^+ cations reside inside the bimolecular G-quadruplex

$[d(G_4T_4G_4)]_2$.[47] Evidence was also presented that each Na^+ resides between two G-quartets, as opposed to being coordinated within the plane of a G-quartet. There was no evidence of Na^+ cations in the thymine loop.[47]

4.4 Measuring the Dynamics of Cation Exchange by NMR Spectroscopy

NMR spectroscopy has been used extensively to measure the dynamics of cations coordinated within G-quadruplexes. By analyzing resonance line-widths in NMR spectra of the quadrupolar nuclei ^{23}Na and ^{39}K, Braunlin and co-workers were able to estimate the rotation times and residence lifetimes of dehydrated cations within G-quadruplexes. These investigators demonstrated that Na^+ and K^+ ions coordinated within the G-quadruplexes $[d(T_2G_4T)]_4$ and $[d(G_4T_4G_4)]_2$ exhibit quadrupolar relaxation effects that reflect the molecular tumbling of the G-quadruplexes, which occurs on the time scale of nanoseconds.[48,49] Thus, cations within these two G-quadruplexes have rotation times that are at least on the order of nanoseconds. The rotational immobilization exhibited by coordinated cations within G-quadruplexes is in distinct contrast to the high rotational mobility of cations associated with the surface of a G-quadruplex. Na^+ ions coordinated within the tetramolecular quadruplex $[d(T_2G_4T)]_4$ were shown to exchange rapidly compared to Na^+ ions coordinated within the bimolecular quadruplex $[d(G_4T_4G_4)]_2$.[49] ^{23}Na NMR relaxation measurements performed as a function of temperature provided an estimate of 250 μs (at 10°C) for the lifetime of Na^+ ions coordinated within $[d(G_4T_4G_4)]_2$.[49] The residence lifetimes determined for Na^+ ions tightly bound within DNA quadruplexes are in distinct contrast to the kinetics of G-quartet base opening, which can be as slow as days to weeks.[50] Thus, cations must move between bulk solution and the cation coordination sites within a G-quadruplex by passing along the axial channel of the quadruplex, rather than by moving through transient side openings in G-quartets. The difference observed between the bound lifetime of Na^+ ions within $[d(T_2G_4T)]_4$ and $[d(G_4T_4G_4)]_2$ suggests that the diagonal loops of $[d(G_4T_4G_4)]_2$ (Figure 3) limit the rate at which Na^+ ions can exchange between the cation coordination sites within this quadruplex and bulk solution.

In addition to serving as a probe for locating cation binding sites in solution, $^{15}NH_4^+$ has also proven to be valuable for monitoring the movement of cations between different coordination sites within G-quadruplexes.[26,27,39] As discussed above, NMR spectroscopy was used to demonstrate that $^{15}NH_4^+$ ions occupy three coordination sites within the bimolecular quadruplex $[d(G_4T_4G_4)]_2$, with two sites being symmetry-related sites (*i.e.*, the outer sites) and one being a unique site at the center of the molecule (*i.e.*, the inner site) (Figure 3).[27] All three binding sites are fully occupied by $^{15}NH_4^+$ ions. ^1H–^{15}N 2D correlational spectra revealed that NH_4^+ ions move along the central axis of the quadruplex between these three sites and the solution (Figure 4), reminiscent of an ion channel. The movement of $^{15}NH_4^+$ ions between the inner and outer coordination sites was shown to result in the $^{15}NH_4^+$ ions having a residence time at

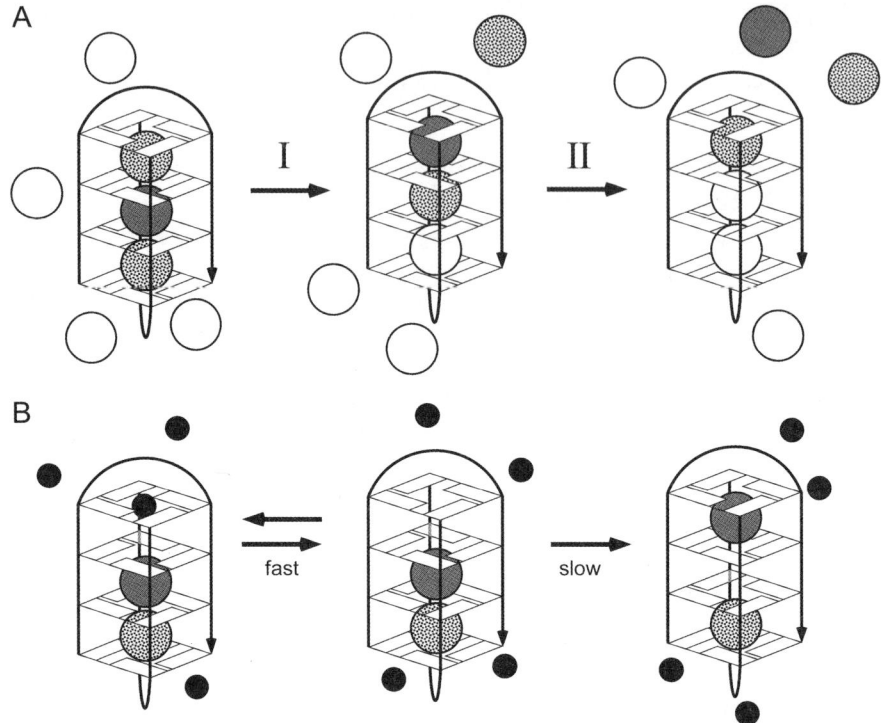

Figure 4 *(A) A schematic representation of* $^{15}NH_4^+$ *movement through the quadruplex* $[d(G_4T_4G_4)]_2$. *Dark gray, light gray, and white spheres represent inner, outer, and bulk* $^{15}NH_4^+$ *ions, respectively, in the initial state. The residence time of* $^{15}NH_4^+$ *ions bound at the inner site was revealed to be approximately 250 ms at 10 °C by 2D heteronuclear NMR exchange spectroscopy.[27] Step I represents a single movement of* $^{15}NH_4^+$ *ions, step II represents a second movement of ions along the same direction through the quadruplex, which was observed by NMR spectroscopy.[27] (B) Illustration of a model for the mechanism by which Na$^+$ ions accelerate the movement of* $^{15}NH_4^+$ *ions through the quadruplex* $[d(G_4T_4G_4)]_2$.[28] Na$^+$ *ions, represented by small black spheres, exchange rapidly between the outer coordination sites of* $[d(G_4T_4G_4)]_2$ *and bulk solution (residence times <1 ms).[49] Thus, movement of an inner* $^{15}NH_4^+$ *to an outer site occupied by Na$^+$ is not impeded by the presence of Na$^+$. The measured residence time for* $^{15}NH_4^+$ *ions bound at the inner site of 36 ms therefore reflects the rate at which* $^{15}NH_4^+$ *can move to an empty outer site. In a solution containing only* $^{15}NH_4^+$ *ions, as shown in A, each movement of a* $^{15}NH_4^+$ *from an inner to an outer site also requires the movement of a* $^{15}NH_4^+$ *from an outer site into bulk solution, which results in a much longer observed residence time for* $^{15}NH_4^+$ *ions bound at the inner site (i.e., 270 ms at 25°C)[28]*

the inner site of 250 ms (at 10°C). Cations were also shown to move from the outer binding sites into solution. The analysis of the cross-peak intensities showed that the number of cations that have moved from the outer to the inner coordination sites, inner to outer coordination sites, and outer coordination sites to bulk are all equal. There was no evidence to support the direct movement of $^{15}NH_4^+$ from the bulk to the inner site. Thus, as indicated above, cations do not enter and exit through the sides the G-quadruplex, but rather move between coordination sites along the central axis of the G-quadruplex and out through the ends. Vacant coordination sites are likely to exist, at least transiently, during $^{15}NH_4^+$ movement. However, the lifetime of these vacancies must be short compared with the lifetime of occupied sites. The relatively large size of $^{15}NH_4^+$ ions in comparison to Na^+ ions is likely to hinder their movement through the quadruplex and provides a rationale for their longer residence time compared with Na^+.[27]

The significant difference in the residence times of NH_4^+ and Na^+ ions bound inside the bimolecular G-quadruplex $[d(G_4T_4G_4)]_2$ has recently been explored within the same experiments by NMR spectroscopy.[28] These experiments have revealed that $^{15}NH_4^+$ movement from the inner to the outer binding sites of $[d(G_4T_4G_4)]_2$ is accelerated by *ca.* 7.5 times in the presence of Na^+ ions. The residence time of $^{15}NH_4^+$ ions within the inner coordination site is reduced from 270 ms in the presence of $^{15}NH_4^+$ ions alone to 36 ms in the presence of both $^{15}NH_4^+$ and Na^+ ions (at 25°C). The addition of Na^+ ions to a solution containing $[d(G_4T_4G_4)]_2$ initially in the NH_4^+ form results in the partial occupancy of cation coordination sites by Na^+ ions. This partial occupation by Na^+ ions is strongly correlated with the acceleration of $^{15}NH_4^+$ movement through the G-quadruplex, whereas bulk ion concentration does not exert a noticeable influence on the rate of $^{15}NH_4^+$ movement from the inner to the outer coordination sites. Even a low concentration (*i.e.*, 5 mM) of Na^+ ions decreases the residence time of $^{15}NH_4^+$ ions bound at the inner coordination site to the minimum value of 36 ms. Further increases in Na^+ concentrations, to the point that one or two coordination sites of $[d(G_4T_4G_4)]_2$ are completely occupied by Na^+ ions, does not lead to further increase in the rate of $^{15}NH_4^+$ movement. Thus, the bound residence time of 36 ms for a $^{15}NH_4^+$ bound at the inner coordination site provides a good estimate for the limiting rate for the movement of a $^{15}NH_4^+$ from the inner to an outer binding site, as Na^+ movement is not rate limiting (Figure 4). The much slower rate of $^{15}NH_4^+$ movement observed when all three coordination sites of $[d(G_4T_4G_4)]_2$ are occupied by $^{15}NH_4^+$ is apparently because of the added restriction that movement of a NH_4^+ to an adjacent site within the G-quadruplex requires a concerted movement of another NH_4^+ out of the adjacent site (*i.e.*, double NH_4^+ movement).[28]

As mentioned above, Tl^+ can also stabilize G-quadruplex structures. The ability for Tl^+ ions to compete with Na^+ ions for coordination within the G-quadruplex $[d(G_4T_4G_4)]_2$ has been verified by solution-state ^1H NMR.[51] The results from these experiments were similar to the K^+–Na^+ titration experiments discussed above for the same G-quadruplex.[51] A recent solution-state NMR structure of $[d(G_4T_4G_4)]_2$ folded in the presence of Tl^+ ions has also confirmed that the Tl^+ form of this quadruplex is similar to that of the K^+ form.[52] The excellent NMR properties of ^{205}Tl, spin 1/2 with a high natural abundance of 70% and relative sensitivity to protons of 0.13, also make ^{205}Tl a promising probe for directly analyzing cation coordination within G-quadru-plexes.[51] Strobel and co-workers have reported the NMR spectra for ^{205}Tl$^+$ ions in a solution containing the tetramolecular G-quadruplex [d(T$_2$G$_4$T$_2$)]$_4$.[16] Owing to the large chemical shift range of ^{205}Tl and its sensitivity to ligands and coordination geometry, resonances of ^{205}Tl$^+$ ions associated with the G-quartet were clearly resolved from the resonance of ^{205}Tl$^+$ ions in bulk solution. Three ^{205}Tl signals were attributed to three ^{205}Tl$^+$ ions bound at distinct sites within the G-quadruplex, presumably the three coordination sites that exist between each adjacent pair of the four G-quartets.[16] Three Tl$^+$ ion binding sites have been established within the G-quadruplex adopted by d(T$_2$G$_4$T$_2$) using ^{205}Tl NMR in solution.[16] Differences in resonance peak line-widths suggested that ions move from one ion coordination site to the other. The central peak was narrower, which was attributed to the slower exchange of the Tl$^+$ ion at the middle coordination site. The outside Tl$^+$ ions exchange with free ions in solution more freely. The residence lifetime of Tl$^+$ ions bound within the G-quadruplex was estimated to be at least 3 μs.[16]

4.5 Thermodynamics of G-Quadruplex Cation-Dependent Stability and Cation Selectivity

4.5.1 Monovalent Cations

The stability of a particular G-quadruplex can exhibit a strong dependence on the species of cation coordinated by its G-quartets. Most cation-dependent stability experiments have focused on the relative stabilities of G-quadruplexes in the presence of K$^+$ *vs.* Na$^+$ ions. In general, the melting temperature (T_m) for a G-quadruplex is higher in the presence of K$^+$ ions than in the presence of Na$^+$ ions. The size of a cation and its energy of hydration both contribute to cation selectivity and the stability of a G-quadruplex. As discussed above, K$^+$ is too large to be coordinated in the plane of G quartet, whereas a Na$^+$ can fit in the center of G-quartet plane. The resulting differences in K$^+$ and Na$^+$ coordination geometries and cation-guanine O6 distances contribute to a difference in the free energy provided by K$^+$ *vs.* Na$^+$ coordination within a G-quadruplex. Of equal importance is the fact that cations differ in their respective energies of hydration. To a good approximation, hydration energies of monovalent ions are inversely proportional to their ionic radii. Thus, the free energy required to dehydrate K$^+$ for coordination within a G-quadruplex is less than that required to dehydrate Na$^+$. The net difference between the free energy of coordination within a G-quadruplex and the free energy of hydration ultimately determines cation selectivity by G-quadruplexes.[3,36,53,54]

The relative free energies of Na$^+$ *vs.* K$^+$ coordination within a G-quadruplex have been determined under equilibrium conditions using ^{1}H NMR spectros-copy.[36] This study used the G-quadruplex formed by d(G$_3$T$_4$G$_3$), a sequence that forms a bimolecular quadruplex with three G-quartets in the presence of either Na$^+$ or K$^+$ ions.[37,38] ^{1}H chemical shifts were followed as a sample of [d(G$_3$T$_4$G$_3$)]$_2$ was converted from its Na$^+$ form to its K$^+$ form by titration with

KCl. Equilibrium constants for the replacement of Na^+ by K^+ were derived by fitting the G-quadruplex ^1H chemical shifts as a function of KCl concentration. Perfect fits were obtained for a model in which two Na^+ ions were replaced by two K^+ ions. Quantitative analyses demonstrated that the preferred coordination of K^+ over Na^+ is actually driven by the greater energetic cost of Na^+ dehydration with respect to K^+ dehydration, whereas the intrinsic free energy of Na^+ coordination within $[d(G_3T_4G_3)]_2$ is actually more favorable than K^+ coordination (Figure 5). Overall, the conversion of the G-quadruplex $[d(G_3T_4G_3)]_2$ from its Na^+ form to K^+ form is associated with a net free energy change ($\Delta\Delta G^\circ$) of -1.7 kcal mol^{-1}.[36] Subsequent calculations have provided additional support for the argument that cation dehydration is the dominant free energy that determines Na^+ *vs.* K^+ selectivity by G-quartets.[55–57]

The extent to which K^+ increases the stability of a G-quadruplex over Na^+ also depends on the nucleotide sequence of the quadruplex, and DT_m can vary from as low as 2°C to over 30°C (Table 1).[13,54,58–60] Structural origins for these sequence-dependent differences are likely because the K^+ and Na^+ forms of some G-quadruplexes are different in structure and/or in the number of coordinated cations. For example, it has been known for over a decade that G-quadruplexes formed from human telomeric repeat oligonucleotide

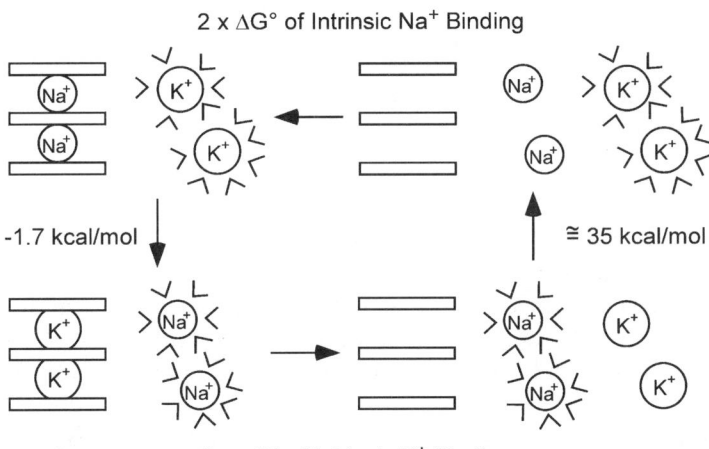

Figure 5 *Free energy cycle for the conversion of $[d(G_3T_4G_3)]_2$ from the Na^+ to the K^+ form in aqueous solution. The free energy difference between the di-Na^+ quadruplex and the di-K^+ quadruplex of -1.7 kcal mol^{-1} was determined under conditions of equilibrium cation exchange by ^1H NMR spectroscopy.[36] The relatively large difference between the free energy of dehydration for two Na^+ ions vs. that for two K^+ ions, with respect to the free energy difference between the Na^+ and K^+ forms of $[d(G_3T_4G_3)]_2$, revealed that Na^+ ions are actually more favorably coordinated by G-quartets than K^+ ions. However, K^+ is selectively bound over Na^+ because of its less unfavorable free energy of dehydration (Reproduced from ref. 36; copyright American Chemical Society)*

Table 1   *Melting temperatures of G-quadruplexes*

	K^+		Na^+		
Sequence	$T_m{}^a$ (°C)	*Salt concentration (mM)*	$T_m{}^a(Na^+)$ (°C)	*Salt concentration (mM)*	*Reference*
Monomolecular G-quadruplexes					
d[GGTTGGTGTGGTTGG]	48	100	~20	100	54
d[AGG(TTAGG)$_3$]	42	100	40	100	54
d[AGGG(TTAGGG)$_3$]	62	100	55	100	54
d[TTAGGG]$_4$	50.2	49	42.4	50	13
d[TTAGGG]$_4$	63	70	49	70	61
d[G$_4$(T$_4$G$_4$)$_3$]	84	100	67	100	58
d[TG$_3$(TTAG$_3$)$_3$]	51.1	10	37.7	10	102
d[TG$_3$(TTAG$_3$)$_3$]	81.8	100	62.8	100	102
d[G$_3$(TG$_3$)$_3$]	100	1	86.2	10	102
d[G$_3$(T$_2$G$_3$)$_3$]	80.0	1	60.3	10	102
d[G$_4$(T$_2$G$_4$)$_3$]	79.1	10	51.9	10	102
d[G$_4$TTG$_4$TGTG$_4$TTG$_4$]	-		86.0	200	121
Bimolecular G-quadruplexes					
d[G$_4$T$_3$G$_4$]	-		44	70	97
d[G$_4$T$_4$G$_4$]	n.r.^{b}		53	100	54
d[G$_4$T$_4$G$_4$]	-		47.0^c/66.6	200	121
d[G$_3$TTAG$_3$]	42	70	31	70	61
Tetramolecular G-quadruplexes					
d[TGGT]	48	110	16	110	60
d[TGGGT]	>90	100	55	100	54
d[TG$_4$T]	-		74.6	200	121
d[TTAGGG]	50	110	17	110	60
d[TTAGGGT]	55	110	24	110	60
d[TTAGGG]	50	110	17	110	60

^{a} T_m values in the table refer to both reversible transitions and apparent melting temperatures ($T_{1/2}$) for the non-reversible processes.
^{b} Non-reversible.
^{c} Premelting process.

sequences (*e.g.*, d(TTAGGG)$_4$) exhibit different loop conformations depending upon whether Na^+ or K^+ ions are present.[61] As discussed below, it is now known that the G-quadruplex formed by some G-rich sequences in the presence of K^+ ions can be dramatically different in strand orientation (*i.e.*, parallel *vs.* antiparallel) and strand number (monomolecular *vs.* tetramolecular) from the G-quadruplex formed in the presence of Na^+ ions. Thus, it is not surprising that a simple relationship does not exist for the difference in T_m for the Na^+ and K^+ forms for all known G-quadruplexes (Table 1).

Other monovalent cations shown to promote the formation of G-quadruplex structures include Rb^+, Cs^+,[12,13] NH_4^+[14,15], and Tl^+.[16] Decades ago Chantot and Guschlbauer measured the melting temperature of gels formed by 8-bromo-guanosine (gels that are now known to contain G-quartets) as a function of monovalent cation species.[29] The resulting order of monovalent cations for their relative ability to stabilize guanosine gels was $K^+ >> Rb^+ > NH_4^+ > Na^+ > Li^+$. This general ordering has proven correct for quadruplex structures that contain G-tetrads. For example, a systematic study on the thermal stability of the 24-mer d(TTAGGG)$_4$, derived from the human telomere repeat sequence, has indicated the following ordering of monovalent cations according to their stabilization of monomolecular G-quadruplex structures: $K^+ > Na^+ \geq Rb^+ > Li^+ > Cs^+$.[13] The selective binding of monovalent cations to G-quartet structures formed by GMP has also been studied by ^{23}Na NMR in the solid-state.[46] A series of ion titration experiments provided quantitative thermodynamic parameters for the relative cation coordination free energy within the central cavity of a G-quadruplex structure. The measured values for the free energy difference ($\Delta\Delta G°$) between the coordination of each cation *vs.* Na^+ are K^+ (-1.9 kcal mol^{-1}), NH_4^+ (-1.8 kcal mol^{-1}), Rb^+ (-0.3 kcal mol^{-1}), and Cs^+ (1.8 kcal mol^{-1}). These experiments produced the following ordering of monovalent cations for their ability to stabilize GMP G-tetrad assemblies: $K^+ > NH_4^+ > Rb^+ > Na^+ > Cs^+ > Li^+$.[46] Together, T_m and cation competition experiments consistently show that K^+ ions have an optimal balance between favorable coordination within a G-quadruplex and not too unfavorable free energy of dehydration. At the other end of this spectrum are Cs^+ and Li^+. It appears that these cations are not nearly as favorable for G-quadruplex stabilization because Cs^+ is too large for optimal coordination by G-quartets within a quadruplex, whereas the free energy required for Li^+ dehydration is much higher than that of other alkali metal ions.

4.5.2 Divalent Cations

G-quadruplexes can also be stabilized by divalent cations, including Sr^{2+},[13,17] Ba^{2+},[18] and Pb^{2+}.[14,19,62] Studies of 8-bromo-guanosine gel melting transition temperatures indicate the relative G-quartet stabilizing propensities of divalent cations as $Sr^{2+} >> Ba^{2+} > Ca^{2+} > Mg^{2+}$. Like the monovalent cation series, this same ordering of divalent cations has generally been observed for G-quadruplex stabilization. However, the effects of divalent cations on G-quadruplex stability have proven to be more complex. Sr^{2+}, with a similar ionic radius to K^+, has been shown in several studies to facilitate formation of G-quadruplexes.[1,17,19] For example, melting studies of the monomolecular G-quadruplex d(TTAGGG)$_4$ has shown that Sr^{2+} is even more stabilizing than K^+.[13] Pb^{2+} was also shown to bind more effectively than K^+ to the thrombin-binding aptamer (TBA), d(GGTTGGTGTGGTTGG), and to induce an monomolecular fold that is similar to the structure formed in the presence of K^+.[19] Similar behavior was reported for other monomolecular G-quadruplexes, but no strong Pb^{2+} binding was observed in the case of hairpin bimolecular G-quadruplexes.[19] Davis and co-workers have shown that, in the presence of Pb^{2+}, lipophilic guanine analogues can associate to form quadruplex-like structures in organic solvents.[63] These results affirmed earlier studies that Pb^{2+} induces a more stable and compact structure compared to K^+, which is evident from the comparison of cation-O6 bond lengths, O6–O6 diagonal distances and inter-quartet separation.[19]

Studies by Hardin and co-workers have shown that Mg^{2+} and Ca^{2+} ions stabilize the tetramolecular parallel-stranded G-quadruplexes formed by the sequences d(CGCG$_3$GCG) and d(TATG$_3$ATA) more than the monovalent alkali cations with similar radii, Li$^+$, and Na$^+$, respectively.[64,65] A CD spectroscopic study by Sugimoto and co-workers has shown that the bimolecular G-quadruplex [d(G$_4$T$_4$G$_4$)]$_2$, formed in the presence of NaCl with antiparallel strands (Figure 1), is actually destabilized by only 1 mM concentrations of Mg^{2+}, Ca^{2+}, Mn^{2+}, Co^{2+}, or Zn^{2+}.[66–69] Thermodynamic parameters revealed that these divalent cations destabilize the bimolecular, antiparallel G-quadruplex [d(G$_4$T$_4$G$_4$)]$_2$ in the order: Zn^{2+} > Co^{2+} > Mn^{2+} > Mg^{2+} > Ca^{2+}. Furthermore, CD spectra indicate that at higher Ca^{2+} concentrations, in the presence of 100 mM NaCl, d(G$_4$T$_4$G$_4$) undergoes a transition from the antiparallel to a parallel-stranded G-quadruplex structure. This transition is complete at 20 mM CaCl$_2$.[66]

Millimolar concentrations of Mg^{2+}, Ca^{2+}, Ba^{2+}, and Sr^{2+} will induce polymeric d(GGA)$_n$ repeat molecules to form quadruplexes.[70] For these structures, the relative stabilizing effects of the divalent alkaline-earth cations follow the order: Sr^{2+} > Ba^{2+} > Ca^{2+} > Mg^{2+}.[64,65,70] The situation with divalent cations again seems more complicated than that of monovalent cations. For example, while Ca^{2+} ions have been shown to fold poly-d(GGA) into quadruplex structures,[14,70] Ca^{2+} and Mg^{2+} do not result in the formation of any ordered structure by the four repeats of human telomere sequence d(TTAGGG)$_4$.[13] On the other hand, triethylene tetraamine has been suggested to stabilize and lead to formation of the G-quadruplex adopted by the 24-mer d(TTAGGG)$_4$.[71]

The more complicated G-quadruplex stabilizing properties of divalent cations may result from the ability for some divalent cations (*e.g.*, Mg^{2+}) to only screen electrostatic repulsions between the backbones of a G-quadruplex that coordinates other cations, whereas other divalent cations (*e.g.*, Sr^{2+}) can stabilize a G-quadruplex both by screening backbone repulsions and by coordination with G-quartets. Qualitatively different behavior may be seen with monovalent cations, as the ability for monovalent cations to screen backbone repulsions is significantly less than that of divalent cations, and monovalent cation stabilization of G-quartets may be less sensitive to G-quadruplex topology and nucleotide sequence.

4.6   Cations and G-Quadruplex Polymorphism

In addition to their central role in stabilizing G-quadruplexes, cations also play an important role in determining G-quadruplex structure. G-quadruplexes are polymorphic with respect to several structural features, including strand orientation (*i.e.*, parallel *vs.* antiparallel), strand number (*e.g.*, bimolecular *vs.* tetramolecular), *syn* and *anti* conformation across guanine nucleoside glycosidic bonds, and loop topology (*e.g.*, lateral loops *vs.* diagonal loops). Oligonucleotides with single G-tracts typically form tetramolecular structures with parallel strands, while oligonucleotides with two G-tracts separated by two or more other bases have a tendency to fold into bimolecular structures with antiparallel strands. Oligonucleotides with four G-tracts separated by two or more non-G bases tend to fold into intramolecular G-quadruplexes. The polymorphism of G-quadruplexes results from a subtle balance between several energetic terms, including the cation coordination, base–base stacking interactions, hydrogen bonding, and hydrophobic effects. In this section we discuss how cation species, particularly K^+ and Na^+, can promote different G-quadruplex structures for the same G-rich sequence. Some of these differences are relatively minor, whereas others are quite dramatic. Analysis of cation-specific structures continues to provide insights into the factors that govern the structure and formation of G-quadruplexes in general.

4.6.1 Minor Cation-Dependent Polymorphisms

The NMR structure of the G-quadruplex formed by $d(G_4T_4G_4)$ in the presence of NaCl was one of the first G-quadruplex structures determined at high resolution.[31,32] This structure revealed that $d(G_4T_4G_4)$ forms a symmetrical bimolecular G-quadruplex with four G-quartets and two dT_4 loops that span diagonally across each end of the quadruplex (Figure 2). This structure was also determined by NMR to have the same topology assumed by $d(G_4T_4G_4)$ in the presence of K^+ ions.[33] A subsequent X-ray crystal structure of $[d(G_4T_4G_4)]_2$ in the Na^+ form, located in a cavity of the *O. nova* single-strand telomere-binding protein complex,[21] revealed the same fold as the NMR structure. A recent X-ray study of the K^+ form of $[d(G_4T_4G_4)]_2$, which crystallizes in two crystal forms,[22] also revealed identical folds to the NMR and the protein-associated structures. These results are likewise in accordance with the conformations deduced by Raman spectroscopy.[72] The same bimolecular, diagonal loop fold is also assumed by the closely related quadruplex $d[(G_3T_4G_3)]_2$.[31,33,38,43,73] However, the precise structure and dynamics of the T_4 loops of both G-quadruplexes are cation-dependent. A comparison of the Na^+, K^+, and NH_4^+ NMR structures of $[d(G_4T_4G_4)]_2$ indicated that the coordination of Na^+ ions within the planes of the outer G-quartets allowed for the participation of the O2 carbonyl of the third loop thymine (T7) in the coordination of these Na^+ ions (Figure 6).[33] In contrast, coordination of K^+ or NH_4^+ ions between the planes of the G-quartets did not allow for coordination by the same loop thymine, which results in a different loop structure (Figure 6). When formed in the presence of K^+ ions, there is also evidence of greater motion in the diagonal loops of the quadruplexes with respect to the Na^+ structure.[31,33,37,43,74] Strahan et al.[43] have shown using a combined multidimensional NMR and molecular dynamics approach that the dynamics can be modeled by exchange between two conformations that differ primarily in the stacking and unstacking of the fourth thymine (T8) residue over the outer G-quartet. The greater motion of the loops of the K^+ form with respect to the Na^+ form is also consistent with the position of the K^+ being located between the outer- and the inner-G-quartets, and hence too far removed from atoms of the loop residues to allow coordination.

As mentioned above, NMR studies indicate that the K^+ form of $[d(G_3T_4G_3)]_2$ is basically the same as the fold adopted in the presence of Na^+

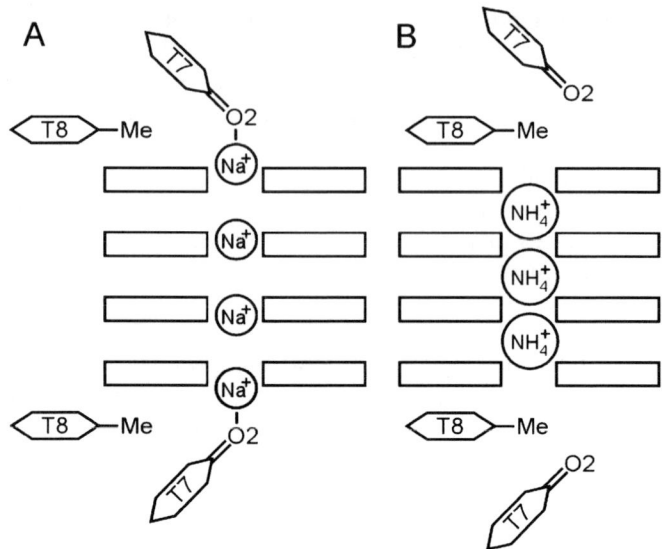

Figure 6   *An illustration of the proposal made by Schulze et al.[33] to explain the different loop conformations of [d($G_4T_4G_4$)]$_2$ in the Na$^+$ form versus the similar NH$_4^+$ and K$^+$ forms. (A) Na$^+$ ions are coordinated in the plane of the outer G-quartets of [d($G_4T_4G_4$)]$_2$, where additional coordination is possible with O2 carbonyls of thymidine loop residues. (B) The larger NH$_4^+$ (right) is coordinated between G-quartets, where interaction with loop resides is not possible. This difference in cation coordination apparently contributes to the loops of the Na$^+$ form being less mobile than in the K$^+$ form. Additionally, the base of loop residue T8 in the NH$_4^+$ and K$^+$ forms, in the absence of steric clash with T7, stacks directly on the base of residue G9*

ions, except for localized structural variations. However, some significant qualitative differences between the K$^+$ and Na$^+$ forms of this G-quadruplex have been reported.[43] Several days after the addition of KCl to a solution of d(G$_3$T$_4$G$_3$), broad resonance lines began to appear in the NMR spectra. The new species has been assigned to a linear, tetramolecular G-quadruplex, which can efficiently stack end to end, giving rise to broad NMR peaks. In contrast, no signs of a linear quadruplex species have been observed in the presence of NaCl. These results were interpreted in terms of kinetic and equilibrium differences. In the presence of NaCl, the free energy difference, $\Delta\Delta G°$, between the diagonally looped, bimolecular and the linear, tetramolecular quadruplex forms is sufficiently large that only the bimolecular quadruplex is detected. In the presence of KCl, however, the $\Delta\Delta G°$ is significantly smaller and detectable amounts of each structure are observed at equilibrium. The true equilibrium conditions were only established after several days at NMR concentrations, since the kinetics of formation of the tetramolecular G-quadruplex are much slower than for the bimolecular structure.[43] The apparent larger size of G-quartets stabilized by K$^+$ ions provides greater loop flexibility, resulting in a weakening of the loop-stabilizing interactions. This may lead to a lower overall stability of the diagonally looped, bimolecular structure in the case of K^+ and, correspondingly, a smaller $\Delta\Delta G^\circ$ between the bimolecular and the tetramolecular quadruplexes in KCl compared to NaCl.[43]

A similar cation-dependent loop polymorphism has been observed in the structure of a G-quadruplex that contains $G \cdot C \cdot G \cdot C$ quartets, along with G-quartets. A cation-induced conformational exchange was observed in the bimolecular, laterally looped fold adopted by $d(G_3CT_4G_3C)$.[75,76] The conformation of thymines in the loop is, again, different between Na^+ and K^+ forms. However, in the case of this quadruplex, the loop conformation is apparently affected by the presence of a K^+ binding site in the T_4 loop, whereas there is no such binding site for Na^+. The loop conformation, in turn, is sensitive to the conformation of the outer G-quartet, which is substantially different in both forms. In the Na^+ form of this G-quadruplex, the first and last thymine in the loop stack over the adjacent bases in the G-quartet, but in the K^+ form, the thymine at the 3' end of the loop is pulled away from this position to create a K^+ binding site.[75,76] A K^+ ion is also postulated to partially penetrate the $G \cdot C \cdot G \cdot C$ quartet at the expense of hydrogen bonding between the $G \cdot C$ base pairs.[75,76] In solutions containing both Na^+ and K^+ ions, the two forms of the $d(G_3CT_4G_3C)$ quadruplex are in exchange at a rate that is slow on the NMR time scale. Equal amounts of the two conformations coexist in a solution of 50 mM NaCl and 7.5 mM KCl (at 10°C).[76]

The two-repeat human telomeric sequence $d(TAG_3T_2AG_3T)$ has been shown to form both parallel and antiparallel G-quadruplex structure in the presence of K^+ ions.[77] Both structures are dimeric and comprise three stacked G-quartets. The two structures can coexist and interconvert in solution. They have different thermodynamic properties and different kinetics of folding and unfolding. The antiparallel structure with lateral loops on the opposite sides of G-quadruplex core is more favorable at low temperatures (<50°C), while parallel structure with double-chain-reversal loops is more favorable at higher temperatures.[77] $d(TG_4T_2G_4T)$, an analogous sequence with two G-for-A substitutions, forms two asymmetric, bimolecular G-quadruplexes in solution with Na^+ ions.[78] Both structures consist of four G-quartets and two loops that span along the edges of the end G-quartets. In one structure both loops are at the same end of the G-quadruplex core (head-to-head arrangement), while in the second loops are located at the opposite ends (head-to-tail arrangement). In both structures adjacent strands are parallel and antiparallel.[78]

4.6.2 Major Cation-Dependent Polymorphisms

The NMR structure of the monomolecular G-quadruplex formed by (almost) four repeats of the human telomere sequence, $d[AGGG(TTAGGG)_3]$, in the presence of Na^+ ions was reported several years ago.[79] It comprises a core of three G-quartets held together by strands in alternating orientations (Figure 7). This results in two lateral and one diagonal d(TTA) loop that are positioned at the G-quartet ends. This monomolecular G-quadruplex is stabilized by three

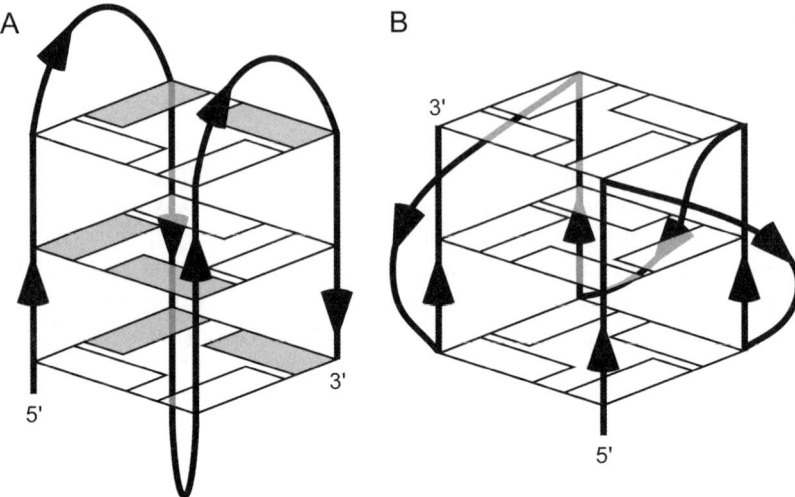

Figure 7   *Schematic diagrams of human telomeric quadruplex folding topologies. (A) The fold determined by NMR spectroscopy of the intramolecular G-quadruplex formed by d(AG₃(T₂AG₃)₃) in solution with only Na⁺ ions. Gray rectangles represent nucleosides with* syn *glycosidic torsion angles. For clarity, only G-bases are represented.[79] (B) The G-quadruplex fold, with parallel GGG elements and external loops, as determined by X-ray crystallography for the same sequence crystallized in the presence of K⁺ ions[23]*

stacked G-quartets with *syn–anti* G-glycosidic bond conformations along individual strands and *syn–anti–anti–syn* conformations around each G-quartet.[79] The crystal structure of the K⁺ form shows a dramatically different fold,[23] with all four GGG segments being parallel (Figure 7). All guanine residues are in the *anti* conformation. All three d(TTA) loops connect the top of one GGG strand with the bottom of the other, and are positioned alongside the grooves rather than at the ends of the G-quartet stack (referred to as "external" loops). The overall structure has a highly asymmetric propeller shape, in contrast to the compact globular shape of the Na⁺ form.

The reasons for the dramatic difference between the solution-state Na⁺ form and the crystal-state K⁺ form of d[AGGG(TTAGGG)₃] are yet to be fully elucidated; major changes in structure have previously been noted in CD studies of the transition,[61] although it had not been possible to assign structures to the individual species. On the other hand, several recent studies have demonstrated that an antiparallel G-quadruplex is formed by the human telomere repeat sequence in solution in the presence of both Na⁺ and K⁺ ions.[80–84] Attempts to determine the high-resolution structure of the 22-mer human telomeric sequence G-quadruplex in the presence of K⁺ ions by NMR in solution have been unsuccessful because of the presence of multiple conformations.[77] The crystal structure of the related bimolecular G-quadruplex formed by the sequence d(TAGGGTTAGGGT) in the presence of K⁺ has the same propeller fold and loop architecture.[23] However, the same sequence was found to form both parallel and antiparallel G-quadruplexes in the presence of K^+ ions in solution.[77] While the parallel structure was similar to that observed in the crystal structure, the antiparallel structure featured two lateral loops on opposite ends of G-quartet stacks. The two structures coexist and interconvert in solution.[77] The oligonucleotide with a single repeat of the human telomere repeat sequence, d(TTAGGG), forms a tetramolecular parallel-stranded G-quadruplex at low K^+ concentrations, which aggregates to form higher order structures when the concentration of K^+ is increased.[85]

The sequence $d(G_4T_4G_3)$, which has one $3'$ guanine missing from the sequence $d(G_4T_4G_4)$ discussed at length above, has been shown to form a G-quadruplex with a significantly different structure and greater cation-dependent polymorphism than $d(G_4T_4G_4)$. It has previously been shown that removal of both $5'$ and $3'$ dG residues from $d[(G_4T_4G_4)]_2$ resulted in the absence of the central G-quartet to create an asymmetric G-quadruplex $[d(G_3T_4G_3)]_2$.[37,43,74] The single-truncation sequence, $d(G_4T_4G_3)$, also forms a bimolecular G-quadruplex, but with a distinctive asymmetric fold involving three stacked G-quartets,[86] as opposed to the four G-quartets of the $[d(G_4T_4G_4)]_2$ structure. The two guanines not involved in the G-quartets are unpaired and are involved in stacking interactions with one G-quartet face and a thymine in the adjacent T_4 loop. A single structure of $[d(G_4T_4G_3)]_2$ was found in the presence of Na^+ ions, while multiple forms were observed in the presence of K^+ or $^{15}NH_4^+$ ions.[86] A more recent study concerning the formation of G-quadruplexes by oligonucleotides with the common sequence motif dG_4-loop-dG_4, where the loop consisted of non-nucleosidic residues, also demonstrated formation of a single structure in the presence of Na^+ ions and several structures in the presence of K^+ or $^{15}NH_4^+$ ions.[87] These bimolecular G-quadruplexes exhibited lower stability in comparison to the parent $[d(G_4T_4G_4)]_2$ structure in the presence of Na^+ ions.[87]

A systematic study of G-rich oligonucleotides containing runs of G residues of various lengths, and with varying numbers intervening residues, revealed a cation-dependent pattern for the G-quadruplex folds adopted by these sequences. It was shown that K^+ ions are required to stabilize "chair" type structures, that is, G-quadruplexes with lateral loops instead of diagonal loops.[88] A minimum of three residues in the central loop and two residues in the two outer loops is required to form an intramolecular chair-type quadruplex, otherwise a mixture of tetramolecular parallel-stranded quadruplexes is favored.[88,89] Longer loops and more G-quartets favor G-quadruplexes with a diagonal loop in the central position.[88] G-quadruplexes with diagonal loops and tetramolecular parallel quadruplexes can be formed in the presence of either K^+ or Na^+, but intramolecular G-quadruplexes that have only lateral loops can form in the presence of K^+ and Na^+ ions, but not Na^+ alone.[88] TBA is an example of a chair-type quadruplex that forms in the presence of K^+ ions. Only one K^+ per DNA strand is required for TBA to adopt its characteristic chair-type fold, but in the presence of two or more K^+ ions per DNA strand there is a minor modification of the central lateral loop structure to create a binding site for a second K^+ between the central loop and the adjacent quartet.[90] This structural change further stabilizes the quadruplex. The presence of purines in the lateral loops prevents the formation of a chair-like quadruplex structure because of steric effects and the lack of a suitable coordinating carbonyl group.

Finally, an even greater sensitivity to cation species is seen in the Fragile X syndrome triplet repeat, $d(TG_2CG_2C)$. In the presence of K^+ ions at neutral pH, $d(TG_2CG_2C)$ forms a tetramolecular parallel quadruplex.[91] Under the same conditions but in the presence of Na^+ ions, $d(TG_2CG_2C)$ forms an antiparallel duplex. An equilibrium between duplex and quadruplex was observed upon lowering pH. When the pH is lowered to 2.2, $d(TG_2CG_2C)$ exists entirely as a G-quadruplex in the presence of Na^+.[91] In contrast, the analogous oligonucleotide $d(TG_3CG_2C)$ forms a tetramolecular quadruplex in solutions containing either Na^+ or K^+.[92]

4.7   Cation-Driven Conformational Switches

The possibility of cation-induced changes in G-quadruplex structure has formed the basis for the hypothesis that DNA G-quadruplexes might function as molecular switches *in vivo*.[93,94] In this section we review experiments related to reports of cation-driven conformational changes by G-quadruplexes. While this proposal is certainly related to the cation-dependent polymorphisms described in Section 4.6, the results presented in this section are more focused on G-quadruplex sequences and transitions that have been proposed as possible cation-dependent molecular switches, particularly those sequences derived from telomere DNA sequences.

Sen and Gilbert were the first to report that K^+ ions preferentially stabilize G-quadruplex structures with folds that differ from those formed in the presence of Na^+ ions.[94] These investigators deduced, by gel electrophoresis, that the interconversion of G-rich DNA sequences between antiparallel and parallel G-quadruplexes could be controlled by Na^+/K^+ concentrations, with K^+ ions favoring formation of the antiparallel structure.[95] It was suggested that changes in cation concentrations within the cell might be used to control the conformation of the single-stranded G-rich overhang of telomeres.[94] Hardin *et al.* subsequently established, by additional physical techniques, that changing counterions from Na^+ to K^+ specifically induces a conformational transition in the sequence $d(T_2G_4)_4$ (four units of telomeric DNA repeats from *Tetrahymena*) from a monomolecular to a multistranded G-quadruplex structure.[96] Similarly, only bimolecular diagonally looped structures are formed by $d(G_3T_4G_3)$ and $d(G_4T_4G_4)$ in the presence of Na^+ ions, while both bimolecular and linear, tetramolecular G-quadruplexes form in the presence of K^+ ions.[43,97]

There is also evidence for the formation of simpler, hairpin structures stabilized by GG base pairs, rather than G-quartets, in oligonucleotides derived from the *O. nova* telomere repeat, $d(T_4G_4)_2$ and $dT_6(T_4G_4)_2$, in the presence of 10–50 mM NaCl.[98] In the presence of 150 mM KCl, these sequences are converted to parallel, tetramolecular quadruplexes. These GG-based hairpins were proposed as intermediates in the formation of bimolecular G-quadruplexes *via* association of two such hairpins. Similar sequence/cation-dependent structural relationships have been reported for other metal ions as well. For example, in the sequences $d(G_4T_2)_4$ and $d(G_4T_4)_4$, Sr^{2+} ions exhibited a preferential stabilization of the linear G-quadruplex forms rather than folded forms,[17] while this divalent cation is very effective at stabilizing the monomolecular quadruplex form of the sequence $d(G_2T_2G_2TGTG_2T_2G_2)$ with no sign of linear, tetramolecular structures.[19] Clearly, the situation is complicated, potentially involving a variety of equilibria among mono-, bi-, and tetramolecular structures.[96] In a biological context, where the concentration of K^+ and Na^+ ions can change with time and precise location within a cell, we may anticipate significant variability in the mixture of G-quadruplex structures potentially adopted by G-rich sequences.

Thomas and co-workers used Raman spectroscopy to generate a phase diagram of antiparallel and parallel G-quadruplex structures as controlled by concentrations of Na^+ and K^+ ions.[93] Both alkali ions facilitate the formation of an antiparallel diagonal-loop G-quadruplex by four units of the telomere repeat sequence of *O. nova*, $d(T_4G_4)_4$, at low concentrations. An extended, parallel-stranded tetramolecular G-quadruplex is formed at higher cation concentrations. However, K^+ is more effective than Na^+ in inducing the parallel strand association. The midpoints of the Na^+ and K^+ concentrations required for the structural transition are 225 and 65 mM, respectively.[93]

Under low salt conditions, the human telomere DNA repeats tend to be in either disordered single-stranded states or parallel tetramolecular G-quadruplex structures.[99] For example, the double repeat sequence $d(TTAGGG)_2$ adopts a disordered, single-stranded structure at 5 mM Na^+ concentration. A mixture of parallel and antiparallel G-quadruplexes are formed upon increase of Na^+ concentrations to 140 mM. CD spectra show evidence of an external loop G-quadruplex in the presence of 100 mM K^+ with either 5 or 140 mM Na^+.[99] Similarly, $d(GGGTTA)_2GGG$, $d(GGGTTA)_3GGG$, $d(TTAGGG)_4$, and $d(GGGTTA)_4$ show evidence of antiparallel G-quadruplex structures at 140 mM Na^+ concentration, whereas there are more than one structural types present at 5 mM Na^+ concentration.

As mentioned above, the switching between antiparallel and parallel G-quadruplexes can also be driven by divalent cations.[66] A continuous structural transition from an antiparallel to a parallel G-quadruplex structure adopted by $d(G_4T_4G_4)$, which spontaneously assembles into G-wires, was observed with Ca^{2+} titration in the presence of 100 mM NaCl. The structural transition was completed by adding 20 mM Ca^{2+}, and $d(G_4T_4G_4)$ has been shown to fold into a parallel G-quadruplex in the presence of Ca^{2+} ions alone. As little as a 1 mM concentration of divalent cations was sufficient to destabilize the bimolecular, antiparallel G-quadruplex formed by $d(G_4T_4G_4)$. Thus, divalent cations might also prove to be a powerful tool for regulating G-quadruplex structures.[66,67]

G-rich sequences that fold into G-quadruplex structures can, of course, form Watson–Crick duplexes with their complementary C-rich strands. The formation of G-quadruplex or duplex structures within the cell will therefore depend on the relative stability of the G-quadruplex and the corresponding duplex.[69,80,100–102] Furthermore, the C-rich strand might have the possibility of forming an intramolecular i-motif that likewise competes with duplex formation, which can also be influenced by cations.[103,104] The human telomere repeat sequence $[d(G_3(TTAGGG)]_3$ forms a duplex in the presence of the complementary C-rich strand under physiological conditions. The *Tetrahymena* sequence $d[G_4(T_2G_4)_3]$, the sequence $d[G_3(T_2G_3)_3]$, and sequences related to regions of the *c-myc* promoter $d(G_4AG_4TG_4AG_4)$ and $d(G_4AG_3TG_4AG_3)$ preferentially adopt the quadruplex form in buffers containing K^+ ions, even in the presence of a 50-fold excess of their complementary C-rich strands. In comparison, these sequences adopt predominantly duplex structures in the presence of Na^+ ions. The HIV integrase inhibitor $d[G_3(TG_3)_3]$ forms an extremely stable quadruplex that is not affected by addition of a 50-fold excess of the complementary C-rich strand in both K^+- and Na^+-containing buffers.[102] Thus, in considering possible cation-driven G-quadruplex switches *in vivo*, it is important to consider whether the G-rich DNA is single stranded, as in the single-stranded overhang telomere DNA, or double stranded, as is the case for the vast majority of all other DNA within a cell.

Finally, an entirely different ion-dependent conformational switch has been observed for $d(G_2AG_2AG)$. At moderate ionic strength, 150 mM NaCl, this oligonucleotide forms a four-stranded quadruplex.[105] The four-stranded dimeric quadruplex comprises a unique $A \cdot (G \cdot G \cdot G \cdot G) \cdot A$ hexad motif that is stabilized by sheared $G \cdot A$ base pairs, in addition to the G-tetrad.[105] At low ionic strength (10 mM NaCl) $d(G_2AG_2AG)$ forms a two-stranded "arrowhead" motif.[106] The v-shaped arrowhead motif is stabilized by cross-strand stacking and mismatch formation. A similar concentration-dependent switch by the same oligonucleotide has been observed as a function of KCl concentration. The cation binding sites associated with the two folds of this oligonucleotide suggest a rationale for the cation-dependent switch. There are two Na^+ binding sites centered between the two stacked hexad planes and adjacent to the phosphate backbone in the hexad motif, in addition to the cation binding sites between adjacent G-quartets.[105] In contrast, there are no such additional cation binding sites in the arrowhead motif, suggesting that cation binding stabilizes the bimolecular hexad motif.

4.8   Possible Applications of Cation-Dependent G-Quadruplex Formation

In addition to their biological implications, the propensity for G-quadruplexes to form cation-specific supramolecular structures has opened the possibility for using guanosine nucleosides and G-rich DNA sequences for applications that range from environmental remediation to organic synthesis and nanotechnology. For example, molecular self-assembly of guanine-containing nucleosides has been shown to be useful for the construction of highly selective ionophores. These self-assembled guanosine-based hexadecamers and isoguanosine-based decamers are excellent $^{226}Ra^{2+}$ selective ionophores even in the presence of excess alkali (Na^+, K^+, Rb^+, and Cs^+) and alkaline-earth (Mg^{2+}, Ca^{2+}, Sr^{2+}, and Ba^{2+}) cations over the pH range 3–11.[107] Picrate anions are needed to provide a neutral assembly, whereas the isoguanosine analogue assembly extracts $^{226}Ra^{2+}$ cations without any such additives. Both guanosine-picrate and isoguanosine assemblies show $^{226}Ra^{2+}$ extraction even at a 0.35×10^6-fold excess of Na^+, K^+, Rb^+, Cs^+, Mg^{2+}, or Ca^{2+} (10^{-2} M) to $^{226}Ra^{2+}$ (2.9×10^{-8} M) and at a 100-fold salt to ionophore excess. In the case of the guanosine-picrate assembly, more competition was observed from Sr^{2+} and Ba^{2+}, as extraction of $^{226}Ra^{2+}$ ceased at an $M^{2+}/^{226}Ra^{2+}$ ratio of 10^6 and 10^4, respectively. With the isoguanosine assembly, $^{226}Ra^{2+}$ extraction also occurred at a $Sr^{2+}/^{226}Ra^{2+}$ ratio of 10^6, but ceased at a 10^6 excess of Ba^{2+}.[107] Self-assembled isoguanosine-based extractants for Ra^{2+} ions from industrially produced water have shown a high $^{226}Ra^{2+}$ selectivity under laboratory conditions.[108]

G-quartets have recently been used to promote the synthesis of specific products. Guanosine hydrazide yields a stable supramolecular hydrogel based on the formation of G-quartets in presence of metal cations. This assembly has been exploited to generate a constitutional dynamic library when mixed with various aldehydes.[109] The process amounts to gelation-driven self-organization with component selection and amplification based on G-quartet formation and reversible covalent connections. The observed self-organization and component selection occurs by means of a multilevel self-assembly involving three dynamic processes, two of a supramolecular and one of a reversible covalent nature. This approach has extended constitutional dynamic chemistry to phase-organization and phase-transition events.[109]

It has been known for sometime that G-rich DNA can form linear nanostructures, termed G-wires and frayed wires, whose structure and stability depend on the relative concentration of cations present in solution.[110,111] The possibility exists for controlling the electronic properties of the G-wire by using different metal ions to stabilize the G-quartet stacks, which could be a powerful tool for tuning the conduction properties of these putative nanowires.[112] In another application inspired by nanotechnology, the addition of cations to a DNA duplex has been shown to trigger G-wire dimerization through G-quartet assembly to yield new nanostructures.[113] $G \cdot G$ mismatches incorporated within DNA duplexes can be used to assemble two such duplexes through G-quartet formation.[114] The parallel and antiparallel assembly of this the so-called "synapsable" DNA shows a dramatic dependence on cation species.[114] The equilibrium between DNA duplex and G-quadruplex structures, as driven by cations present in solution, has even been proposed as a means to create nanomotors that are capable of extension and shrinking movements.[115,116]

4.9 Summary

Our knowledge of possible G-quadruplex structures is far from complete and it is not currently possible to delineate general rules governing the folding of (human) telomeric and non-telomeric[117–119] repeats, and more specifically, the role of cations in this process. A few guidelines are, however, now apparent. In general, G-quadruplexes tend to have all strands parallel when associated with K^+ ions, regardless of the number of separate strands involved. Different metal ions can induce conformational changes, which can be dramatic for some G-quadruplexes. In some cases, a simple change in cation (salt) concentration has been found to result in significant structural transitions.[13,90,120] In addition to the continued interest in the possible biological roles of G-quadruplexes, and in determining the physical origins of G-quadruplex topological sensitivity to cation concentration and species, there are now several reports where such structural changes are being exploited for the design of nanometer-scale devices.

Acknowledgment

Financial support from NATO Collaborative Programs Section (CLG grant 979520) to the authors is gratefully acknowledged.

References

1. W. Guschlbauer, J.F. Chantot and D. Thiele, *J. Biomol. Struct. Dyn.*, 1990, **8**, 491.
2. D. Sen and W. Gilbert, *Methods Enzymol.*, 1992, **211**, 191.
3. J.R. Williamson, *Annu. Rev. Biophys. Biomol. Struct.*, 1994, **23**, 703.
4. D.E. Gilbert and J. Feigon, *Curr. Opin. Struct. Biol.*, 1999, **9**, 305.
5. C.C. Hardin, A.G. Perry and K. White, *Biopolymers*, 2001, **56**, 147.
6. T. Simonsson, *Biol. Chem.*, 2001, **382**, 621.
7. M.A. Keniry, *Biopolymers*, 2001, **56**, 123.
8. S. Neidle and G.N. Parkinson, *Curr. Opin. Struct. Biol.*, 2003, **13**, 275.
9. J.T. Davis, *Angew. Chem. Int. Ed. Engl.*, 2004, **43**, 668.
10. K. Phillips, Z. Dauter, A.I.H. Murchie, D.M.J. Lilley and B. Luisi, *J. Mol. Biol.*, 1997, **273**, 171.
11. D. Sen and W. Gilbert, *Curr. Opin. Struct. Biol.*, 1991, **1**, 435.
12. M.M. Cai, X.D. Shi, V. Sidorov, D. Fabris, Y.F. Lam and J.T. Davis, *Tetrahedron*, 2002, **58**, 661.
13. A. Wlodarczyk, P. Grzybowski, A. Patkowski and A. Dobek, *J. Phys. Chem. B*, 2005, **109**, 3594.
14. J.S. Lee, *Nucleic Acids Res.*, 1990, **18**, 6057.
15. N. Nagesh and D. Chatterji, *J. Biochem. Biophys. Meth.*, 1995, **30**, 1.
16. S. Basu, A. Szewczak, M. Cocco and S.A. Strobel, *J. Am. Chem. Soc.*, 2000, **122**, 3240.
17. F.M. Chen, *Biochemistry*, 1992, **31**, 3769.
18. E.A. Venczel and D. Sen, *Biochemistry*, 1993, **32**, 6220.
19. I. Smirnov and R.H. Shafer, *J. Mol. Biol.*, 2000, **296**, 1.
20. G. Laughlan, A.I.H. Murchie, D.G. Norman, M.H. Moore, P.C.E. Moody, D.M.J. Lilley and B. Luisi, *Science*, 1994, **265**, 520.
21. M.P. Horvath and S.C. Schultz, *J. Mol. Biol.*, 2001, **310**, 367.
22. S. Haider, G.N. Parkinson and S. Neidle, *J. Mol. Biol.*, 2002, **320**, 189.
23. G.N. Parkinson, M.P.H. Lee and S. Neidle, *Nature*, 2002, **417**, 876.

24. S.L. Forman, J.C. Fettinger, S. Pieraccini, G. Gottareli and J.T. Davis, *J. Am. Chem. Soc.*, 2000, **122**, 4060.

25. J.P. Deng, Y. Xiong and M. Sundaralingam, *Proc. Natl. Acad. Sci. USA*, 2001, **98**, 13665.

26. N.V. Hud, V. Sklenar and J. Feigon, *J. Mol. Biol.*, 1999, **286**, 651.

27. N.V. Hud, P. Schultze, V. Sklenar and J. Feigon, *J. Mol. Biol.*, 1999, **285**, 233.

28. P. Sket, M. Crnugelj, W. Kozminski and J. Plavec, *Org. Biomol. Chem.*, 2004, **2**, 1970.

29. J.F. Chantot and W. Guschlbauer, *FEBS Lett.*, 1969, **4**, 173.

30. A. Caceres, G. Wright, C. Gouyette, G. Parkinson and J.A. Subirana, *Nucleic Acids Res.*, 2004, **32**, 1097.

31. P. Schultze, F.W. Smith and J. Feigon, *Structure*, 1994, **2**, 221.

32. F.W. Smith and J. Feigon, *Nature*, 1992, **356**, 164.

33. P. Schultze, N.V. Hud, F.W. Smith and J. Feigon, *Nucleic Acids Res.*, 1999, **27**, 3018.

34. M. Borzo, C. Detellier, P. Laszlo and A. Paris, *J. Am. Chem. Soc.*, 1980, **102**, 1124.

35. C. Detellier and P. Laszlo, *J. Am. Chem. Soc.*, 1980, **102**, 1135.

36. N.V. Hud, F.W. Smith, F.A.L. Anet and J. Feigon, *Biochemistry*, 1996, **35**, 15383.

37. F.W. Smith, F.W. Lau and J. Feigon, *Proc. Natl. Acad. Sci. USA*, 1994, **91**, 10546.

38. M.A. Keniry, G.D. Strahan, E.A. Owen and R.H. Shafer, *Eur. J. Biochem.*, 1995, **233**, 631.

39. N.V. Hud, P. Schultze and J. Feigon, *J. Am. Chem. Soc.*, 1998, **120**, 6403.

40. P. Sket, M. Crnugelj and J. Plavec, *Nucleic Acids Res.*, 2005, **33**, 3691.

41. P. Sket, M. Cmugelj and J. Plavec, *Bioorg. Med. Chem.*, 2004, **12**, 5735.

42. M. Crnugelj, P. Sket and J. Plavec, *J. Am. Chem. Soc.*, 2003, **125**, 7866.

43. G.D. Strahan, M.A. Keniry and R.H. Shafer, *Biophys. J.*, 1998, **75**, 968.

44. A. Wong, J.C. Fettinger, S.L. Forman, J.T. Davis and G. Wu, *J. Am. Chem. Soc.*, 2002, **124**, 742.

45. G. Wu, A. Wong, Z.H. Gan and J.T. Davis, *J. Am. Chem. Soc.*, 2003, **125**, 7182.

46. A. Wong and G. Wu, *J. Am. Chem. Soc.*, 2003, **125**, 13895.

47. G. Wu and A. Wong, *Biochem. Biophys. Res. Commun.*, 2004, **323**, 1139.

48. Q.W. Xu, H. Deng and W.H. Braunlin, *Biochemistry*, 1993, **32**, 13130.

49. H. Deng and W.H. Braunlin, *J. Mol. Biol.*, 1996, **255**, 476.

50. F.W. Smith and J. Feigon, *Biochemistry*, 1993, **32**, 8682.

51. J. Feigon, S.E. Butcher, L.D. Finger and N.V. Hud, *Methods Enzymol.*, 2001, **338**, 400.

52. M.L. Gill, S.A. Strobel and J.P. Loria, *J. Am. Chem. Soc.*, 2005, **127**, 16723.

53. W.S. Ross and C.C. Hardin, *J. Am. Chem. Soc.*, 1994, **116**, 6070.

54. B. Sacca, L. Lacroix and J.L. Mergny, *Nucleic Acids Res.*, 2005, **33**, 1182.

55. J.D. Gu and J. Leszczynski, *J. Phys. Chem. A*, 2000, **104**, 6308.

56. J.D. Gu and J. Leszczynski, *J. Phys. Chem. A*, 2002, **106**, 529.

57. F.C. Meng, W.R. Xu and C.B. Liu, *Chem. Phys. Lett.*, 2004, **389**, 421.

58. J.L. Mergny, A.T. Phan and L. Lacroix, *FEBS Lett.*, 1998, **435**, 74.

59. A. Risitano and K.R. Fox, *Nucleic Acids Res.*, 2004, **32**, 2598.

60. J.L. Mergny, A. De Cian, A. Ghelab, B. Sacca and L. Lacroix, *Nucleic Acids Res.*, 2005, **33**, 81.

61. P. Balagurumoorthy and S.K. Brahmachari, *J. Biol. Chem.*, 1994, **269**, 21858.

62. I.V. Smirnov, F.W. Kotch, I.J. Pickering, J.T. Davis and R.H. Shafer, *Biochemistry*, 2002, **41**, 12133.

63. F.W. Kotch, J.C. Fettinger and J.T. Davis, *Org. Lett.*, 2000, **2**, 3277.

64. C.C. Hardin, T. Watson, M. Corregan and C. Bailey, *Biochemistry*, 1992, **31**, 833.

65. C.C. Hardin, M. Corregan, B.A. Brown and L.N. Frederick, *Biochemistry*, 1993, **32**, 5870.

66. D. Miyoshi, A. Nakao, T. Toda and N. Sugimoto, *FEBS Lett.*, 2001, **496**, 128.

67. D. Miyoshi, A. Nakao and N. Sugimoto, *Nucleic Acids Res.*, 2003, **31**, 1156.

68. D. Miyoshi, S. Matsumura, W. Li and N. Sugimoto, *Nucleosides Nucleotides Nucleic Acids*, 2003, **22**, 203.

69. W. Li, D. Miyoshi, S. Nakano and N. Sugimoto, *Biochemistry*, 2003, **42**, 11736.

70. S.W. Blume, V. Guarcello, W. Zacharias and D.M. Miller, *Nucleic Acids Res.*, 1997, **25**, 617.

71. F. Yin, J.H. Liu and X.J. Peng, *Bioorg. Med. Chem. Lett.*, 2003, **13**, 3923.

72. C. Krafft, J.M. Benevides and G.J. Thomas, *Nucleic Acids Res.*, 2002, **30**, 3981.

73. M.A. Keniry, E.A. Owen and R.H. Shafer, *Nucleic Acids Res.*, 1997, **25**, 4389.

74. G.D. Strahan, R.H. Shafer and M.A. Keniry, *Nucleic Acids Res.*, 1994, **22**, 5447.

75. A. Kettani, S. Bouaziz, A. Gorin, H. Zhao, R.A. Jones and D.J. Patel, *J. Mol. Biol.*, 1998, **282**, 619.

76. S. Bouaziz, A. Kettani and D.J. Patel, *J. Mol. Biol.*, 1998, **282**, 637.

77. A.T. Phan and D.J. Patel, *J. Am. Chem. Soc.*, 2003, **125**, 15021.

78. A.T. Phan, Y.S. Modi and D.J. Patel, *J. Mol. Biol.*, 2004, **338**, 93.

79. Y. Wang and D.J. Patel, *Structure*, 1993, **1**, 263.

80. L.M. Ying, J.J. Green, H.T. Li, D. Klenerman and S. Balasubramanian, *Proc. Natl. Acad. Sci. USA*, 2003, **100**, 14629.

81. S. Redon, S. Bombard, M.A. Elizondo-Riojas and J.C. Chottard, *Nucleic Acids Res.*, 2003, **31**, 1605.

82. Y.J. He, R.D. Neumann and I.G. Panyutin, *Nucleic Acids Res.*, 2004, **32**, 5359.

83. J.Y. Qi and R.H. Shafer, *Nucleic Acids Res.*, 2005, **33**, 3185.

84. J. Li, J.J. Correia, L. Wang, J.O. Trent and J.B. Chaires, *Nucleic Acids Res.*, 2005, **33**, 4649.
85. Y. Kato, T. Ohyama, H. Mita and Y. Yamamoto, *J. Am. Chem. Soc.*, 2005, **127**, 9980.
86. M. Crnugelj, N.V. Hud and J. Plavec, *J. Mol. Biol.*, 2002, **320**, 911.
87. M. Cevec and J. Plavec, *Biochemistry*, 2005, **44**, 15238.
88. V.M. Marathias and P.H. Bolton, *Biochemistry*, 1999, **38**, 4355.
89. I. Smirnov and R.H. Shafer, *Biochemistry*, 2000, **39**, 1462.
90. V.M. Marathias and P.H. Bolton, *Nucleic Acids Res.*, 2000, **28**, 1969.
91. P.K. Patel, N.S. Bhavesh and R.V. Hosur, *Biochem. Biophys. Res. Commun.*, 2000, **278**, 833.
92. P.K. Patel, N.S. Bhavesh and R.V. Hosur, *Biochem. Biophys. Res. Commun.*, 2000, **270**, 967.
93. T. Miura, J.M. Benevides and G.J. Thomas Jr., *J. Mol. Biol.*, 1995, **248**, 233.
94. D. Sen and W. Gilbert, *Nature*, 1990, **344**, 410.
95. D. Sen and W. Gilbert, *Nature*, 1988, **334**, 364.
96. C.C. Hardin, E. Henderson, T. Watson and J.K. Prosser, *Biochemistry*, 1991, **30**, 4460.
97. P. Balagurumoorthy, S.K. Brahmachari, D. Mohanty, M. Bansal and V. Sasisekharan, *Nucleic Acids Res.*, 1992, **20**, 4061.
98. L. Laporte and G.J. Thomas, *J. Mol. Biol.*, 1998, **281**, 261.
99. I.N. Rujan, J.C. Meleney and P.H. Bolton, *Nucleic Acids Res.*, 2005, **33**, 2022.
100. A.T. Phan and J.L. Mergny, *Nucleic Acids Res.*, 2002, **30**, 4618.
101. W. Li, P. Wu, T. Ohmichi and N. Sugimoto, *FEBS Lett.*, 2002, **526**, 77.
102. A. Risitano and K.R. Fox, *Biochemistry*, 2003, **42**, 6507.
103. K. Halder, V. Mathur, D. Chugh, A. Verma and S. Chowdhury, *Biochem. Biophys. Res. Commun.*, 2005, **327**, 49.
104. V. Mathur, A. Verma, S. Maiti and S. Chowdhury, *Biochem. Biophys. Res. Commun.*, 2004, **320**, 1220.
105. A. Kettani, A. Gorin, A. Majumdar, T. Hermann, E. Skripkin, H. Zhao, R. Jones and D.J. Patel, *J. Mol. Biol.*, 2000, **297**, 627.
106. A. Kettani, S. Bouaziz, E. Skripkin, A. Majumdar, W.M. Wang, R.A. Jones and D.J. Patel, *Structure*, 1999, **7**, 803.
107. F.W.B. van Leeuwen, W. Verboom, X.D. Shi, J.T. Davis and D.N. Reinhoudt, *J. Am. Chem. Soc.*, 2004, **126**, 16575.
108. F.W.B. Van Leeuwen, C.J.H. Miermans, H. Beijleveld, T. Tomasberger, J.T. Davis, W. Verboom and D.N. Reinhoudt, *Environ. Sci. Technol.*, 2005, **39**, 5455.
109. N. Sreenivasachary and J.M. Lehn, *Proc. Natl. Acad. Sci. USA*, 2005, **102**, 5938.
110. T.C. Marsh and E. Henderson, *Biochemistry*, 1994, **33**, 10718.
111. E. Protozanova and R.B. Macgregor, *Biochemistry*, 1996, **35**, 16638.
112. A. Calzolari, R. Di Felice, E. Molinari and A. Garbesi, *J. Phys. Chem. B*, 2004, **108**, 2509.

113. E.A. Venczel and D. Sen, *J. Mol. Biol.*, 1996, **257**, 219.
114. R.P. Fahlman and D. Sen, *J. Mol. Biol.*, 1998, **280**, 237.
115. J.W.J. Li and W.H. Tan, *Nano Lett.*, 2002, **2**, 315.
116. P. Alberti and J.L. Mergny, *Proc. Natl. Acad. Sci. USA*, 2003, **100**, 1569.
117. A.K. Todd, M. Johnston and S. Neidle, *Nucleic Acids Res.*, 2005, **33**, 2901.
118. S. Rankin, A.P. Reszka, J. Huppert, M. Zloh, G.N. Parkinson, A.K. Todd, S. Ladame, S. Balasubramanian and S. Neidle, *J. Am. Chem. Soc.*, 2005, **127**, 10584.
119. P. Hazel, J. Huppert, S. Balasubramanian and S. Neidle, *J. Am. Chem. Soc.*, 2004, **126**, 16405.
120. Y. Wang and D.J. Patel, *Biochemistry*, 1992, **31**, 8112.
121. L. Petraccone, E. Erra, V. Esposito, A. Randazzo, L. Mayol, L. Nasti, G. Barone and C. Giancola, *Biochemistry*, 2004, **43**, 4877.

CHAPTER 5

DNA Quadruplex–Ligand Recognition: Structure and Dynamics

MARK S. SEARLE AND GRAHAM D. BALKWILL

Centre for Biomolecular Sciences, School of Chemistry, University of Nottingham, University Park, Nottingham, NG7 2RD, UK

5.1 Telomeric DNA and Cell Immortalisation

The telomeric ends of chromosomes are repetitive non-coding DNA sequences that protect the cell from recombination, end-to-end fusion and nuclease degradation.[1,2] In human cells the telomeric DNA typically consists of 5–8 kilobases of a double-stranded tandem repeat of the guanine-rich sequence TTAGGG with a single-stranded 3′-end overhang of 100–200 bases necessary to ensure complete chromosomal DNA replication. Telomeres shorten by 50–200 base pairs with each cell division because synthesis of the lagging strand of DNA is unable to replicate the 3′-end overhang. Telomere shortening to a critical length stops normal growth and the cell enters a state of senescence where end-to-end fusion and chromosomal instability leads to cell death. Tumour cells evade this ultimate fate and become immortalised through activation of the enzyme telomerase that stabilises the length of the telomeres.[3–6]

The ribonucleoprotein telomerase is responsible for maintaining telomere length by adding TTAGGG repeats to the 3′-ends of chromosomes. Telomerase consists of several components including an endogenous RNA template (hRNA) of 11 nucleotides and a reverse transcriptase (hTERT) that catalyses the addition of telomeric repeats. A prerequisite for telomerase activity is the availability of the single-stranded telomere as a primer.[7–9] All human somatic cells contain the hRNA template but lack the hTERT;[8] however, telomerase has been shown to be active in 85–90% of all human tumour cells.[6] The consequence of the activation of the telomerase mechanism is that tumour cells have unlimited proliferative potential. Consequently, telomerase has become a high-profile target for the development of novel anti-cancer agents.[10–13]

The endogenous RNA template region of the telomerase requires an extended single-stranded, telomeric DNA primer for the effective addition of the telomeric repeats.[7–9] However, earlier investigations of guanine-rich telomeric sequences demonstrated the ability to assemble into quadruplex structures consisting of the alignment of four strands of nucleic acid to facilitate the formation of guanine tetrads stabilised by monovalent cations (Na^+ and K^+) (Figure 1).[14–17]

The investigations of Zahler and co-workers[18] established that telomerase activity and the extent of telomere elongation could be negatively regulated *in vivo* through G-quadruplex stabilisation by K^+ ions. Subsequently, small molecules that bind and stabilise DNA quadruplex structures appear to be potent telomerase inhibitors, with lower levels of general cytotoxicity than associated with ligand binding to duplex DNA. The elucidation of solution NMR and X-ray crystallographic structures of G-quadruplex DNA has revealed the distinct geometrical features associated with the different parallel, anti-parallel and mixed strand alignments (intramolecular folded assemblies, intermolecular linear structures and dimeric hairpins; Figure 1). These include four distinct grooves arising from combinations of *syn* and *anti* deoxyguanosine conformations, a channel of negative electrostatic potential and the possibility

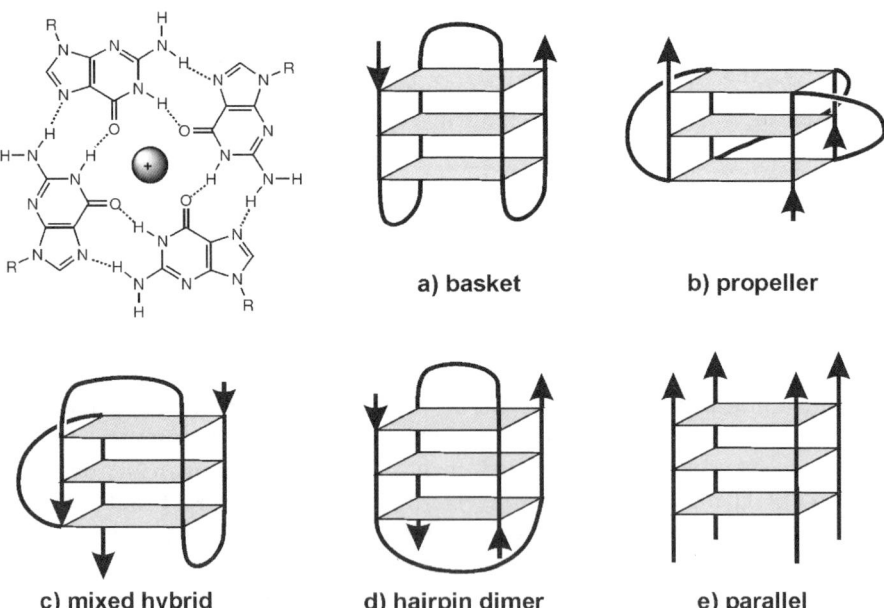

a) basket   b) propeller

c) mixed hybrid   d) hairpin dimer   e) parallel

Figure 1   *Structure of a G-tetrad showing complexation around a monovalent cation (Na⁺ or K⁺), and five possible G-quadruplex structures derived from intramolecular folding (a) basket, (b) propeller and (c) mixed hybrid structure; (d) an intermolecular hairpin dimer structure and (e) an intermolecular parallel-stranded structure, with the arrows showing the strand directions from 5′ to 3′.*

of recognition motifs within the flexible loop regions, all of which make them attractive targets for drug design.[15–17,19–23] The high anti-tumour chemotherapeutic potential of ligands with the capability to stabilise G-quadruplex has focused much research activity on structure-based design approaches to the development of molecules that interact specifically with G-quadruplexes. Recent developments in elucidating structural details of ligand–quadruplex interactions and the dynamics of these systems are the subject of this chapter.

5.2 Structural Diversity of Quadruplex-Targeted Ligands

A number of recent reviews have captured the range of quadruplex-targeted ligands derived from structure-based design approaches that turn out to be potent telomerase inhibitors.[10–13,23] Typical of DNA intercalators in general (as illustrated for ethidium bromide **1**), many of these ligands are polyaromatic heterocycles designed for extensive π-stacking interactions with G-tetrads offering the possibility for either intercalation within the stack of G-tetrads or through stacking on the end of the G-quadruplex. Some of the diversity of structures that bind to quadruplexes and inhibit telomerase are illustrated in Figure 2.

The anthraquinone derivative (**2**) represents the first example from 1997, with numerous others following in rapid succession, including the dibenzophenanthroline derivatives[24] and tri-substituted acridines (**3**),[25] which were reported to inhibit telomerase action in tumour cell lines with IC_{50} values of up to 28 and 60 nM, respectively. The tri-substituted acridines (**3**) were developed from the simple acridine (**4**) on the basis of structure-based design principles to maximise the quadruplex binding affinity. Thus, inhibition of telomerase by these compounds appears to be correlated to selective stabilisation of the human DNA quadruplex structure. Tetra-(*N*-methyl-4-pyridyl)-porphyrins (**5**),[26,27] 2,7-disubstituted fluorenones[28,29] and a dicationic perylene tetra-carboxylic diimide derivative (**6**) (abbreviated PIPER)[30] have also been reported to inhibit telomerase activity with small IC_{50} values.

A novel fluorinated polycyclic quinoacridinium cation RHPS4 (**7**) demonstrates enhanced binding to higher ordered DNA structures (triplex/quadruplex) and has been shown to induce telomere shortening with an IC_{50} value of 0.33 μM, while decreasing tumour cell proliferation of breast 21NT cells at concentrations as low as 0.2 μM.[31] RHPS4 is weakly cytotoxic (mean GI_{50} value in the NCI 60 human tumour cell panel is 13.18 μM), giving a therapeutic index (GI_{50}/IC_{50}) of 40.[32] This activity does not appear to be associated with Taq polymerase and topoisomerase II inhibition, strongly suggesting that RHPS4 is an inhibitor of telomerase function. The dimeric macrocycle (**8**),[33] in contrast to the related monomeric forms, selectively binds quadruplex DNA as demonstrated by competition dialysis experiments, while the triazine derivatives (**9**) have also been shown to induce oligonucleotides to fold into higher order quadruplex structures, with nano-molar inhibition of telomerase.[34] Other

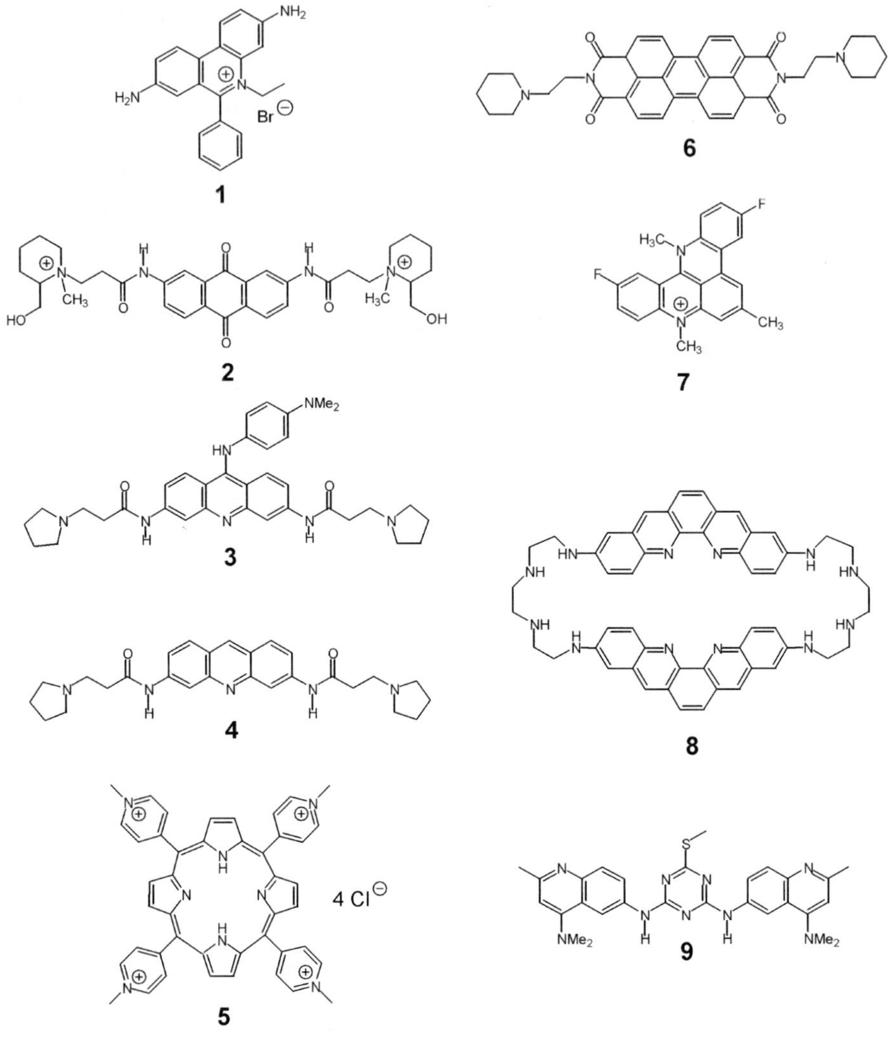

Figure 2 *Panels showing the diverse range of small molecule G-quadruplex binding ligands (1–16).*

new ligands are constantly emerging based upon biarylpyrimidines,[35] porphyrin-aminquinoline conjugates[36] or indoloquinoline templates.[37]

More recently, a novel family of carbocyanine–peptide conjugates has been identified through a combinatorial selection approach[38] building on earlier studies by Chen *et al.*[39] in identifying a carbocyanine dye (**10**) that bound to hairpin quadruplex structures. Schouten *et al.*[38] used the heterocyclic hemicyanine (HC) core motif and a tetrapeptide library to identify three peptide conjugates (**11**) (HC-XXXX-CONH$_2$, where XXXX = RKKV, KRSR and FRHR, one-letter amino acid code) with K_d values < 50 µM and a high quadruplex/duplex selectivity.[38] The peptide recognition motifs were

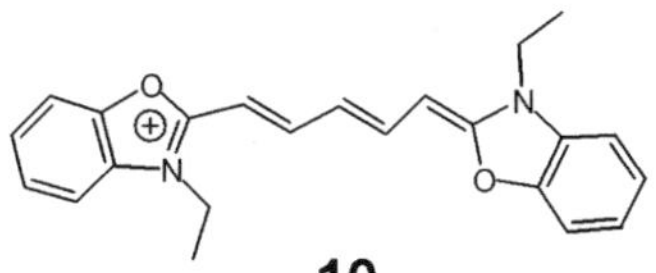

10

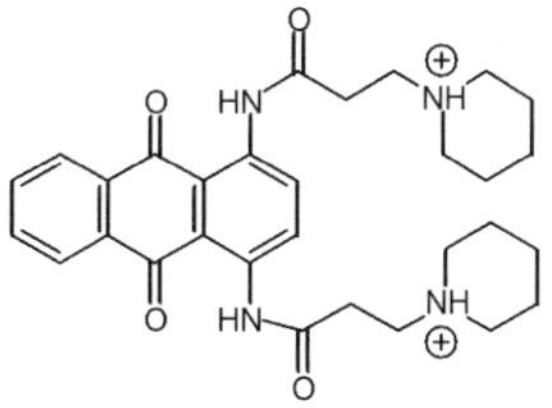

14

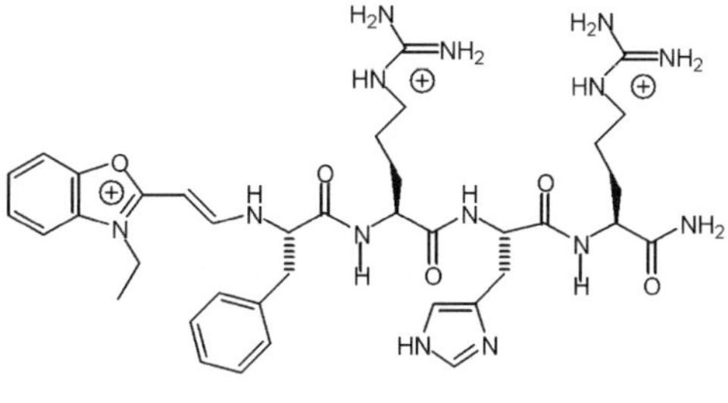

11

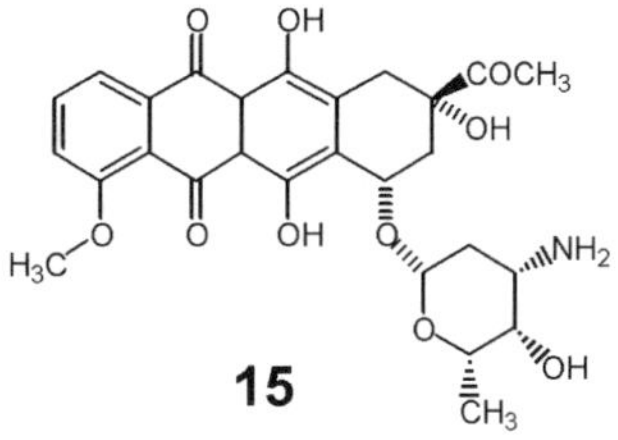

15

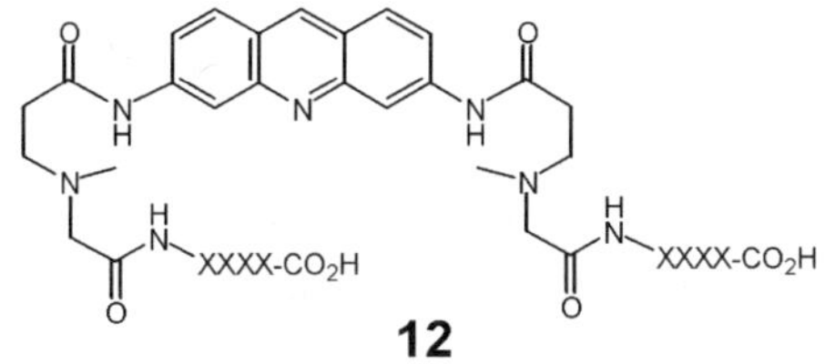

12

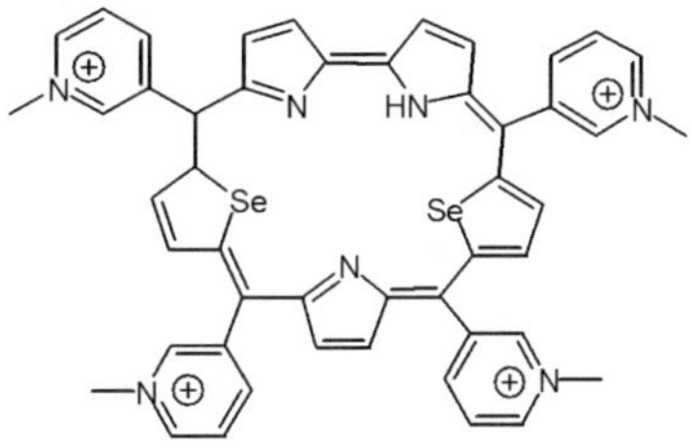

16

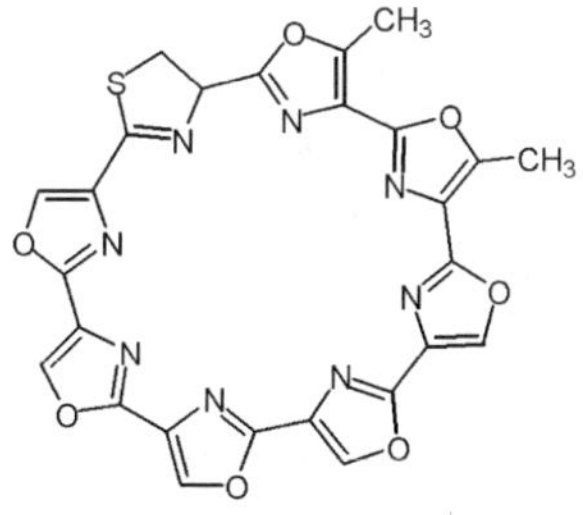

13

Figure 2 *(continued)*.

subsequently tethered to acridine (**12**) and acridone platforms to generate bis-conjugates that also strongly interacted with G-quadruplex structures.[40]

The versatility and modular nature of zinc finger DNA binding proteins, combined with phage display technology, have previously been used to select for high-affinity binding to novel DNA targets. Recently, a three-finger library has successfully identified clones that bind in a sequence-dependent and structure-specific manner to the single-stranded human telomeric repeat sequence when folded into a G-quadruplex structure, although the full structural details of the recognition process are yet to be described.[41] By analogy, ribosomal display using the Human Combinatorial Antibody Library has also identified single-chain antibody fragments that recognise G-quadruplexes with high affinity ($K_d = 125$ pM), and which demonstrate at least 1000-fold specificity between parallel and anti-parallel conformations.[42]

Last but not least, the most potent natural product known to act as a selective telomerase inhibitor is telomestatin (**13**), isolated from *Streptomyces anulatus*.[43] This novel macrocycle has been shown to bind quadruplex structures with high specificity and indeed influence the type of G-quadruplex fold formed in solution (see below).[44] We now examine some of these systems in more detail.

5.3 Structural Studies of Ligand–Quadruplex Interactions: Intercalation or End-Stacking?

Whether quadruplex-targeted ligands interact externally by stacking on the ends of G-quadruplexes or by intercalating between G-tetrads has been a key issue in ligand design (Figure 3). One early study suggested intercalation of tetra-(*N*-methyl-2-pyridyl) porphyrin between the G-tetrads on the basis of calorimetric (ITC) and spectroscopic data, with the conclusion that the neighbouring site exclusion principle, widely observed in ligand binding to duplex structures, may be violated in the context of G-quadruplexes [Figure 3(a)].[45] The same porphyrin derivative was examined by an electrophoretic photo-cleavage assay,[26] NMR and UV spectroscopy,[27] but these data were consistent with the ligand bound externally to the G-tetrads. Molecular modelling studies also suggested that ligands are able to form stable complexes by either intercalating or end-stacking with G-tetrads adding weight to this controversy.[45] The general paucity of detailed structural studies on drug–quadruplex interactions in solution may be explained by the observation of largely intractable NMR spectra arising from extensive drug-induced line broadening and multiple binding site occupancy, demonstrating in many cases the dynamic nature of the binding interaction. A combined NMR and modelling study of the complex of PIPER (**6**) with a DNA quadruplex showed that the drug forms either a sandwich complex bound between the blunt ends of a quadruplex dimer formed from d(TTAGGG)$_4$ [Figure 3(b)] or intercalates by end-stacking primarily at the GpT step of d(TAGGGTTA)$_4$ [Figure 3(c)].[30] None of the data supported ligand insertion within the stack of G-tetrads.

5.4 Modelling Ligand–Quadruplex Interactions

Several studies have shown that structural information derived from the modelling of quadruplex–ligand interactions can lead to new and more effective

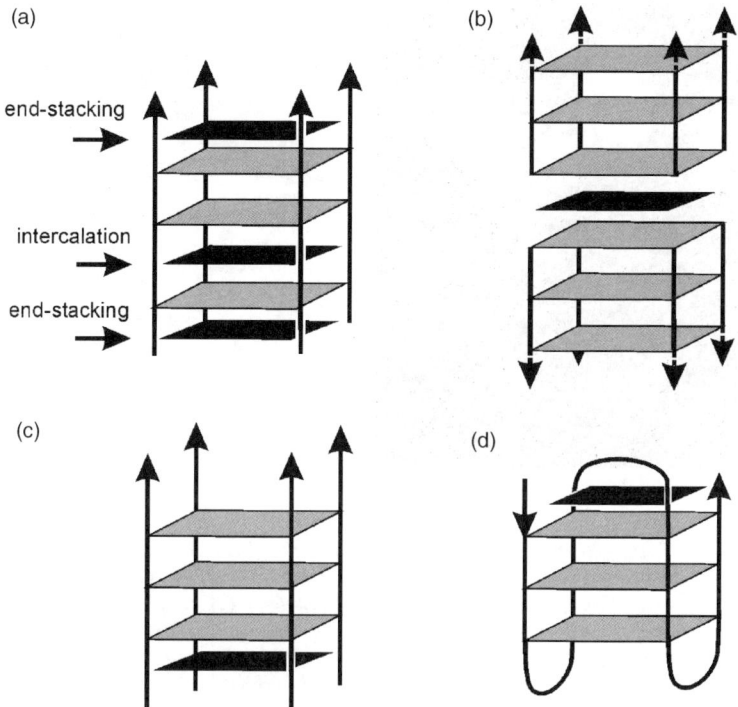

Figure 3 *(a) Parallel-stranded quadruplex structure showing possible intercalation sites corresponding to end-stacking on the terminal G-tetrads, and intercalation between G-tetrads that requires the displacement of a bound monovalent cation. (b) Sandwich model in which a drug molecule (PIPER, 6) is bound between the blunt-ends of two parallel-stranded quadruplexes d(TTAGGG)₄ and (c) when end-stacking at the GpT step of d(TAGGGTTA)₄, and (d) end-stacking and binding within the TTA loop region of the intramolecular human telomeric basket structure.*

telomerase inhibitors. One investigation with a 1,4-disubstituted amidoanthraquinone ligand (14) was able to define a plausible model for the ligand-binding interaction by combining low-resolution X-ray fibre diffraction data with molecular dynamics simulations to distinguish between internally intercalated and externally stacked binding models.[46] The inability of the complex of d(TGGGGT) with the bound ligand to give diffracting single crystals suggested inherent disorder indicative of conformational flexibility and/or multiple bound orientations of the ligand. However, the diffraction data were consistent with a repeating unit of four G-tetrads with two stacked ligand molecules, which could be modelled in more detail. During 1 ns MD simulations, the intercalation model (TG·GG·GT;·=ligand) led to rapid structural distortions of the core quadruplex, while end-stacking models (T·GGGG·T) remained very stable, consistent with the NMR models for the PIPER complex. Not only are the end-stacked structures likely to be more kinetically accessible,

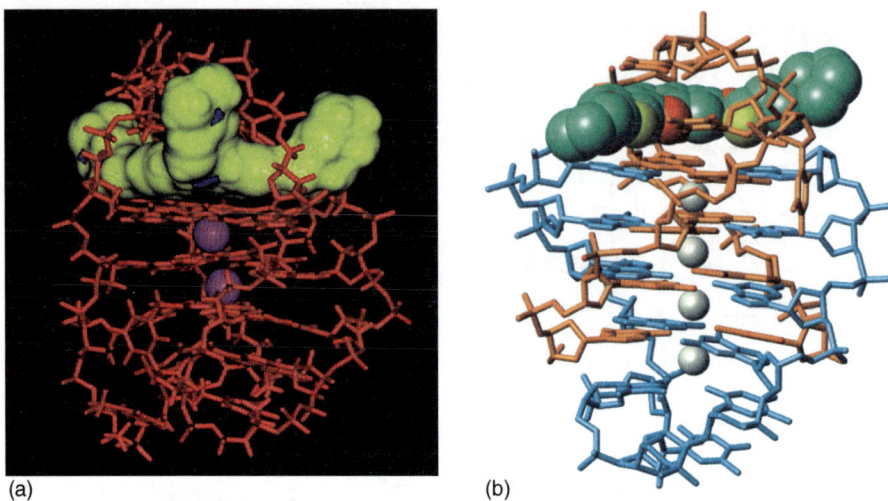

(a)   (b)

Figure 4   *Structural model of the complex of the 3,6,9-trisubstituted acridine ligand (3) with the intramolecular basket quadruplex structure of the human telomeric sequence. (a) End-stacking with the terminal G-tetrad permits the substituent at the 9-position to interact with the flexible TTA loop structure in a specific manner. (b) X-ray structure of the complex of a di-substituted aninoalkylamido acridine (4) with the dimeric hairpin quadruplex structure derived from the Oxytricha nova sequence d(GGGGTTTTGGGG)₂. The drug end-stacks and interacts extensively with the thymine loop ((a) Adapted from ref 25 and (b) from RCSB Protein Data Bank co-ordinates 1L1H (ref 49)).*

but also the core internally stabilising monovalent cations are undisturbed by the binding interaction.

Extending this approach to the design of 3,6,9-trisubstituted acridine ligands (**3**), binding to the intramolecular quadruplex formed by the human telomeric sequence suggested that a substituent at the 9-position would lie in the groove of the quadruplex with enhanced selectivity for the higher order structure over duplex DNA.[25] The acridine ligand was modelled by end-stacking on the quadruplex "basket" structure of Wang and Patel[17] forming interactions within the TTA loop region. Substitution at the 9-position with an anilino derivative suggested favourable interactions within the quadruplex groove and with the loop structure [Figure 4(a)]. Surface plasmon resonance (SPR) experiments were used to estimate binding constants to duplex and quadruplex structures by immobilising the DNA substrates. The two tri-substituted anilino compounds, compared to the simpler di-substituted analogue, bound to human quadruplex DNA 30–40 times more strongly than to duplex, with a quadruplex affinity 10-fold higher than for the simple di-substituted acridine. The rationally designed anilino-acridines were subsequently shown to be potent telomerase inhibitors in TRAP assays ($EC_{50} < 0.1$ µM), with lower cytotoxicities presumably reflecting their reduced affinity for duplex DNA. These data, together with earlier results, seem to strongly correlate telomerase inhibition with quadruplex binding energy.

5.5 Structure and Dynamics of a Quino-Acridinium Cation Bound to a G-Quadruplex

The NMR structure of the parallel-stranded DNA quadruplex d(TTAGGGT)$_4$, containing the human telomeric repeat, was used as the drug target for complexation studies with a fluorinated pentacyclic quino[4,3,2-*kl*]acridinium cation (Figure 2; RHPS4, **9**).[47,48] RHPS4 has been identified as a potent inhibitor of telomerase at submicromolar levels (IC$_{50}$ value of 0.33 ± 0.13 µM), exhibiting a wide differential between telomerase inhibition and acute cellular toxicity. NMR (^{1}H and ^{19}F) investigations of RHPS4 show that the ligand forms a dynamic interaction with d(TTAGGGT)$_4$ with fast exchange observed between binding sites at 318 K. The dependence of ^{1}H imino proton line widths on ligand concentration demonstrates classical fast exchange characteristics broadening considerably at a drug/DNA ratio of ~ 0.3, but sharpening as the fully bound state is reached (Figure 5). The line width dependence as a function of the fraction of bound ligand gives an estimated off-rate of 2800 (± 1000) s^{-1} at 318 K. Perturbations to ^{1}H NMR DNA chemical shifts and 24 intermolecular NOEs identified the 5′-ApG and 5′-GpT steps as the principle intercalation sites and allowed a structural model to be refined using NOE-restrained molecular dynamics (Figure 6a and b).

The central G-tetrad core remains intact with drug molecules stacking at the ends of the G-quadruplex, suggesting that the energetic cost of displacing the core K$^+$ ions is too great and not sufficiently compensated by drug–DNA π-stacking interactions. In both ApG and GpT intercalation sites, the drug is seen to adopt a similar orientation in which the π-system of the drug overlaps primarily with two bases of each G-tetrad [Figure 6(b)]. The drug is held in place primarily through stacking interactions with the G-tetrads; however, there is some evidence for a more dynamic weakly stabilised A-tetrad that partially stacks on top of the drug at the 5′-end of the sequence. Molecular dynamics simulations demonstrate larger amplitude fluctuations in the conformation of the adjacent A-tetrad than for the core G-tetrads but suggest nevertheless that the A-tetrad may be contributing to the stabilisation of the binding interaction [Figure 6(c)]. Together the interactions of RHPS4 increase the T_m of the quadruplex by $\sim 20°$C. There is no evidence for drug intercalation within the G-quadruplex; however, the structural model driven by experimental restraints strongly supports end-stacking interactions with the terminal G-tetrads as demonstrated in the complex of PIPER,[30] and in other ligand-interaction studies.

5.6 High-Resolution X-Ray Analysis of Drug–Quadruplex Interactions

Subsequently, two X-ray structures have been reported: one of a drug-bound dimeric anti-parallel G-quadruplex containing the *Oxytricha nova* sequence d(GGGGTTTTGGGG) in which a di-substituted aminoalkylamido acridine drug (**4**, see Figure 2) end-stacks with a terminal G-tetrad and interacts with one of the loops.[49] In the other structure a daunomycin (**14**) trimer stacks on

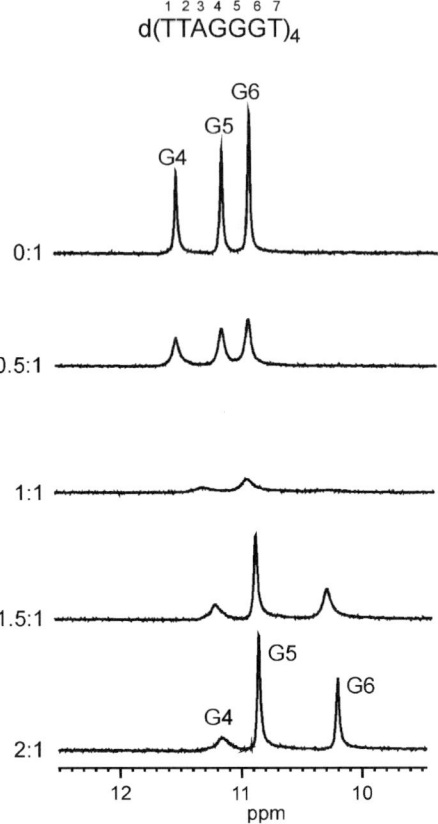

Figure 5 *¹H NMR titration of the pentacyclic acridinium cation RHPS4 (7) with the intermolecular quadruplex d(TTAGGT)₄ monitoring the changes to the imino proton shifts and line widths of the three core G-tetrads, demonstrating fast-exchange characteristics as the signals broaden and then sharpen as the binding sites at the ApG and GpT steps are saturated (Adapted from ref 48).*

the terminal G-tetrad of a parallel intermolecular quadruplex.[50] The former demonstrates that the diagonal thymine loop can act as a flexible recognition motif stabilising the complex by specific hydrogen-bonding interactions [Figure 4(b)]. Its intrinsic flexibility suggests that alternative conformations are highly likely to accommodate other ligands with a different arrangement of interacting functional groups. Similarly, loops of different length, base composition and geometry seem likely to be key determinants in ligand binding specificity in the context of both telomeric and non-telomeric quadruplexes. The key structural feature to emerge from this study appears to be the active role played by the flexible loop region in binding and recognition, together with further support for earlier NMR models that favour end-stacking rather than interior interca-lation. To probe the stability of the X-ray structure, molecular dynamics simulations were run for 2 ns and indeed demonstrated that the arrangement

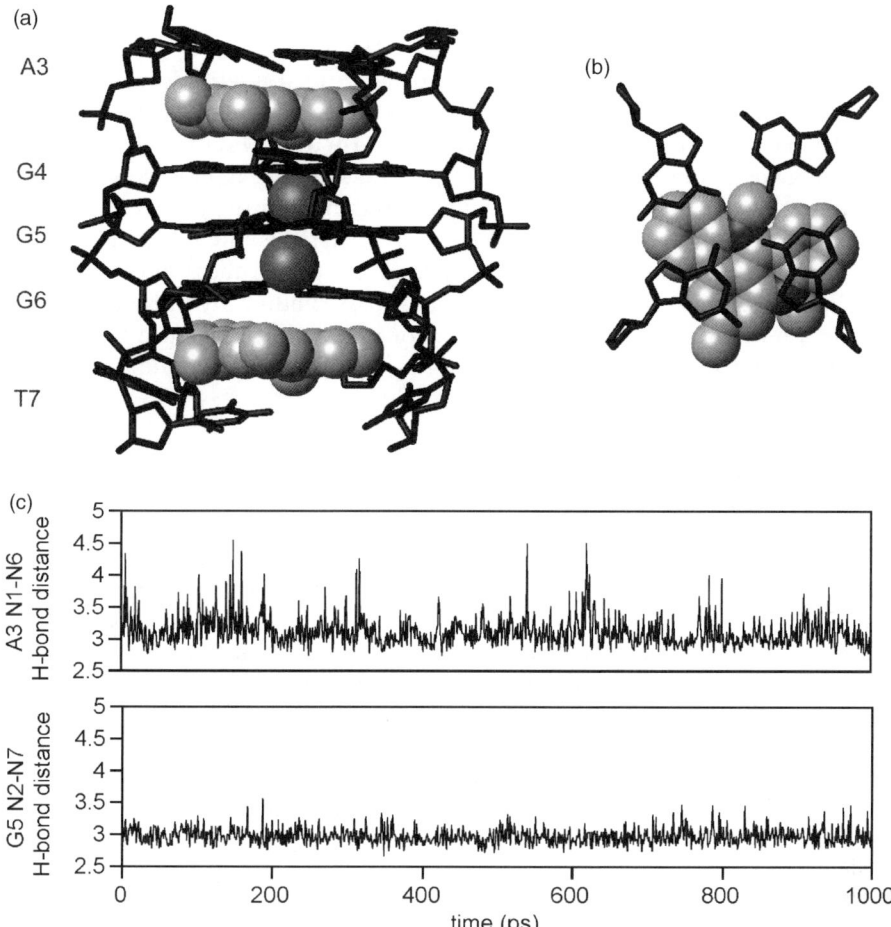

Figure 6 *(a) NOE-derived structure of the 2:1 complex of RHPS4 with the parallel-stranded G-quadruplex d(TTAGGGT)₄ with drug molecules end-stacked at the ApG and GpT steps. (b) Off-set π-stacking with the G4 tetrad with the 13-N of the acridine positioned above the centre of the G-tetrad in the region of high negative charge density probably occupied by a sodium ion in the ligand-free state. (c) Dynamic fluctuations in the conformation of purine (G or A) tetrads measured from the N1–N6 distance of the A-tetrad (top) and N2–N7 distance of the G-tetrad of G5 (bottom) during 1000 ps of unrestrained molecular dynamics simulation of the 2:1 RHPS4−d(TTAGGGT)₄ complex at 300 K in explicit solvent ((b) Adapted from ref 47 and (c) from ref 48).*

was quite stable on this timescale. The NMR data reported above for the binding of RHPS4 (**9**)[48] and PIPER (**8**)[30] to model quadruplexes emphasise the dynamic nature of the interaction, reflecting the relatively simple architectural features of the ligands so far investigated, with ligands exchanging on a timescale >2000 s^{-1} (lifetime for the bound state <0.5 ms). Thus, although the barriers to association and dissociation are relatively small, the timescales involved are still largely inaccessible to the MD simulations.

In the second X-ray structure, the parallel-stranded intermolecular quadruplex d(TGGGGT)$_4$ was shown to interact with a self-assembled daunomycin (15) trimer by stacking with the terminal G-tetrad (Figure 7).[50] In such an arrangement the trimer is able to optimise stacking interactions with the G-tetrad allowing the daunosamine sugars to dangle in the grooves and form stabilising hydrogen-bonding interactions. Whether this interaction is an artefact of the crystallisation process is unclear; however, others have suggested that ligand aggregation can enhance quadruplex binding by presumably presenting a more extensive binding surface for interaction. Evidence is available to suggest that the anthracycline antibiotics doxorubicin and daunomycin are able to interact with telomeric DNA.[51]

5.7   Quadruplex-Targeted Small Molecule-Peptide Conjugates

A novel family of carbocyanine–peptide conjugates has been identified through a combinatorial selection approach by combining binding elements that recognise distinct structural features of the G-quadruplex.[38] Earlier studies by Chen et al.[39] identified a carbocyanine dye (10) that bound to hairpin quadruplex structures in preference to duplex DNA. These ligands bind with the induction of unique spectroscopic features in the CD spectrum that appear to be quadruplex specific, identifying an important role for the loop sequences (in particular thymine bases) in determining the bound orientation. Fluorescence energy transfer experiments suggest that at least some partial stacking interaction adjacent to the thymine-rich loops. In the work of Schouten et al.[38] the heterocyclic HC core motif, derived from the carbocyanine dye structures described by Chen et al.,[39] was combined with a tetrapeptide library using a subset of 13 amino acids and was screened using the radiolabelled single-stranded human telomeric quadruplex. Subsequently, three peptide conjugates (11) (HC-XXXX-CONH$_2$, where XXXX = RKKV, KRSR and FRHR, one-letter amino acid code) were selected from the potential pool of >25,000 and characterised by SPR methods revealing K_d values better than 50 μM and a high quadruplex/duplex selectivity of ~40.[38] The molecular basis for the high level of structural discrimination observed for this novel class of ligands remains to be established. The authors subsequently extended this approach to using an acridine "platform" to generate a 3,6-bis-peptide conjugate (12) with the ligand properties shown to be highly peptide sequence-dependent with the FRHR conjugate showing considerable specificity for quadruplex over duplex as evident from K_d values estimated from SPR sensograms (Figure 8).[40] The molecular basis for this high specificity was probed using molecular modelling approaches suggesting that in an end-stacking model with the parallel-stranded intramolecular quadruplex (Figure 1b, propeller model), the tetrapeptides are readily accommodated through interactions with the TTA loop pockets with some evidence that the Phe side chain may

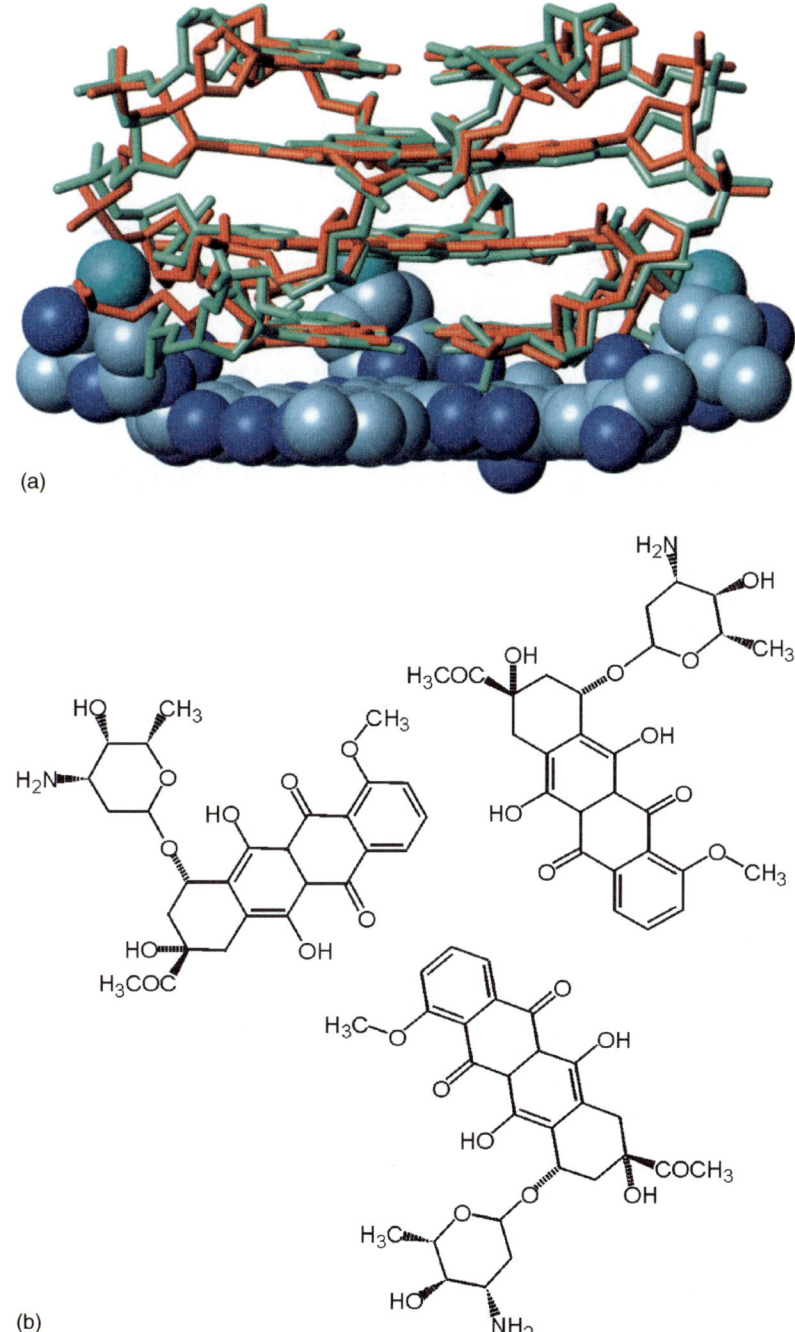

Figure 7 *(a) Structure of the intermolecular parallel quadruplex d(TGGGGT)₄ with a self-assembled daunomycin trimer (b) stacked on the terminal G-tetrad. The structure in (a), with the drug trimer in space filling representation, is overlayed with the quadruplex structure of the ligand-free structure showing little drug-induced change in quadruplex conformation (Adapted from ref 50 and PDB co-ordinates 100 K).*

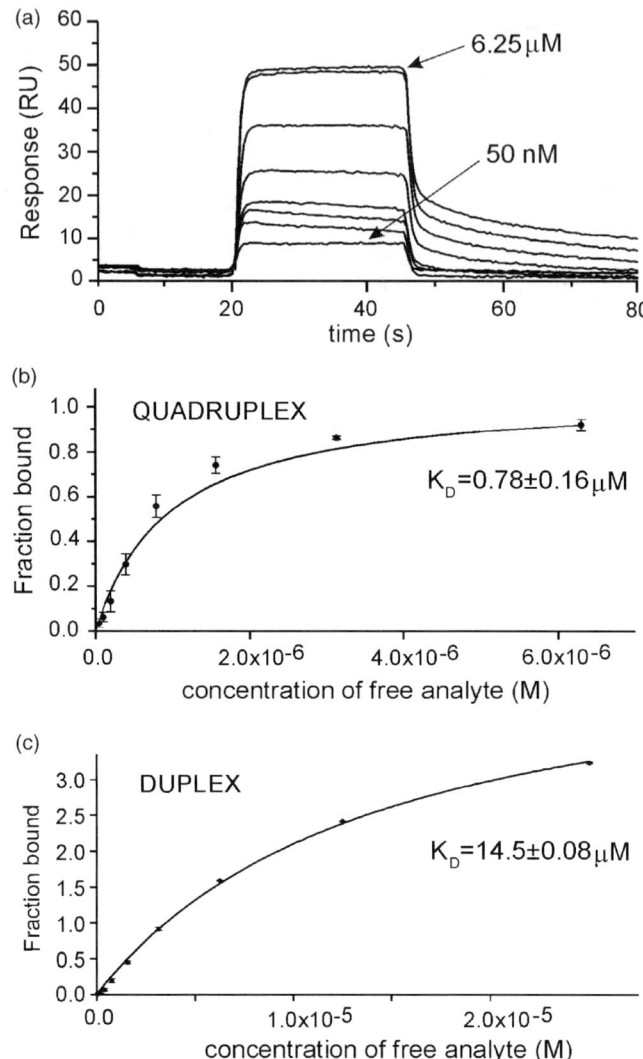

Figure 8   *(a) SPR sensograms for the binding of the acridine-peptide conjugate (**12**) to the human intramolecular telomeric quadruplex over the drug concentration range 0.05–6.25 µM (bottom to top). Binding curves (fraction bound vs. free) used to determine K_d values for the interaction with the quadruplex (b) and duplex (c) (Adapted from ref 40).*

contribute to the stability of the complex through π-stacking interactions with the terminal G-tetrad. The Arg and His residues contribute through strong electrostatic interactions. A large electrostatic component of binding is evident for all three high-affinity binding peptide sequences (FRHR, RKKV and KRSR). Similar model building studies and MD simulations with the peptide conjugates threaded through a B-DNA helix suggest that the observed differences in K_d values stem from the incompatibility of the substituted acridines with the molecular dimensions of the B-DNA helix that precludes the peptide chains from being optimally accommodated in the grooves. In particular the Phe side chain of the FRHR conjugate (specificity > 50-fold for quadruplex) is left solvent exposed in the B-DNA model.

5.8 Quadruplex Recognition by Distamycin: In the Grooves or End-Stacking?

The polypyrrole ligands distamycin A and netropsin, which have been extensively developed as duplex DNA minor groove binding ligands, have been shown to present a versatile template for the design of mixed DNA sequence recognition using side by side dimerisation to infer sequence binding specificity. These ligands have also been shown to interact in a dimeric anti-parallel fashion with the quadruplex grooves all involved in the formation of a dynamic short-lived complex.[52] NMR titration studies suggest a 4:1 drug/quadruplex interaction with the symmetrical d(TGGGGT)$_4$ structure in which the dyad symmetry is retained in the bound state (Figure 9). The initially proposed model suggested that distamycin dimers can bind symmetrically in two of the grooves of the quadruplex, but a fully saturated complex with all grooves occupied was not observed, suggesting that the occupation of one groove may distort the width of adjacent grooves preventing further interaction. The ability of the distamycin A dimer to expand the minor groove of duplex DNA is well-documented suggesting analogous effects on the quadruplex structure as a possibility. However, a subsequent analysis using chemical shift perturbation data and intermolecular NOEs, particularly from the drug methyl resonances, seems to suggest that distamycin A binds in much the same way as other ligand complexes by end-stacking with the terminal G-quartets but with two molecules lying side by side flat on the terminal G-tetrad.[53] A distamycin derivative has recently been shown to inhibit telomerase activity suggesting a new lease of life for the polypeptide distamycin/netropsin motif in the design of quadruplex-specific therapeutics.[54]

Selectivity, binding stoichiometry and mode of binding of distamycin, a PIPER-like ligand (Tel01) and diethylthiocarbocyanine (DTC) have also been investigated by electrospray ionisation–mass spectrometry (ESI–MS) to observe non-covalent interactions in the gas phase.[55] Collisionally activated dissociation of these complexes produces distinct fragmentation patterns for these ligands suggesting different binding modes. While the Tel01–quadruplex complex undergoes facile dissociation to drug and single strands of DNA, the distamycin and DTC complexes show evidence for dissociation to drug-bound single strands from which the authors concluded that the latter ligands interact through quadruplex groove binding rather than end-stacking.

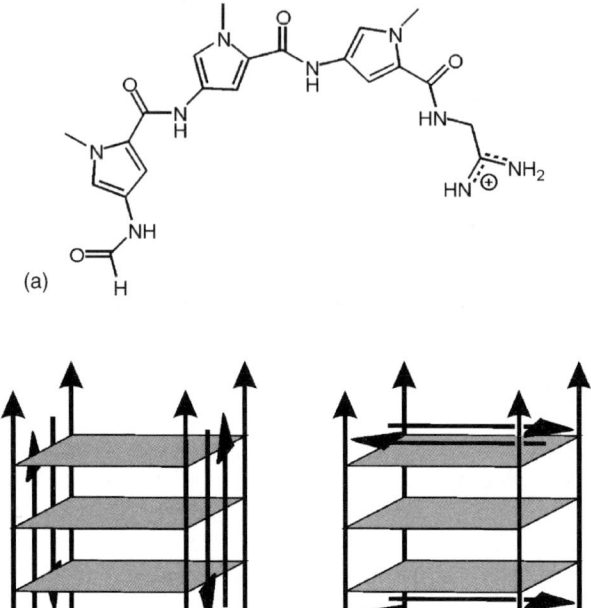

Figure 9 *(a) Structure of distamycin and schematic representations of two 4:1 drug/*
quadruplex complexes with the symmetrical d(TGGGGT)₄ parallel quadruplex
structure in which the overall dyad symmetry of the complex is retained in the
bound state, (b) drug molecules are bound as anti-parallel side-by-side dimers
in the grooves with extensive drug–drug interactions, and (c) the drug molecules
end-stack with the terminal G-tetrads with π-stacking primarily with the
guanine bases.

5.9 Selectivity of Telomestatin for Different G-Quadruplex Isoforms

The human telomeric repeat sequence has been reported to fold into a number
of possible intramolecular G-quadruplex structures depending on the incubation conditions and the nature of the stabilising cation (Na^+ or K^+). The two
key intramolecular structures are the "basket" and "propeller" forms both of
which are stabilised by the same G-tetrad core but differ in the orientation of
the connecting loops and strand alignment (see Chapter 4) (Figure 1). The
Wang and Patel basket structure, which is stabilised in the presence of Na^+
ions, has one diagonal and two lateral loops at the ends of the structure. In
contrast, the crystal structure of the propeller form stabilised in the presence of
K^+ ions, forms a homogeneous parallel alignment of strands with the loops
lying to the sides (propeller loops) rather than forming caps at the ends of the
stack of tetrads (Figure 1). A number of alternate structures derived from non-
human telomeric sequences have also been reported, including a mixed parallel/
antiparallel hybrid structure (Figure 1). The natural product telomestatin binds

strongly and specifically to the human intramolecular quadruplex and, unusually, was shown to induce quadruplex formation in the absence of monovalent cations.[44] In parallel studies with the synthetic diselena sapphyrin chloride [Se2SAP (**16**), Figure 2], CD studies have been used to demonstrate that these ligands are capable of converting one form to another.[56] Telomestatin binds preferentially to the basket form and induces such a structure in the preformed mixed hydrid, while Se2SAP (**16**) converts the basket form to the mixed hybrid. Telomestatin also induces the basket form in the random coil sequence in the absence of significant concentrations of added cations. The CD changes identified are significant with each of the basket and hybrid species having characteristic spectroscopic features. The ability of different ligands to promote the folding of G-rich sequences found in telomeres and certain promoter regions of important oncogenes (such as *c-myc*)[57,58] suggests that the biological consequences extend beyond their ability to stabilise a preformed structure and inhibit telomerase activity.

5.10 Porphyrin Interaction with the *c-myc*-Derived Quadruplex

The *c-myc* promoter region regulates the transcription of the *MYC* oncogene an important element of which is the nuclease hypersensitivity element III (NHE III) that controls ~90% of total *MYC* transcription. The 27-nucleotide purine-rich strand of NHE III (Pu27) contains six guanine tracts and is known to assemble to form G-quadruplex structures, the stabilities of which appear to influence the level of *MYC* transcription. In particular, extensive investigations of the interaction of the cationic porphyrin (**5**; see Figure 2) with Pu27 show that it forms a quadruplex-stabilising interaction that suppresses transcription.[57,58] Phan *et al.*[59] have recently shown that the Pu27 sequence in K^+ solution gives NMR spectra indicative of multiple conformations. However, truncation of Pu27 and substitution at key positions to minimise conformational heterogeneity, enables a major intramolecular conformer to be isolated in solution. The Pu24I sequence [Figure 10(a)] gives high-quality spectra that yield to detailed structural analysis by model-independent resonance assignment methods using site-specific low-enrichment ^{13}C and ^{15}N labelling.[60] The structure of Pu24I shows an unusual G-quadruplex fold consisting of three guanine tetrads one of which contains a snapback 3′-end *syn* guanine (see Figure 10b, adapted from ref. 59). Subsequently, ligand binding studies with the cationic porphyrin (**5**), and also the classic minor groove binder Hoechst 33258 and the DNA intercalators daunomycin (**15**) and ethidium (**1**) resulted in subtly different effects. Monitoring imino proton chemical shifts and line widths shows that all but the porphyrin are in fast exchange between free and bound state interacting through stacking with the terminal G-tetrad. In contrast, the tetramethylpyridylporphyrin (**5**) binds in slow exchange with signals from the free and bound states evident in the presence of half an equivalent of ligand, with magnetisation transfer between the two forms evident in 2D NOESY spectra.[59] Large chemical shift changes to the imino proton resonances are

(a)

Pu27: TGGGGAGGGTGGGGAGGGTGGGGAAGG

Pu24I: TGAGGGTGGIGAGGGTGGGGAAGG

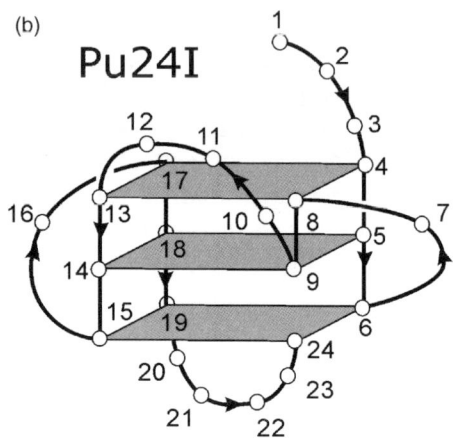

Figure 10 *(a) DNA sequence for the Pu27 c-myc NHE III promoter region, and the modified Pu24I sequence used in structural studies; (b) schematic representation of the folded structure of Pu24I showing the core G-tetrad structure (Adapted from ref 59).*

consistent with a well-defined stacking orientation on the top tetrad (G4, G8, G13 and G17), again consistent with the previously proposed ligand–quadruplex binding models presented above. The Pu24I sequence provides a viable model system, with a novel intramolecular fold, for the development of potential anti-cancer drugs targeted at the NHE III element of the *c-MYC* promoter region that regulates the level of aberrant expression of human *c-MYC* transcription in the progression of a range of cancers.

5.11 Conclusions

The emergence of a range of structural motifs and novel folds formed by multi-guanine-tract DNA sequences found in telomeres and oncogenic promoter regions presents a challenge for structure determination, but at the same time attractive targets for anti-cancer drug design. This chapter has summarised some of the structural details that have so far been reported on drug–quadruplex recognition. An emerging feature has been the important role of the non-guanine-rich connecting loop sequences in defining ligand binding sites. The stabilising role of monovalent cations between stacked G-tetrads appears to preclude the possibility of ligand molecules intercalating within the stack of tetrads because of the high energetic penalty associated with cation displacement. Consequently, ligand binding studies to date have identified end-stacking as the principal mode of interaction. By forming π–π interactions between an aromatic ligand and the terminal G-tetrad the interconnecting loop sequences at the ends of the quadruplex are brought into play in a sequence-dependent manner to stabilise ligand binding interactions. To date, the emphasis has been on electrostatic contributions to ligand–quadruplex stability through π-stacking with largely heterocyclic aromatic ligands; however, other structural features such as sequence- and fold-dependent groove widths and novel loop structures generated by different fold topologies are yet to be widely exploited in rational ligand design, presenting plenty of future challenges for medicinal chemists and chemical biologists.

Acknowledgements

G.B. is grateful for the support provided by the EPSRC (UK). We also acknowledge support from the University of Nottingham.

References

1. E.H. Blackburn, Telomeres – no end in sight, *Cell*, 1994, **77**, 621–623.
2. V.A. Zakian, Telomeres – beginning to understand the end, *Science*, 1995, **270**, 1601–1607.
3. T. Delange, Activation of telomerase in a human tumour, *Proc. Natl. Acad. Sci. USA*, 1994, **91**, 2882–2885.
4. T.M. Bryan and T.R. Cech, Telomerase and the maintenance of chromosome ends, *Curr. Opin. Cell Biol.*, 1999, **11**, 318–324.
5. C.B. Harley and B. Villeponteau, Telomeres and telomerase in aging and cancer, *Curr. Opin. Genet. Dev.*, 1995, **5**, 249–255.
6. N.W. Kim, M.A. Piatyszek, K.R. Prowse, C.B. Harley, M.D. West, P.L.C. Ho, G.M. Coviello, W.E. Wright, S.L. Weinrich and J.W. Shay, Specific association of human telomerase activity with immortal cells and cancer, *Science*, 1994, **266**, 2011–2015.
7. J. Lingner, T.R. Hughes, A. Shevchenko, M. Mann, V. Lundblad and T.R. Cech, Reverse transcriptase motifs in the catalytic subunit of telomerase, *Science*, 1997, **276**, 561–567.
8. T.M. Nakamura, G.B. Morin, K.B. Chapman, S.L. Weinrich, W.H. Andrews, J. Lingner, C.B. Harley and T.R. Cech, Telomerase catalytic subunit homologs from fission yeast and human, *Science*, 1997, **277**, 955–959.
9. J.L. Feng, W.D. Funk, S.S. Wang, S.L. Weinrich, A.A. Avilion, C.P. Chiu, R.R. Adams, E. Chang, R.C. Allsopp, J.H. Yu, S.Y. Le, M.D. West, C.B. Harley, W.H. Andrews, C.W. Greider and B. Villeponteau, The RNA component of human telomerase, *Science*, 1995, **269**, 1236–1241.
10. J.L. Mergny, P. Mailliet, F. Lavelle, J.F. Riou, A. Laoui and C. Helene, The development of telomerase inhibitors: The G-quartet approach, *Anti-Cancer Drug Des.*, 1999, **14**, 327–339.

11. P.J. Perry and T.C. Jenkins, Recent advances in the development of telomerase inhibitors for the treatment of cancer, *Exp. Opin. Invest. Drugs*, 1999, **8**, 1981–2008.

12. E. Raymond, J.C. Soria, E. Izbicka, F. Boussin, L. Hurley and D.D. Von Hoff, DNA G-quadruplexes telomere-specific proteins, and telomere-associated enzymes as potential targets for new anticancer drugs, *Invest. New Drugs*, 2000, **18**, 123–137.

13. S. Neidle and G. Parkinson, Telomere maintenance as a target for anti-cancer drug discovery, *Nat. Reviews, Drug Disc.*, 2002, **1**, 383–393.

14. J.R. Williamson, Guanine quartets, *Curr. Opin. Struct. Biol.*, 1993, **3**, 357–362.

15. J. Feigon, K.M. Koshlap and F.W. Smith, ^{1}H NMR spectroscopy of DNA triplexes and quadruplexes, *Nucl. Magn. Reson. Nucleic Acids*, 1995, **261**, 225–255.

16. R.H. Shafer, Stability and structure of model DNA triplexes and quadruplexes and their interactions with small ligands, *Progr. Nucleic Acid Res. Mol. Biol.*, 1998, **59**, 55–94.

17. D.J. Patel, S. Bouaziz, A. Ketttani and Y. Wang, *Oxford Handbook of Nucleic Acid Structure*, Oxford University Press, New York, 1998, 389–453.

18. A.M. Zahler, J.R. Williamson, T.R. Cech and D.M. Prescott, Inhibition of telomerase by G-quartet DNA structures, *Nature*, 1991, **350**, 718–720.

19. K. Phillips, Z. Dauter, A.I.H. Murchie, D.M.J. Lilley and B. Luisi, Crystal structure of a parallel-stranded guanine tetraplex at 0.95 Å resolution, *J. Mol. Biol.*, 1997, **273**, 171–182.

20. M.P. Horvath and S.C. Schultz, DNA quartets in a 1.86 Å resolution structure of an *Oxytricha nova* telomeric protein-DNA complex, *J. Mol. Biol.*, 2001, **310**, 367–377.

21. G.N. Parkinson, M.P.H. Lee and S. Neidle, Crystal structure of parallel quadruplexes from human telomeric DNA, *Nature*, 2002, **417**, 876–880.

22. S. Neidle and G.N. Parkinson, The structure of telomeric DNA, *Curr. Opin. Struct. Biol.*, 2003, **13**, 275–283.

23. J.T. Davis, G-quartets 40 years later: From 5'-GMP to molecular biology and supramolecular chemistry, *Angew. Chemie. Int. Ed.*, 2004, **43**, 668–698.

24. J.L. Mergny, L. Lacroix, M.P. Teulade-Fichou, C. Hounsou, L. Guittat, M. Hoarau, P.B. Arimondo, J.P. Vigneron, J.M. Lehn, J.F. Riou, T. Garestier and C. Helene, Telomerase inhibitors based on quadruplex ligands selected by a fluorescence assay, *Proc. Natl. Acad. Sci. USA*, 2001, **98**, 3062–3067.

25. M. Read, R.J. Harrison, B. Romagnoli, F.A. Tanious, S.H. Gowan, A.P. Reszka, W.D. Wilson, L.R. Kelland and S. Neidle, Structure-based design of selective and potent G-quadruplex-mediated telomerase inhibitors, *Proc. Natl. Acad. Sci. USA*, 2001, **98**, 4844–4849.

26. F.X.G. Han, R.T. Wheelhouse and L.H. Hurley, Interactions of TMPyP4 and TMPyP2 with quadruplex DNA. Structural basis for the differential effects on telomerase inhibition, *J. Am. Chem. Soc.*, 1999, **121**, 3561–3570.

27. R.T. Wheelhouse, D.K. Sun, H.Y. Han, F.X.G. Han and L.H. Hurley, Cationic porphyrins as telomerase inhibitors: The interaction of tetra-(*N*-methyl-4-pyridyl)porphine with quadruplex DNA, *J. Am. Chem. Soc.*, 1998, **120**, 3261–3262.

28. S. Neidle, R.J. Harrison, A.P. Reszka and M.A. Read, Structure-activity relationships among guanine-quadruplex telomerase inhibitors, *Pharmacol. Ther.*, 2000, **85**, 133–139.

29. P.J. Perry, M.A. Read, R.T. Davies, S.M. Gowan, A.P. Reszka, A.A. Wood, L.R. Kelland and S. Neidle, 2,7-disubstituted amidofluorenone derivatives as inhibitors of human telomerase, *J. Med. Chem.*, 1999, **42**, 2679–2684.

30. O.Y. Fedoroff, M. Salazar, H.Y. Han, V.V. Chemeris, S.M. Kerwin and L.H. Hurley, NMR-based model of a telomerase-inhibiting compound bound to G-quadruplex DNA, *Biochemistry*, 1998, **37**, 12367–12374.

31. S.M. Gowan, R.A. Heald, M.F.G. Stevens and L.R. Kelland, Potent inhibition of telomerase by small-molecule pentacyclic acridines capable of interacting with G-quadruplexes, *Mol. Pharmacol.*, 2001, **60**, 981–988.

32. R.A. Heald, C. Modi, J.C. Cookson, I. Hutchinson, C.A. Laughton, S.M. Gowan, L.R. Kelland and M.F.G. Stevens, Antitumor polycyclic acridines. 8. Synthesis and telomerase-inhibitory activity of methylated pentacyclic acridinium salts, *J. Med. Chem.*, 2002, **45**, 590–597.

33. M.-P. Teulade-Fichuo, C. Carrasco, L. Guittat, C. Bailly, P. Alberti, J.-L. Mergny, A. David, J.-M. Lehn and W.D. Wilson, Selective recognition of G-quadruplex telomeric DNA by a bis(quinacridine) macrocycle, *J. Am. Chem. Soc.*, 2003, **125**, 4732–4740.

34. J.F. Riou, A. Laoui, P. Mailliet, O. Petitgenet, F. Koeppel, P.B. Arimondo, D. Labit, C. Helene and J-L. Mergny, Ethidium derivatives bind to G-quartets, inhibit telomerase and act as fluorescent probes for quadruplexes, *Nucleic Acids Res.*, 2001, **29**, 1087–1096.

35. P.M. Murphy, V.A. Phillips, S.A. Jennings, N.C. Garbett, J.B. Chaires, T.C. Jenkins and R.T. Wheelhouse, Biarylpyrimidines: A new class of ligand for high-order DNA recognition, *Chem. Comm.*, 2003, **1160–1161**.

36. A. Maraval, S. Franco, C. Vialas, G. Pratviel, M.A. Blasco and B. Meunier, Porphyrin-aminoquinoline conjugates as telomerase inhibitors, *Org. Biomol. Chem.*, 2003, **1**, 921–927.

37. B. Guyen, C.M. Schultes, P. Hazel, J. Mann and S. Neidle, Synthesis and evaluation of analogues of 10H-indolo[3,2-b] quinoline as G-quadruplex stabilising ligands and potential inhibitors of the enzyme telomerase, *Org. Biomol. Chem.*, 2004, **2**, 981–988.

38. J.A. Schouten, S. Ladame, S.J. Mason, M.A. Cooper and S. Balasubramanian, G-quadruplex-specific peptide-hemicyanine ligands by partial combinatorial selection, *J. Am. Chem. Soc.*, 2003, **125**, 5594–5595.

39. Q. Chen, I.D. Kuntz and R.H. Shafer, Spectroscopic recognition of guanine dimeric hairpin quadruplexes by a carbocyanine dye, *Proc. Natl. Acad. Sci. USA*, 1996, **93**, 2635–2639.

40. S. Ladame, J.A. Schouten, J. Stuart, J. Roldan, S. Neidle and S. Balasu-bramanian, Tetrapeptides induce selective recognition for G-quadruplexes when conjugated to a DNA-binding platform, *Org. Biomol. Chem.*, 2004, **2**, 2925–2931.
41. M. Isalan, S.D. Patel, S. Balasubramanian and Y. Choo, Selection of zinc fingers that bind single-stranded telomeric DNA in the G-quadruplex conformation, *Biochemistry*, 2001, **40**, 830–836.
42. C. Schaffitzel, I. Berger, J. Postberg, J. Hanes, H.J. Lipps and A. Pluckt-hun, *In vitro* generated antibodies specific for telomeric guanine-quadru-plex DNA react with *Stylonychia lemnae* macronuclei, *Proc. Natl. Acad. Sci. USA*, 2001, **98**, 8572–8577.
43. K. Shin-ya, K. Wierzba, K. Matsuo, T. Ohtani, Y. Yamada, K. Furihata, Y. Hayakawa and H. Seto, Telomestatin, *a novel telomerase inhibitor from Streptomyces anulatus, J. Am. Chem. Soc.*, 2001, **123**, 1262–1263.
44. M.-Y. Kim, H. Vankayalapati, K. Shin-ya, K. Wierzba and L.H. Hurley, Telomestatin, a potent telomerase inhibitor that interacts quite specifically with the human telomeric intramolecular G-quadruplex, *J. Am. Chem. Soc.*, 2002, **124**, 2098–2099.
45. I. Haq, J.O. Trent, B.Z. Chowdhry and T.C. Jenkins, Intercalative G-tetraplex stabilization of telomeric DNA by a cationic porphyrin, *J. Am. Chem. Soc.*, 1999, **121**, 1768–1779.
46. M.A. Read and S. Neidle, Structural characterization of a guanine-quad-ruplex ligand complex, *Biochemistry*, 2000, **39**, 13422–13432.
47. E. Gavathiotis, R.A. Heald, M.F.G. Stevens and M.S. Searle, Recognition and stabilisation of quadruplex DNA by a potent new telomerase inhibitor: NMR studies of the 2:1 complex of a pentacyclic methylacridinium cation with d(TTAGGGT)$_4$, *Angew. Chemie. Int. Ed.*, 2001, **40**, 4749–4751.
48. E. Gavathiotis, R.A. Heald, M.F.G. Stevens and M.S. Searle, Drug recognition and stabilisation of a parallel stranded DNA quadruplex d(TTAGGT)$_4$ containing the human telomeric repeat, *J. Mol. Biol.*, 2003, **334**, 25–36.
49. S.M. Haider, G.N. Parkinson and S. Neidle, Structure of a G-quadruplex-ligand complex, *J. Mol. Biol.*, 2003, **326**, 117–125.
50. G.R. Clark, P.D. Pytel and C.J. Squire, Structure of the first parallel DNA quadruplex-drug complex, *J. Am. Chem. Soc.*, 2003, **125**, 4066–4067.
51. L.W. Elmore, C.W. Rehder, X. Di, P.A. McChesney, C.K. Jackson-Cook, D.A. Gewitz and S.E. Holt, Adriamycin-induced senescence in breast tumor cells involves functional p53 and telomere dysfunction, *J. Biol. Chem.*, 2002, **277**, 35509–35515.
52. A. Randazzo, A. Galeone and L. Mayol, ^{1}H NMR study of the interaction of distamycin A and netropsin with the parallel stranded tetraplex [d(TGGGGT)]$_4$, *Chem. Commun.*, 2001, **1030–1031**.
53. M.J. Cocco, L.A. Hanakahi, M.D. Huber and N. Maizels, Specific inter-actions of distamycin with G-quadruplex DNA, *Nucleic Acids Res.*, 2003, **31**, 2944–2951.

54. N. Zaffaroni, S. Lualdi, R. Villa, D. Bellarosa, C. Cermele, P. Felicetti, C. Rossi, L. Orlandi and M.G. Daidone, Inhibition of telomerase activity by a distamycin derivative: Effects on cell proliferation and induction of apoptosis in human cancer cells, *Eur. J. Cancer*, 2000, **38**, 1792–1801.
55. W.M. David, J. Drodbelt, S.M. Kerwin and P.W. Thomas, Investigation of quadruplex oligonucleotide-drug interactions by electrospray ionisation mass spectrometry, *Anal. Chem.*, 2002, **74**, 2029–2033.
56. E.M. Rezler, J. Seenisamy, S. Bashyam, M.-Y. Kim, E. White, W.D. Wilson and L.H. Hurley, Telomestatin and diseleno sapphyrin bind selectively to two different forms of the human telomeric G-quadruplex structure, *J. Am. Chem. Soc.*, 2005, **127**, 9439–9447.
57. A. Rangan, O.Y. Federoff and L.H. Hurley, Induction of duplex to G-quadruplex transition in the c-*myc* promoter region by a small molecule, *J. Biol. Chem.*, 2001, **276**, 4640–4646.
58. A. Siddiqui-Jain, C.L. Grand, D.J. Bearss and L.H. Hurley, Direct evidence for a G-quadruplex in a promoter region and its targeting with a small molecule to repress c-*MYC* transcription, *Proc. Natl. Acad. Sci. USA*, 2002, **99**, 11593–11598.
59. A.T. Phan, V. Kuryavyi, H.Y. Gaw and D.J. Patel, Small-molecule interaction with a five-guanine-tract G-quadruplex structure from the human MYC promoter, *Nature Chem. Biol.*, 2005, **1**, 167–173.
60. A.T. Phan and D.J. Patel, A site-specific low-enrichment ^{13}C, ^{15}N isotope-labeling approach to unambiguous NMR spectral assignments, *J. Am. Chem. Soc.*, 2002, **124**, 1160–1161.

CHAPTER 6

Quadruplex Ligand Recognition: Biological Aspects

JEAN-FRANÇOIS RIOU, DENNIS GOMEZ, HAMID MORJANI AND CHANTAL TRENTESAUX

Laboratoire d'Onco-Pharmacologie, JE 2428, UFR de Pharmacie, Université de Reims Champagne Ardenne, 51 rue Cognacq-Jay, 51096, Reims, France

6.1 Introduction

The extreme ends of chromosomes represent a fascinating field of investigation for scientists because of the difference existing between telomeres and the other parts of the chromosome.

At their normal state, chromosomes are protected from any fusion with another extremity of chromosomes. Research has demonstrated that alterations of the telomere structure are responsible for the genomic instability associated with cancer transformation.[1] In addition, telomeres represent a mechanism of control in the lifespan of normal cells and their lengthening is associated with the immortalization required for the proliferation of cancer cells.[2] From these paradigms, telomere and telomerase have emerged as interesting therapeutic targets for cancer therapy, allowing restoration of a replicative senescence to cancer cells.[3–5] During the past few years, we have observed the parallel design of small molecules able to counteract telomerase activity and the determination of the structure of the nucleoprotein complex that forms telomeres. Some of the initial hypothesis concerning the mechanism of action of these agents has been validated and new others have been elucidated or refined by the use of molecular tools derived from the discovery of telomere-binding proteins.

The objective of this chapter is to present the latest view on the cellular mechanism(s) of action of G-quadruplex ligands and to discuss their potential use in therapy.

6.2 G-Quadruplex Ligands as Telomerase Inhibitors

The telomeres of human cells range in size from 3 to 15 kb and are comprised of tandem repeats of the sequence (5'-TTAGGG-3') with a 3' overhang of the

G-strand extending 150–250 bases beyond the C-strand[2] [Figure 1(A)]. Since DNA replication machinery is unable to completely replicate the extreme end of linear DNA molecules, telomere sequences are shortened at each round of normal cell division, a situation that leads to proliferation arrest of normal somatic cells when a critical size is reached.

Two mechanisms have been described in immortal and tumor cells to maintain the telomere length.

(i) A specialized enzyme called telomerase is able to copy, as a reverse transcriptase, the short TTAGGG motif at the end of telomere. Telomerase is composed of a catalytic subunit, hTERT, associated with an RNA containing the template of the telomere repeat unit, hTR.[6] Telomerase is overexpressed in a large number of tumors (about 85%) and is involved in the capping of telomere ends.[7]

(ii) Telomerase activity is absent in about 15% of tumors and telomere lengthening is maintained by recombination mechanisms between telomeres, known as Alternative Lengthening of Telomere (ALT).[8,9]

Telomere length is heterogeneous, varying from long telomeres (>20 kb) in ALT cells to a shorter size (4–15 kb) in telomerase positive cells. Telomere length reflects a homeostasis between the telomere lengthening mechanisms and the replicating degradation at each round of division, and is controlled by a complex association of telomere-binding proteins that regulates the accessibility

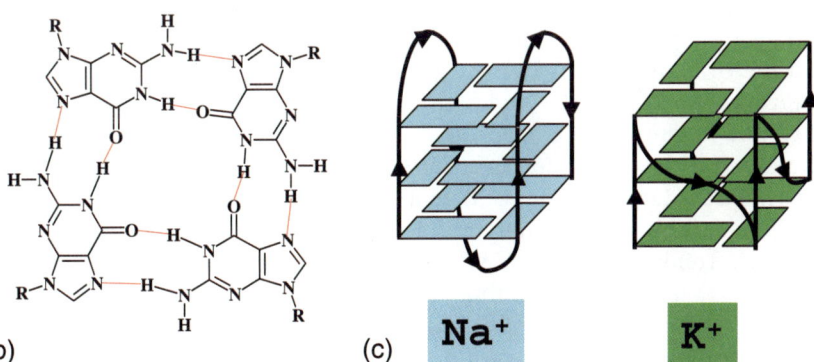

Telomeric duplex   **G-overhang**

5'-TTAGGGTTAGGGTTAGGGTTAGGGTTAGGGTTAGGGTTAGGG**TTAGGG**(**TTAGGG**)$_n$**TTAGGG**-3'
3'-AATCCCAATCCCAATCCCAATCCCAATCCCAATCCCAATC-5'

(a)

(b)   (c)   **Na⁺**   **K⁺**

Figure 1 *Formation of G-quadruplex at the single-stranded G-overhang of human telomeric DNA. (A) Schematic representation of the telomeric end. (B) Structure of a G-quartet involving four coplanar guanines. (C) Two possible conformations of the intramolecular G-quadruplex formed by human telomeric DNA in the presence of Na⁺ or K⁺*

of telomerase to the G-overhang.[10] Although identical telomere-binding proteins are qualitatively found in ALT cells, the exact mechanisms involved in the recombination at telomeres are poorly understood. In ALT cells, telomeric sequences and telomeric proteins are associated in large nuclear complexes to form APBs (ALT-associated PML bodies) that also contain recombination factors.[11,12]

Owing to the repetition of guanines, the G-overhang is prone to form four-stranded DNA structures (G-quadruplex), in which blocks of repeated guanines are engaged in series of quartet structures stabilized by Hoogsteen bond at N7 position of guanines[13,14] [Figure 1(B)]. The telomeric G-overhang can fold in several intramolecular quadruplexes that differ by the position of the adjacent loop regions[15–17] [Figure 1(C)]. Optimal telomerase activity requires the non-folded single-stranded telomere overhang and G-quadruplex formation has been shown to inhibit telomerase elongation *in vitro*.[18] Therefore, the search for small molecules that interact with the telomeric G-quadruplex and subsequently inhibit telomerase activity has been initiated by several groups and represents an alternative strategy to catalytic inhibitors of telomerase in order to alter the telomere homeostasis in cancer cells.

Several classes of ligands have been described such as porphyrins,[19,20] perylenes,[21] amidoanthracene-9,10-diones,[22] 2,7-disubstituted amidofluorenones,[23] acridines,[24,25] ethidium derivatives,[26,27] disubstituted triazines,[28] fluoroquinoanthroxazines,[29] indoloquinolines,[30] dibenzophenanthrolines,[31] bisquinacridines,[32] pentacyclic acridinium,[33] telomestatin,[34] and more recently 2,6-pyridine-dicarboxamide derivatives[35] (for a review see ref 36 and 37). The chemical structure of some of these ligands is presented in Figure 2.

There are fundamental differences between the targeting of the telomeric G-overhang using specific ligands and the inhibition of the catalytic activity of telomerase. Telomeres exist in the absence of telomerase activity and G-quadruplex ligands are expected to have an effect against ALT cells but also on normal dividing cells. In contrast, catalytic inhibitors will take the advantage of the very low expression of telomerase activity in normal cells and should not dramatically affect their growth.

According to the initial paradigm of the lifespan control by telomerase activity and telomere length, G-quadruplex ligands were first evaluated as telomerase inhibitors. Key experiments using hTERT-dominant negatives or oligonucleotides targeting the RNA template of telomerase have demonstrated that the inhibition of the catalytic telomerase function leads to progressive telomere shortening and to the induction of growth arrest accompanied with the features of the senescence induction.

For G-quadruplex ligands, this paradigm is partially true since functional telomerase inhibition was observed in cell lines treated for several weeks with a sub-toxic dosage of the compound. For example, long-term treatment of human A549 lung carcinoma cells at sub-toxic concentrations of disubstituted triazines derivatives (12459, 115405) produces telomere shortening that correlates with the induction of senescence (large morphology of cells and β-galactosidase activity)[28] (Figure 3). Similar findings were observed with

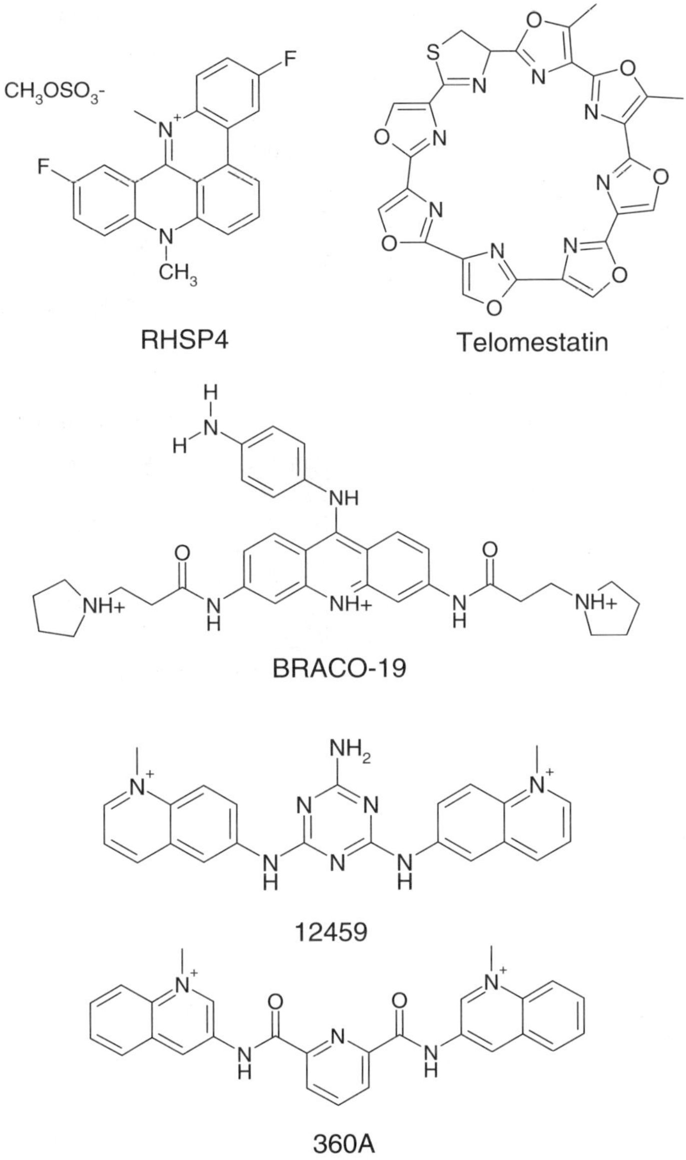

Figure 2 *Chemical formula of some G-quadruplex ligands*

telomestatin against various cell lines.[38-40] However, modification of telomere length was not detectable using either cell lines with short telomeres with triazine derivatives[28] or other G-quadruplex ligands such as BRACO-19, RHSP4, and 307A.[41-43]

All these derivatives were interestingly able to down-regulate telomerase activity in treated cell lines.[28,33,38,41] This effect was also observed with

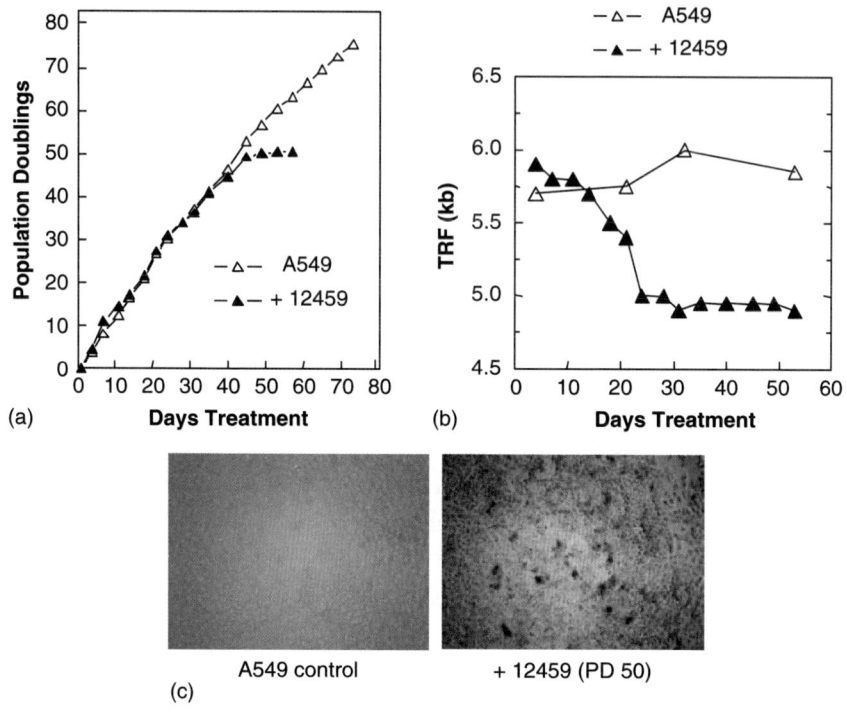

Figure 3   *G-quadruplex ligands behave as telomerase inhibitors.*[28] *(A) Long-term treatment of A549 human cell line with a sub-toxic concentration of 12459 (0.04
μM) leads to the cell-culture growth arrest after several weeks. (B) TRF
decrease induced by 12459 in A549 cells. (C) Senescence induced by 12459,
revealed by SA-b galactosidase activity at PD 50*

TMPyP4, a porphyrin derivative.[44] The mechanism for such effect was not
because of an interaction of the ligand with hTERT but was attributed to
different causes. A significant reduction in *c-myc* gene expression that controls
the hTERT promoter was described for the TMPyP4 porphyrin derivative[45]
while action on the alternative splicing of hTERT was reported for the 12459
derivative.[46] In this case, G-rich sequences that control the splicing have been
found in non-coding regions of hTERT and their stabilization as RNA quadruplexes by 12459, to interfere with the splicing machinery, has been proposed.

For BRACO-19, an important decrease in the nuclear hTERT, together with
the formation of cytoplasmic hTERT bound to ubiquitin may explain the
telomerase activity down-regulation.[47] Other antitumor agents have been also
found to down-regulate telomerase activity and hTERT expression has been
reported as a marker of cell proliferation.[48,49] Thus, the simplest explanation
for the effect of the G-quadruplex ligands to down-regulate telomerase activity
is related to their antiproliferative activity.

6.3 Telomere Dysfunction Induced by G-Quadruplex Ligands

It was also observed earlier that G-quadruplex ligands induced more rapid effects on cell growth than that initially expected for telomerase inhibition alone. Apoptosis and short-term response were observed with triazine derivatives (12459, 115405), telomestatin, and more recently with the pyridine dicarboxamide derivatives (307A, 360A).[28,38,43] Telomestatin induced the activation of ATM and Chk2 that corresponded to an activation of the DNA damage response.[38] 12459 induced apoptosis through the mitochondrial pathway and also provoked the early activation of P53.[50]

Short-term and massive apoptosis were also observed from the interference of the telomere-capping function of telomerase when hTERT or hTR are modified by mutations.[51] The observation that BRACO-19 causes chromosome end-to-end fusions associated with the appearance of p16-associated senescence led to the proposal that G-quadruplex ligands mostly act to disrupt the telomere structure.[52] Such telomeric dysfunction was also observed in cell lines treated with RHSP4 or with 307A and in cell lines resistant to 12459 with typical images of telophase bridges[42–43,53] (Figure 4).

These studies suggest that the direct target of these ligands is rather the telomere than telomerase activity.

The evidence that the antiproliferative effect of G-quadruplex ligands is independent of the presence of telomerase activity also comes from two series of experiments.

(i) Overexpression of hTERT or a dominant negative of hTERT in a telomerase-positive cell line did not modify the antiproliferative effect of the triazine derivative 12459.[53]

(ii) All described ligands were also found to block the proliferation of ALT cell lines.[28,33,52,54]

On the other hand, the re-introduction of hTERT in ALT cell lines only produces a partial protection to the antiproliferative effects of telomestatin and 307A. In that case, it is proposed that hTERT acts to protect the integrity of telomeres (T. Lemarteleur, PhD Thesis, University of Reims, 2005).

6.4 Resistance to G-Quadruplex Ligands

As any potential therapeutic agents, the biological effect of G-quadruplex ligands may be overcome by the appearance of acquired-resistance phenotypes or to be lowered by the intrinsic expression of proteins already known to be associated with resistance to anticancer agents. The selection as well as the analysis of resistant cells is a well-known strategy to obtain information on the mechanisms of drug inactivation and on their target proteins, if a modification of their expression, or a mutation can be related to the resistance phenotype.

The proteins involved in the multidrug-resistance phenotype, including MDR1, MRPs, or BCRP genes products, are the first potential candidates to

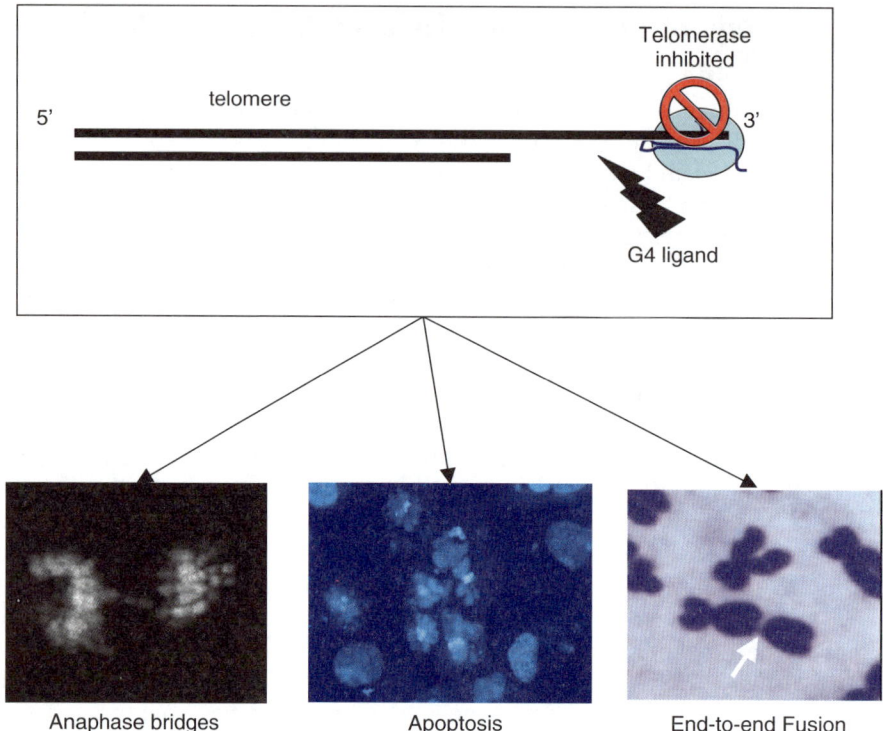

Figure 4   *G-quadruplex ligands induced a telomere dysfunction that is characterized by*
the appearance of anaphase bridges and telomere end-to-end fusions in treated
cells. The induction of short-term apoptosis is also frequently observed. The
telomere dysfunction may appear before any evidence of telomere shortening by
TRF analysis

counteract G4 ligand effects. This family of trans-membrane glyco-
proteins reduces the intracellular levels of numerous pharmacological anticancer agents and therefore decreases interactions with their intracellular
targets.

An easy way to determine the involvement of multidrug-resistance proteins
for G-quadruplex ligands is to determine their cross-resistance using cell lines
models transfected with these genes and/or to analyze the expression of these
genes in cell lines selected for resistance to these ligands. Only a limited number
of studies have been reported to date and concern the triazine and pyridodi-
carboxamide derivatives 12459 and 360A. Results indicated that these compounds are recognized at a low level by these efflux proteins (H. Morjani, J.
Macadré and L. Eddabra, unpublished results). In addition, cell lines resistant
to 12459 did not overexpress multidrug-resistance genes.[39,53]

This situation argues that 12459-resistant cell lines may present potential
modifications of intracellular targets involved in the control of the drug-
induced senescence and/or apoptosis.

6.4.1 Resistance and Telomere Component Expression

Owing to the dual effects of these ligands that induced short- or long-term effects against cancer cells, two strategies to select resistance to these ligands have been used (Figure 5).

The first strategy was to obtain A549 cells lines resistant to the senescence induced by 12459 by a progressive increase of the drug concentration.[39] The resistant cell line obtained, named JFA2 was found resistant to the senescence induction by 12459 and cross-resistant to telomestatin. The telomere shortening induced by these ligands in resistant cells line was abolished. Interestingly, the JFA2 cell line also showed cross-resistance at short-term exposure for 12459, but not for BRACO-19 or telomestatin. This suggests that the mechanism of action of this compound might have some differences compared with other G4 ligands. At short-term exposure, a 4-fold resistance to 12459 was also observed but not to other anticancer agents such as etoposide, camptothecin, or doxorubicin. Thus, this cell line presented a resistance phenotype that seems to be directed against G4 ligands but not to other anticancer agents.

The second strategy was to obtain 12459-resistant A549 clones after short-term exposure to high concentrations of the ligand, together with EMS

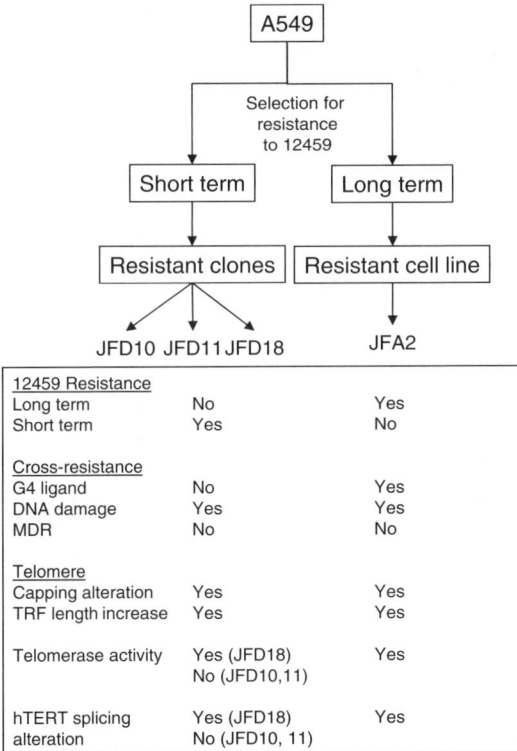

12459 Resistance	JFD10 JFD11 JFD18	JFA2
Long term	No	Yes
Short term	Yes	No
Cross-resistance		
G4 ligand	No	Yes
DNA damage	Yes	Yes
MDR	No	No
Telomere		
Capping alteration	Yes	Yes
TRF length increase	Yes	Yes
Telomerase activity	Yes (JFD18)	Yes
	No (JFD10,11)	
hTERT splicing	Yes (JFD18)	Yes
alteration	No (JFD10, 11)	

Figure 5 *Selection for resistance to 12459 and some of the characteristics of the resistant cell lines obtained[39,46,53]*

mutagenesis.[53] About 15 stable clones have been analyzed and they all have a resistance index between 2 and 8 to the 4-day effect of 12459. Several of these clones, JFD10, JFD11, and JFD18 have been more extensively analyzed. For JFD18, the cross-resistance pattern indicated a mild cross-resistance to the triazine derivative 115405 and to mitomycin C but not to BRACO-19 or telomestatin. In contrast, only a mild resistance effect was observed with long-term treatment with 12459. Thus, in this cell line the resistance phenotype is restricted to the short-term effect of 12459 or a close derivative and may also concern other DNA-damaging agents.

For these two strategies, the up-regulation of MDR efflux was not involved and our results indicated that these resistant models displayed interesting differences for the short- and long-term effects of 12459.

Since telomerase was initially established as one of the potential targets of these ligands, we have determined whether hTERT expression, telomerase activity, and telomere length are modified in these resistant models. Both transcriptional, enzymatic activity and telomere length were increased in JFA2 and in the majority of the JFD clones (8/15), including JFD18 and JFD10, suggesting that the up-regulation of telomerase activity and telomere length increase could play a role in the resistance to this ligand. Important alterations of hTERT splicing were also observed in these resistant cells, in order to overcome down-regulation of hTERT active transcript induced by 12459.[46]

The interpretations of these initial findings become more complex in the light of experiments in which increased telomerase activity was re-introduced into the parental cell line[53] (Figure 6). In this case, no resistance to 12459 was found, using either short- or long-term treatment by the ligand, in A549 cells overexpressing hTERT. Therefore, increased telomerase activity appears to be not sufficient *per se* to confer the resistance phenotype. Since these resistant cell lines displayed an increased number of mitotic alterations (anaphase bridges) that traduced important telomere capping alterations, we have proposed that hTERT overexpression may serve to stabilize or protect the telomere extremities following changes induced by the resistance acquisition. Indeed, the transfection of a dominant-negative hTERT in JFD18 cells partially restores the sensitivity to the short-term treatment with 12459.[53]

The up-regulation of hTERT expression and the increase of telomere length were also reported in a HCT116-resistant cell line established against the cyclin kinase inhibitor flavopiridol.[55] In addition, overexpression of POT1 mRNA was also described for this cell line, suggesting that alteration of the shelterin complex (see paragraph 6) might be associated to the resistance phenotype. Despite the link between these telomere alterations and the mechanisms underlying the flavopiridol resistance is still unknown, the G-quadruplex ligand BRACO-19 showed a synergistic long-term effect with flavopiridol to overcome the resistance. The resistant cell line is also hypersensitive to BRACO-19 alone.

Owing to the link established between hTERT and the protection of the single-stranded G-overhang in normal cells and because of the role of the shelterin complex that includes both single- and double-stranded telomere-binding proteins to protect telomere extremities, a careful examination of this

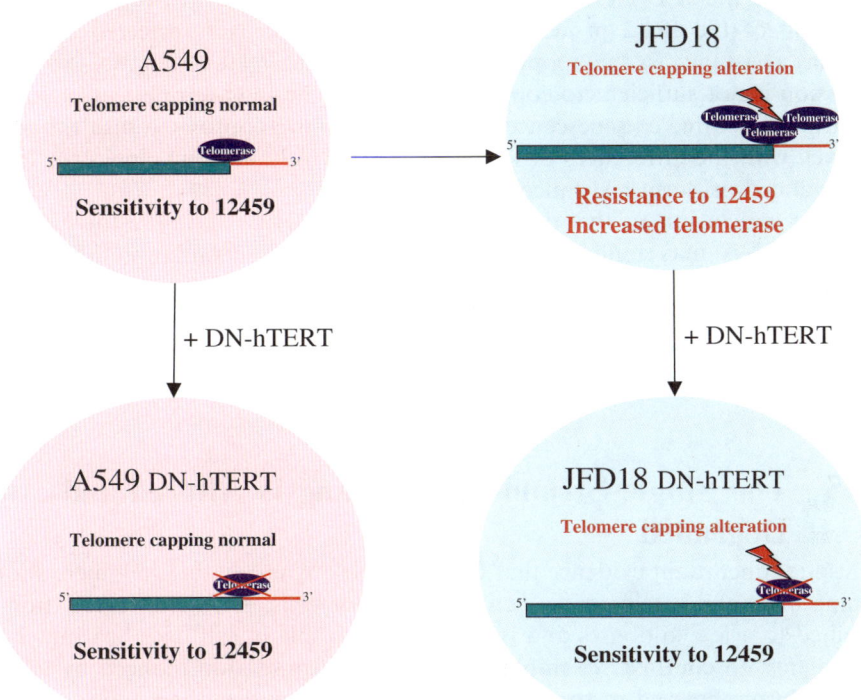

Figure 6 *Proposed role of hTERT overexpression in resistant JFD18 cell line to control telomere capping.[53] The telomere length increase and hTERT overexpression stabilize telomere-capping alterations induced by 12459 in JFD18 resistant cells. Transfection of JFD18 by DN-hTERT reduces the telomere length and restores sensitivity to the short-term effects of 12459. In parental cells, DN-hTERT transfection does not change the sensitivity to 12459, indicating that the short effect of the G-quadruplex ligand is independent of telomerase activity*

protein complex in G-quadruplex ligand resistant cells would give some clues in the future.

Altogether, these studies and the previous finding that G4 ligands are also active against telomerase negative cell lines, strongly suggest that factors other than telomerase are involved in the mechanism of action of these ligands.

6.4.2 Resistance and Apoptosis

For most of the G4 ligands studied so far, apoptotic cell death could be achieved after several cell cycles in tumor-derived cell lines. The triazine ligand 12459 activates the mitochondrial cell death pathway through alteration of the Bax/Bcl-2 balance, which leads to caspase 3 activation. At short-term, it could be noticed that apoptosis predominates over the appearance of senescent cells for this ligand.[50]

Some of the JFD clones selected for resistance to 12459 also show overexpression of the Bcl-2 protein. In addition, A549 cells transfected by Bcl-2 display resistance to the apoptotic action of 12459.[50] However, Bcl-2 overexpression is not sufficient to confer resistance to the long-term effect of 12459. Thus, 12459-directed senescence is uncoupled from apoptosis, a result that fits in well with the differences observed between JFA2 and JFD clones for long-term and short-term resistance studies.

As a conclusion to this part, the modest selectivity of 12459 for G4 over duplex DNA may render questionable some of these findings, that is, the relationship between short-term effects and its molecular action against telomeres. Future work to obtain resistant cell lines from other sources and with more selective G-quadruplex ligands, such as telomestatin or pyridine dicarboxamide derivatives will aim to confirm some of the views presented here.

6.5 The Single-Strand G-Overhang Is Altered and Degraded

Despite concordant evidence that G-quadruplex ligands induce telomere shortening and inhibit telomerase activity in tumor cells, telomere degradation is limited to a few kilobases and is often undetectable in cell lines bearing short telomeres. In contrast, an important telomere degradation induced by telomestatin was observed in some cell lines that arises earlier than expected for a single mechanism involving telomerase inhibition.[38,39] These data, together with reports that G-quadruplex ligands also impaired the growth of ALT cell lines lacking telomerase activity,[28,54] have suggested that additional mechanisms may explain the biological activity of the ligands in tumor cell lines.

The major characteristic of the telomere extremity is the presence of a G-rich 3′ extension (G-overhang) averaging 150–250 bases in length.[56,57] The G-overhangs are present on all chromosomal ends and are formed during S-phase through a complex mechanism that involves cleavage of the C-strand.[56–58] G-overhangs have been implicated in the structure of the telomere extremities to create the T-loop that protects chromosome ends from fusion and their degradation has been associated with the onset of replicative senescence and more recently to the deprotection of telomeres though inactivation of proteins from the shelterin complex.[59]

Since the G-overhang is the only part of the telomere that could exist in a native single-stranded conformation outside the S phase, the hypothesis that G-quadruplex ligands preferentially act to modify G-overhang conformation or induce its degradation has emerged and has been experimentally evaluated.[60]

This approach was evaluated by two techniques. The telomeric oligonucleotide ligation assay (T-OLA) uses the ligation of short oligonucleotides hybridized to the G-overhang and measures the G-overhang length distribution.[61] This assay is somehow difficult to quantify. Another assay, corresponding to the solution hybridization of short oligonucleotides complementary to the G-overhang, is simpler to set up and also gives quantitative information.[39,61]

The amount of oligonucleotides hybridized to the G-overhang is directly proportional to its mean length but does not provide information about its size homogeneity (Figure 7).

By using these two techniques our group has analyzed the effect of several series of G-quadruplex ligands on the behavior of the G-overhang in human cell lines.

In human A549 cells, telomestatin is able to significantly reduce the signal of the G-overhang after a 24–48 h delay.[62] Surprisingly, the G-overhang signal loss was found earlier than expected from the cell-survival experiments and we hypothesized that such an effect could be due either to rapid degradation or to modification of the G-overhang conformation that impairs further hybridization reaction. *In vitro* experiments designed to analyze the effect of telomestatin on the hybridization reaction on the G-overhang from purified genomic DNA and competition with another G-quadruplex sequence derived from the *c-myc* promoter allowed us to conclude that telomestatin remains attached to the G-overhang from treated cells and prevents hybridization of the oligonucleotide probe (Figure 7B and C).[62] The tight and specific binding of telomestatin to the telomeric G-overhang quadruplex is also in agreement with results of melting experiments with the telomeric quadruplex that were not modified in the presence of a huge excess of duplex genomic DNA. Further experiments established that a real degradation of the G-overhang occurred after 8–12 days of telomestatin treatment onto A549 cells. At this time G-overhang degradation was estimated to about 40–50% and was found associated with growth arrest of the cells (Figure 7D and E).

In vivo protection experiments with DMS that is able to transfer a methyl group to the N7 position of guanines enabled us to demonstrate that telomestatin engages the telomeric overhang *in vivo* in a DNA structure resistant to action of DMS.[62] These experiments represent the first proof of the existence of G-quadruplexes at the telomeric G-overhang under the effect of the treatment by a G-quadruplex ligand.

Such rapid and dramatic effects of telomestatin against the conformation of the G-overhang have been extended to other cell lines models such as EcR293 immortalized cells and the ALT cell lines WI38-VA13 and MRC5-V1 (D. Gomez and T. Lemarteleur, unpublished results).

Figure 7   *Schematic explanation of solution hybridization experiments in the presence of telomestatin.[62] (A) C-strand probe ((CCCTAA)₄) hybridizes to the telomeric G-overhang in untreated cells. (B) Treatment with telomestatin (24 h) provokes the formation of stable telomestatin-quadruplex complexes at the G-overhang that further impairs the hybridization reaction. (C) Competition with a G-quadruplex-forming sequence (Pu22myc) displaces the telomestatin-quadruplex complexes and allows the hybridization of the probe to the G-overhang. (D) Treatment with telomestatin (12 days) induces a G-overhang shortening that is masked by the complete inhibition of the hybridization reaction. (E) Competition with Pu22myc allows to detect the G-overhang shortening induced by telomestatin treatment of cells*

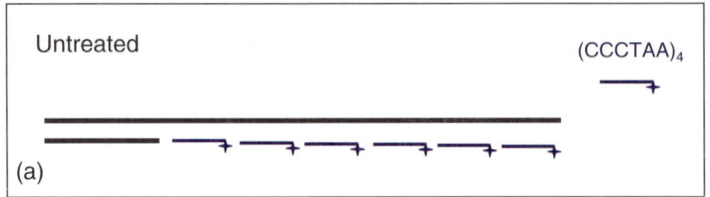

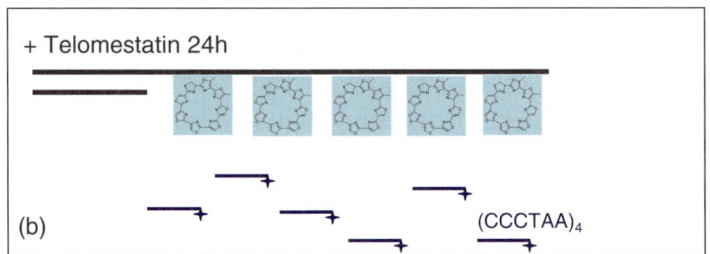

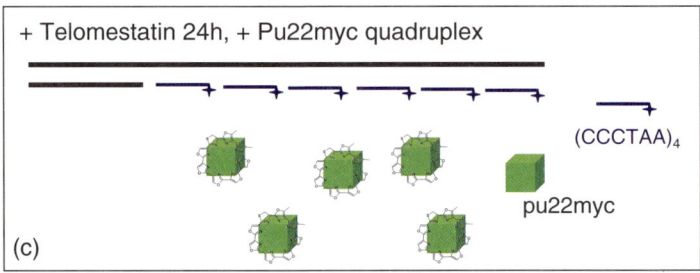

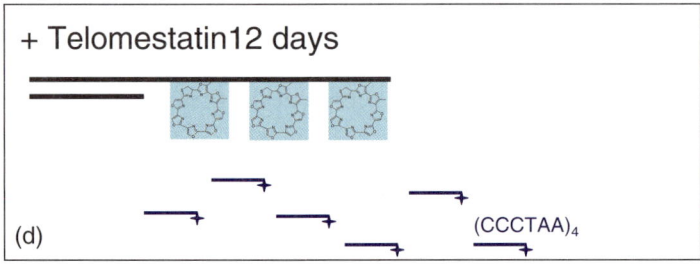

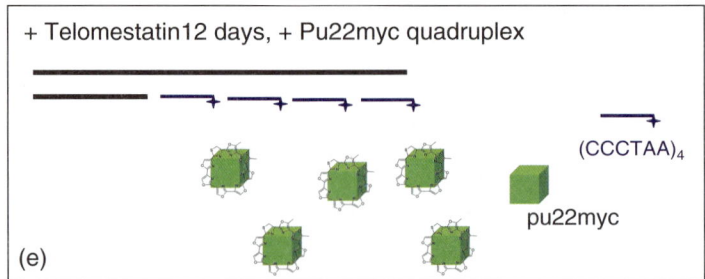

Other G-quadruplex ligands, such as the triazine derivative 12459 and the pyridine dicarboxamide derivatives 360A and 307A showed different effects, as compared to telomestatin, on the conformation and the integrity of the telomeric G-overhang.[43,50]

In vitro, against purified genomic DNA, the derivative 12459 is poorly active and requires a 100 μM concentration to completely inhibit the hybridization reaction on the G-overhang.[50] Derivatives 360A and 307A have an IC_{50} of 1 μM which is comparable to telomestatin, inhibit the hybridization at the G-overhang in Na^+ and K^+ conditions. However, no inhibition was observed with 360A and 307A when Mg^{2+} was added to the hybridization reaction. This distinguishes these ligands from telomestatin and suggests that the stability of the G-quadruplex formed is less than that formed by telomestatin. In agreement with this, we did not observe any dramatic reduction of the signal of the G-overhang in A549 cells treated by 307A or 360A after a 48–72 h delay. However, for longer treatment (10 days) in T98G cells, a 30% decrease of the G-overhang signal was observed for 1 μM 307A, a concentration that induces apoptosis and growth arrest of the cell line after 12 days.[43]

In contrast to the pyridine dicarboxamide derivatives, the triazine derivative 12459 is able to induce a rapid decrease of the telomeric G-overhang signal in A549 cells after short-term treatment (24–48 h).[50] Competition experiments with the myc quadruplex did not show the reversion observed with telomestatin. Therefore, we concluded that the G-overhang signal loss in 12459-treated A549 cells corresponds to effective degradation of the telomeric G-overhang. For this ligand, the G-overhang degradation seems to correlate with the appearance of apoptosis in A549 cells but is not modified in Bcl-2 transfected cells resistant to the apoptotic process. Since the antitumor drug camptothecin does not induce the G-overhang degradation at concentrations able to trigger apoptosis in this cell line, we proposed that G-overhang degradation is an early event uncoupled from the apoptotic processes and selectively induced by 12459. We do not know however whether this degradation precedes or is important to trigger the apoptotic response since other early events such as P53 induction and ROS activation are also induced by 12459.[50]

The difference in effects between these ligands (telomestatin, 307A, 12459) on telomeric G-quadruplex in the same cellular model suggest several remarks and questions (Figure 8).

Is the mode of interaction of the ligand with the telomeric G-quadruplex and its stability responsible for the rapid onset of the G-overhang degradation?

From our *in vitro* and cellular experiments it appears that rapid degradation of the G-overhang is inversely proportional to the effect and the stability of the ligand. One can imagine that effective stabilization of a G-quadruplex at the G-overhang could present an "overcapping" of the telomere that protects it from nucleolytic degradation. On the other hand the effect of the G-overhang length and the number of quadruplexes that could potentially be stabilized has not been investigated. A compound that induced one or two stable G-quadruplexes for a G-overhang mean size of 300b will only have a small inhibitory effect (10–15%) *in vitro* toward the hybridization reaction but may sufficiently destabilize

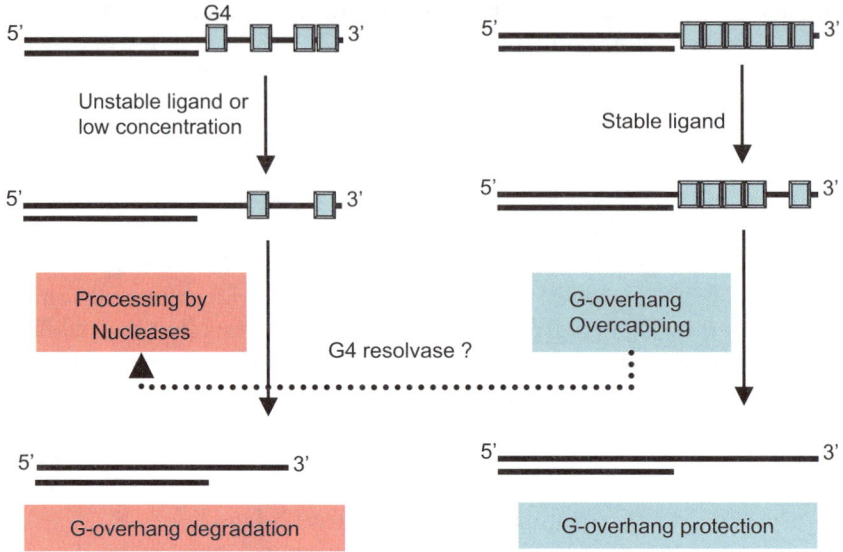

Figure 8   *Model to explain the differential effects of G-quadruplex ligands to induce G-overhang degradation or overcapping. G-quadruplex stabilization using an unstable ligand or with a low concentration is expected to induce partial modification of the G-overhang that might be processed after signalization of damage to get a G-overhang degradation (left part). A ligand that induced stable G-quadruplexes on most of the G-overhang will induce an overcapping that protects against the degradation by nucleases. The overcapped telomere might be solved by specific events such as G4 resolvases during replication to generate further degradation of the G-overhang*

the telomeric end to induce a nucleolytic response in the non-G-quadruplex region (the G-overhang that remains single-stranded). From these remarks we can hypothesize that strongly binding ligands, such as telomestatin, could protect the G-overhang from the intracellular activation of exonucleases but will induce defects during telomere replication, or later, following removal of the G-quadruplex by an active mechanism.

6.6   Deregulation of Telomere-Binding Proteins

The proteins that protect telomere extremities have been identified during the last decade and comprise a complex termed shelterin.[59] This complex is composed of three proteins that directly bind to the telomeric repeats (TRF1, TRF2, and POT1), TIN2, which binds to TRF1 and TRF2, TPP1, which binds to TIN2 and POT1, and Rap1, which binds to TRF2 (Figure 9).[59] TRF1 and TRF2 have a Myb-type DNA binding domain for the recognition of 5′-YTAGGGTTR-3′ in duplex DNA,[63] while POT1 has two OB-folded domains and has a strong preference for the single-stranded 5′-(T)TAGGTTAG-3′ sequence.[64] Thus, shelterin appears to make the connection between duplex telomeric DNA and the 3′-G-overhang.

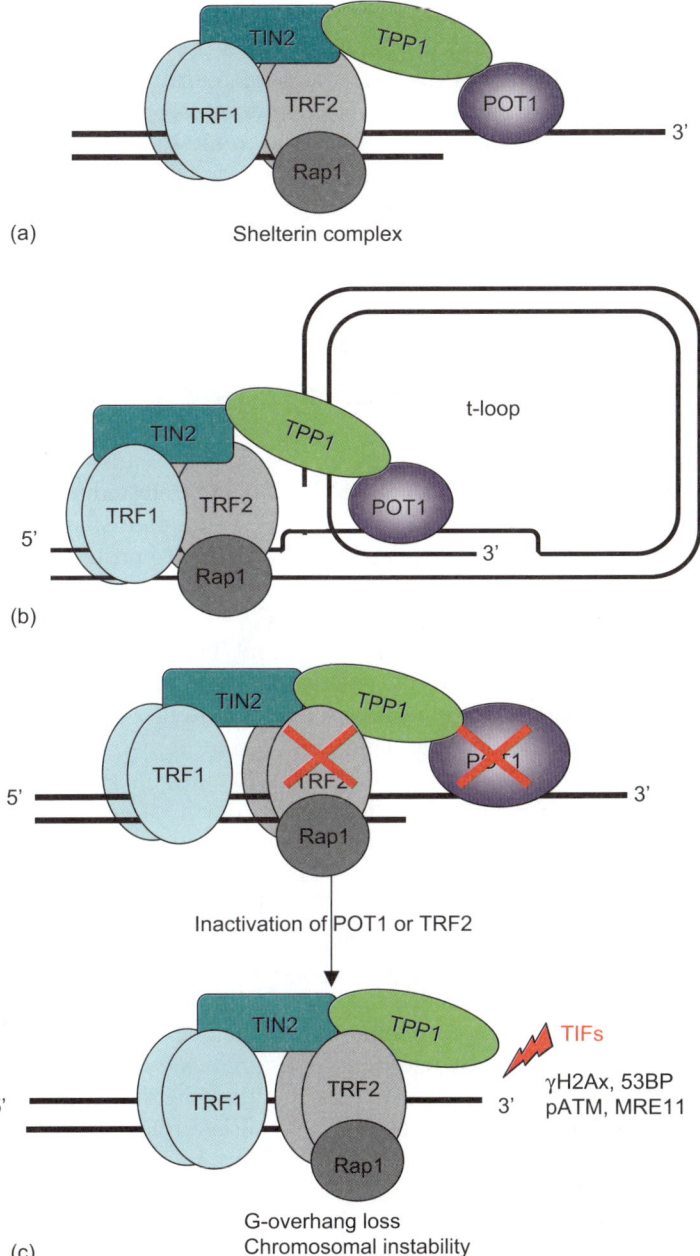

Figure 9 *Schematic representation of the shelterin complex that binds and protects the telomere ends.[59] (A) Shelterin components represented on an open telomere structure accessible to telomerase. (B) t-loop structure forming a closed and protected telomere structure. The 3' G-overhang invades the telomeric duplex to form a hypothetical three-stranded structure stabilized by POT1 and by TRF2 at the junction with the duplex. (C) Inactivation of POT1 by RNA interference or TRF2 by dominant negatives lead to the instability of the shelterin complex, the appearance of telomere dysfunction induced foci (TIFs), chromosomal instability, and G-overhang degradation*

Shelterin is involved in the structure of telomeric DNA and affects the structure of the 3'end. The G-overhang has been proposed to invade double-stranded telomeric DNA, to displace the base pairing with the C-strand in order to form a t-loop, as shown by electron microscopy[65,66] (Figure 9). The TRF2 component of the shelterin complex was found to have a crucial function in the formation and the stabilization of the t-loop.[65,66]

As shelterin also associates with several proteins involved in recombinational repair (Mre11/Rad50/Nbs1, ERCC1/XPF, WRN, BLM, DNA-PK, PARP-2, TANK1) (for a review, see ref 59), it was suggested that these factors could contribute to the maintenance and the formation of the t-loop. When either TRF2 or POT1 are inactivated,[67,68] the overall amount of the single-stranded G-overhang is diminished by 30–50% and a specific DNA damage response is induced at most telomere ends.[69] After TRF2, TIN2, or POT1 inactivation or when telomeres become critically short, 53BP1, γ-H2AX, the Mre11 complex, and phosporylated ATM, accumulate at chromosome ends.[67,70,71] The structures formed by these DNA damage factors are referred to as Telomere dysfunction Induced Foci (TIFs).[72] These studies are consistent with the view that telomere ends are involved in a peculiar structure to protect their integrity. The protein complex shelterin is able to actively change its architecture and to control its detection by DNA-damage factors.

Interestingly, the inactivation of TRF2 and POT1 cause cellular effects analogous to those reported with G-quadruplex ligands, that is, chromosomal instability, loss of the telomeric G-overhang, followed by the appearance of apoptotic and/or senescent cells. Thus, one recent activity of our group has been to examine the effect of G-quadruplex ligands on the function of the shelterin complex. We have investigated the effect of telomestatin on the binding of POT1 to the telomeric G-overhang *in vitro* and in human cells using a GFP–POT1 fusion protein (Gomez, D., *et al.* The G-quadruplex ligand telomestatin inhibits POT1 binding to telomeric sequences in vitro and induces GFP–POT1 dissociation from telomeres in human cells. *Cancer Res.* 2006, in press). Electrophoretic Mobility Shift assays on a short oligonucleotide that reconstitutes the double-stranded telomere with a short 3′ overhang was performed in the presence of telomestatin. These experiments showed that G-quadruplex stabilization by telomestatin dramatically impairs the binding of POT1 to the telomeric G-overhang *in vitro*. On the other hand, telomestatin has no effect on the binding of TRF2 to the telomeric duplex.

The treatment of EcR293 and HT1080 cell lines by telomestatin provokes a major reduction in GFP–POT1 at its normal telomere sites. Telomestatin induced an accumulation of POT1 in cytoplasmic and nucleolus foci that were consistent with the inhibition of POT1 binding through the stabilization of G-quadruplexes at the telomeric G-overhang (Figure 10).

An examination of the time-course of the GFP–POT1 delocalization in living cells also revealed that this event coincides with the alteration of the G-overhang structure in EcR293 cells and is an early event occurring before any evidence of apoptosis or antiproliferative effects of the ligand. Several anticancer agents including doxorubicin, etoposide, and vinblastin, used as

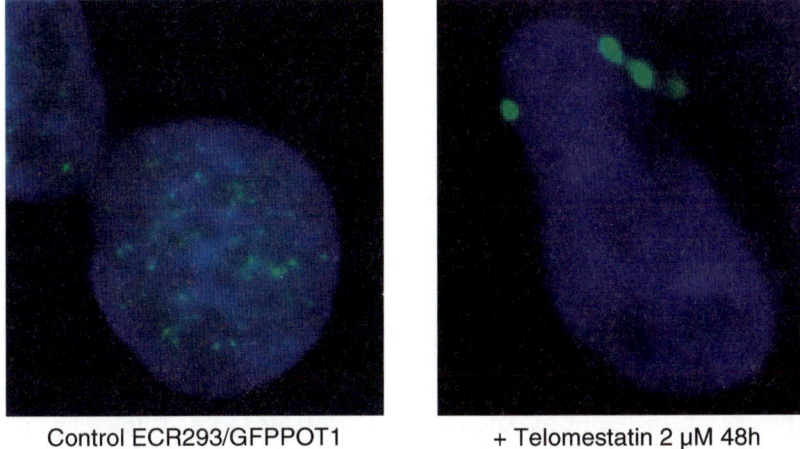

Control ECR293/GFPPOT1   + Telomestatin 2 μM 48h

Figure 10   *G-quadruplex stabilization by telomestatin impairs the binding of GFP-POT1*
to its normal telomere sites and induces cytoplasmic foci of POT1 in EcR293
cells

control, did not show any evidence of effects on POT1 localization, suggesting
that telomestatin has a selective effect on POT1.

Telomestatin treatment (24 h) also induced the formation of phosphorylated
γ-H2AX foci in EcR293 and HT1080 cells that partially colocalize with TRF2.
The analysis of cells in metaphase showed that the γ-H2AX response is
observed at the distal part of chromosomes. Thus, telomestatin is able to
induce an early DNA damage response associated with telomeres, persisting
during mitosis, which corresponds to TIFs.

It has been proposed that G-quadruplexes may act as caps to protect the $3'$
G-overhang from degradation, but these structures have to be resolved during
replication to allow telomere extension.[18,73] In humans, POT1 in stoichiometric
conditions is able to restore normal telomerase activity on telomeric oligonucleotides which efficiently adopt a G-quadruplex conformation and which are
poor primers for telomerase.[74] The authors suggest that POT1 traps and
maintains the oligonucleotide in an unfolded form accessible to telomerase.
Therefore, the mode of action of telomestatin could be seen as a deleterious
"overcapping" of telomeres that triggers a DNA damage response and blocks
telomerase accessibility.

Interestingly, recent work showed that the antitumor response to the
G-quadruplex ligand BRACO-19 parallels the loss of nuclear hTERT protein
expression.[47] A cytoplasmic hTERT expression that co-localized with ubiquitin
was observed in immunostaining of xenograft tissues, suggesting enhanced
destruction of hTERT because of BRACO-19 treatment. Since the telomerase
complex binds and acts at the telomeric G-overhang, these results are in good
agreement with the notion that G-quadruplex ligands will impair or inactivate
the function of single-stranded telomere-binding proteins.

RNA interference inactivation of POT1 in human cells was shown to modify the structure of 3′ and 5′ ends of the telomere and to reduce by twofold the TRF2 binding by ChIP.[67,75]

Despite telomestatin having no direct effect *in vitro* against the binding of TRF2 to telomeric duplex, it is possible that it results in the delocalization of a small fraction of TRF2 that further destabilizes the t-loop. In EcR293 cells, ChIP experiments and fluorescence microscopy did not reveal any significant alteration of TRF2 binding to telomeres. However, the situation seems more complex, since other data indicates that telomestatin provokes the release of TRF2, associated with rapid degradation of duplex telomeric DNA in other cell lines models (K. Shin-ya, personal communication). The reason for these differences is still unknown and may be related to the composition of the shelterin complex and associated proteins that greatly differ between cell lines.

Future works will aim to determine the effect of G-quadruplex ligands on other components of the shelterin complex and associated factors, such as WRN or BLM helicases that cooperate with POT1 for telomere sequence unwinding.[76]

6.7   Importance of Telomere Ends and Replication Processes

One may argue that the short-term effects of G-quadruplex ligands corresponded to side-effects of the ligands because of other targets than the telomere. That was certainly true for the first generation of ligands including cationic porphyrins and some triazine derivatives (115405A). Specificity for G-quadruplex DNA over duplex DNA was greatly improved in trisubstituted acridines, pyridine dicarboxamides, and telomestatin and is >50-fold in biochemical assays.[35,43,77,78]

The selectivity also concerns other G-quadruplex sequences. Since all ligands reported to stabilize telomeric G-quadruplex also display potent activities against the *c-myc* quadruplex *in vitro*,[35] their cellular effects might be due to non-selective stabilization of other quadruplexes in the genome, controlling the function of other genes.[79,80] Although *c-myc* down-regulation was well established for the porphyrin derivative TMPyP4 in tumor cell lines,[81] telomestatin, which has nearly equal activity *in vitro* against *c-myc* and telomeric G-quadruplexes does not affect *c-myc* gene expression in several cell line models (T. Lemarteleur, PhD Thesis, University of Reims, 2005). In addition, a gene expression profiling study revealed that only 51 genes were affected after 7 days of treatment using a gene array chip representing 33,000 genes.[82] None of these genes plays any significant role in DNA recombination, cell cycle control, and apoptosis. A porphyrin analogue, Se2SAP with an increased selectivity for *c-myc* over telomeric quadruplex, has been synthesized[83] but no results concerning its biological activity are yet available.

One of the most interesting demonstration that telomere is a significant target for these ligands has been provided by a study using a radiolabelled G-quadruplex ligand. The localization of the chromosomal binding sites of a tritiated pyridine dicarboxamide derivative: [³H]360A has been performed in metaphase spreads from cell lines bearing different telomere length.[84] The result showed that [³H]360A preferentially binds to the terminal parts of chromosomes in mitosis. Binding sites in internal chromosomal regions were also observed but were significantly lower than those at the end of chromosomes. This ligand displays an interesting *in vitro* binding preference for G-quadruplex over non-G-quadruplex DNA sequences but does not discriminate telomeric repeats from other G-quadruplex-forming sequences, and could be considered as a "non-discriminating" G-quadruplex ligand.[35,84] We also observe preferential *in vitro* binding of the ligand to the telomeric G-overhang from purified genomic DNA. At first view, these results suggest a greater biological importance for telomeric ends than other parts of the genome. The prevalence of the G-overhang to form G-quadruplex, as compared to other G-rich sequences, which are in a duplex context during most of the cell cycle, is a logical explanation. However, the technical design of these experiments represents *per se* a bias and further work to examine the binding of the ligand in other part of the cell cycle, including the S phase, will be of considerable interest.

Several proteins have been described to act as G-quadruplex resolvases. Human RecQ helicases, Bloom's syndrome[85,86] and Werner's syndrome DNA helicases[87] have been shown to unwind intra and intermolecular G-quadruplex formed by telomeric sequences or other G-rich sequences. BLM has been proposed to maintain telomeres in ALT cells through recombination mechanisms.[88] WRN belongs to the DNA polymerase δ replication complex and defective WRN cells are unable to fully replicate telomeres because of a defect in lagging strand synthesis.[89] The DEXH helicase encoded by DHX36 was also identified as the major source of G-quadruplex resolvase in HeLa extracts.[90] Furthermore, Rtel, a murine gene encoding a DNA helicase homologue to dog-1 is required for telomere elongation.[91] Dog-1 was initially identified in the nematode *C. elegans* as maintaining the stability of long poly dG-poly dC runs.[92] Its putative role might be to resolve G-quadruplexes associated with the poly dG repeats and formed during replication, in order to maintain genome integrity. More recently, a study in ciliates has demonstrated that G-quadruplex formation at telomeres is regulated through the cell cycle. TEBPα–β proteins tethers the telomere to the nuclear matrix and stabilizes G-quadruplex formation.[73] During S phase, the phosphorylation of TEBPβ results in the release of telomeric DNA and the dissociation of G-quartets. Then, the chromosome ends are accessible for replication and extension of the G-overhang by telomerase. These observations and the redundancy of G-quadruplex resolvases indicate that mammalian cells require mechanisms for the removal of G-quadruplex during replication.

G-quadruplex ligands from the trisubstituted acridine series were also found to inhibit *in vitro* the unwinding by the RecQ helicases, BLM and WRN.[93] Therefore, because of their additional effects on RecQ helicases, these ligands should also disrupt telomere synthesis (Figure 11). In addition, their action against other G-quadruplex resolvases during replication might provoke DNA

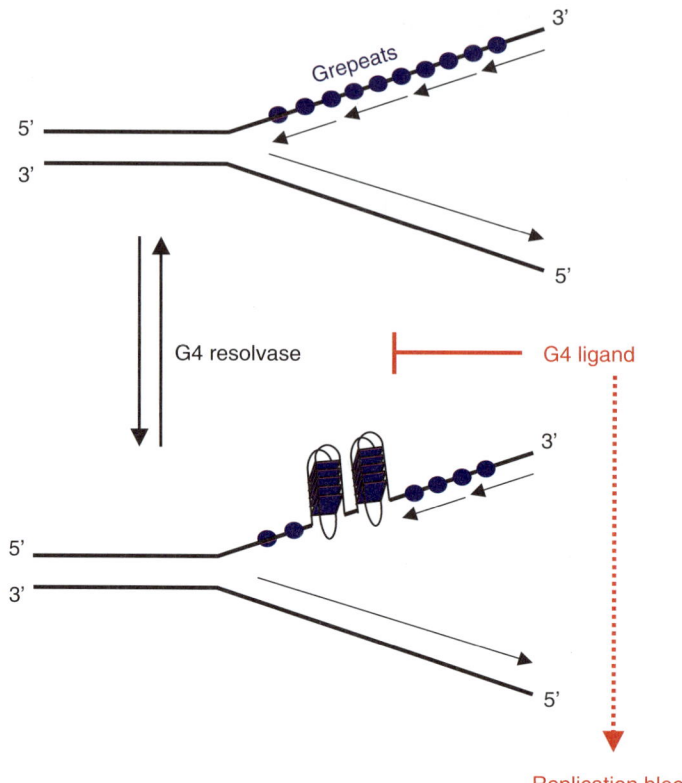

Figure 11   *Schematic model presenting the possible effect of a G-quadruplex ligand on G-quadruplex forming elements during replication. Stabilization of G-quadruplex might block G4 resolvases and further lead to a replication block with deleterious consequences*

synthesis defects in other G-rich regions of the genome (replication stalling, recombination).

It would be interesting to determine whether such action will be a benefit for antitumor activity. A possible way to determine this is to examine the effect of the ligands in cells defective for these resolvases. A replication block at telomere would be expected to rapidly induce larger deletions of telomeric sequence than in wild-type cells. These results may orient the strategy to select future G-quadruplex ligand or to combine the actual ligands with specific inhibitors of these resolvases.

6.8   Concluding Remarks

G-quadruplex interacting agents appear to represent a valuable strategy to target telomere biology. These compounds interact with the telomeric G-overhang to inhibit telomerase activity and to destabilize the telomere shelterin complex. Some possible other biological effects of these ligands have also emerged or needed to be explored, that is, to inhibit the *c-myc* gene transcription and to block resolvases.

In order to design effective therapies, the end point is to achieve antitumor activity and to keep toxicity at a low value. The first criterion was recently obtained for the BRACO-19 derivative that presented antitumor activity as a single agent in human xenografts.[47] Since several other G-quadruplex ligands were reported to present selectivity for tumor cell lines toward normal progenitors, primary astrocytes, or normal cell lines in culture,[38,43,52] the second criterion to achieve a therapeutic index seems possible. These cellular data also suggest that the toxicity of these agents is considerably less important than those observed for "classical" antitumor therapy.Since telomeres are also present in normal cells, the reason for their tumor selectivity is still unknown and would be of great interest to be determined to optimize the future clinical development of these agents. Finally, the rapid telomere delocalization of components of the shelterin complex, including POT1, induced by these ligands could also provide valuable surrogate markers to evaluate their biological activity during clinical development.

Acknowledgment

We thank all the members of the laboratory for fruitful discussions to conceive this chapter, C. Morrison for critical reading of the manuscript, and J.L. Mergny, P. Mailliet, E. Mandine, F. Boussin, A. Londono-Vallejo, K. Shin-ya, M.P. Teulade-Fichou, M.F. O'Donohue for scientific collaborations. This work was supported by the "Association pour la Recherche contre le Cancer," grants #3644 & 4491, by the "Ligue Nationale Contre le Cancer, Equipe Labelliseè 2006" and by an "Action Concertée Incitative Médicament et Cibles Thérapeutiques" from the Ministère de la Recherche et de la Technologie.

References

1. S.A. Stewart and R.A. Weinberg, *Semin. Cancer Biol.*, 2000, **10**, 399.
2. M.J. McEachern, A. Krauskopf and E.H. Blackburn, *Annu. Rev. Genet.*, 2000, **34**, 331.
3. F. Lavelle, J.F. Riou, A. Laoui and P. Mailliet, *Crit. Rev. Oncol. Hematol.*, 2000, **34**, 111.
4. J.L. Mergny, P. Mailliet, F. Lavelle, J.F. Riou, A. Laoui and C. Hélène, *Anticancer Drug Des.*, 1999, **14**, 327.
5. J.L. Mergny, J.F. Riou, P. Mailliet, M.P. Teulade-Fichou and E. Gilson, *Nucleic Acids Res.*, 2002, **30**, 839.
6. C.I. Nugent and V. Lundblad, *Genes Dev.*, 1998, **12**, 1073.
7. J.W. Shay and W.E. Wright, *Cancer Cell*, 2002, **2**, 257.
8. A.A. Neumann and R.R. Reddel, *Nat. Rev. Cancer*, 2002, **2**, 879.
9. R.R. Reddel and T.M. Bryan, *Lancet*, 2003, **361**, 1840.
10. A. Smogorzewska and T. De Lange, *Annu. Rev. Biochem.*, 2004, **73**, 177.

11. T.R. Yeager, A.A. Neumann, A. Englezou, L.I. Huschtscha, J.R. Noble and R.R. Reddel, *Cancer. Res.*, 1999, **59**, 4175.
12. W.Q. Jiang, Z.H. Zhong, J.D. Henson, A.A. Neumann, A.C. Chang and R.R. Reddel, *Mol. Cell. Biol.*, 2005, **25**, 2708.
13. J.T. Davies, *Angew. Chem. Int. Edit.*, 2004, **43**, 668.
14. M. Mills, L. Lacroix, P.B. Arimondo, J.L. Leroy, J.C. Francois, H. Klump and J.L. Mergny, *Curr. Med. Chem. Anti-Canc. Agents*, 2002, **2**, 627.
15. T. Simonsson, *Biol. Chem.*, 2001, **382**, 621.
16. G.N. Parkinson, M.P. Lee and S. Neidle, *Nature*, 2002, **417**, 876.
17. Y. Wang and D.J. Patel, *J. Mol. Biol.*, 1993, **234**, 1171.
18. A.M. Zahler, J.R. Williamson, T.R. Cech and D.M. Prescott, *Nature*, 1991, **350**, 718.
19. H. Han, D.R. Langley, A. Rangan and L.H. Hurley, *J. Am. Chem. Soc.*, 2001, **123**, 8902.
20. H. Han, C.L. Cliff and L.H. Hurley, *Biochemistry*, 1999, **38**, 6981.
21. O.Y. Fedoroff, M. Salazar, H. Han, V.V. Chemeris, S.M. Kerwin and L.H. Hurley, *Biochemistry*, 1998, **37**, 12367.
22. P.J. Perry, S.M. Gowan, A.P. Reszka, P. Polucci, T.C. Jenkins, L.R. Kelland and S. Neidle, *J. Med. Chem.*, 1998, **41**, 3253.
23. P.J. Perry, M.A. Read, R.T. Davies, S.M. Gowan, A.P. Reszka, A.A. Wood, L.R. Kelland and S. Neidle, *J. Med. Chem.*, 1999, **42**, 2679.
24. M.A. Read, A.A. Wood, J.R. Harrison, S.M. Gowan, L.R. Kelland, H.S. Dosanjh and S. Neidle, *J. Med. Chem.*, 1999, **42**, 4538.
25. R.J. Harrison, J. Cuesta, G. Chessari, M.A. Read, S.K. Basra, A.P. Reszka, J. Morrell, S.M. Gowan, C.M. Incles, F.A. Tanious, W.D. Wilson, L.R. Kelland and S. Neidle, *J. Med. Chem.*, 2003, **46**, 4463.
26. F. Koeppel, J.F. Riou, A. Laoui, P. Mailliet, P.B. Arimondo, D. Labit, O. Petitgenet, C. Hélène and J.L. Mergny, *Nucleic Acids Res.*, 2001, **29**, 1087.
27. F. Rosu, E. De Pauw, L. Guittat, P. Alberti, L. Lacroix, P. Mailliet, J.F. Riou and J.L. Mergny, *Biochemistry*, 2003, **42**, 10361.
28. J.F. Riou, L. Guittat, P. Mailliet, A. Laoui, E. Renou, O. Petitgenet, F. Megnin-Chanet, C. Hélène and J.L. Mergny, *Proc. Natl. Acad. Sci. USA*, 2002, **99**, 2672.
29. J.H. Kim, J.H. Kim, G.E. Lee, S.W. Kim and I.K. Chung, *Biochem. J.*, 2003, **373**, 523.
30. V. Caprio, B. Guyen, Y. Opoku-Boahen, J. Mann, S.M. Gowan, L.M. Kelland, M.A. Read and S. Neidle, *Bioorg. Med. Chem. Lett.*, 2000, **10**, 2063.
31. J.L. Mergny, L. Lacroix, M.P. Teulade-Fichou, C. Hounsou, L. Guittat, M. Hoarau, P.B. Arimondo, J.P. Vigneron, J.M. Lehn, J.F. Riou, T. Garestier and C. Hélène, *Proc. Natl. Acad. Sci. USA*, 2001, **98**, 3062.
32. M.P. Teulade-Fichou, C. Carrasco, L. Guittat, C. Bailly, P. Alberti, J.L. Mergny, A. David, J.M. Lehn and W.D. Wilson, *J. Am. Chem. Soc.*, 2003, **125**, 4732.
33. S.M. Gowan, R. Heald, M.F. Stevens and L.R. Kelland, *Mol. Pharmacol.*, 2001, **60**, 981.

34. K. Shin-ya, K. Wierzba, K. Matsuo, T. Ohtani, Y. Yamada, K. Furihata, Y. Hayakawa and H. Seto, *J. Am. Chem. Soc.*, 2001, **123**, 1262.
35. T. Lemarteleur, D. Gomez, R. Paterski, E. Mandine, P. Mailliet and J.F. Riou, *Biochem. Biophys. Res. Commun.*, 2004, **323**, 802.
36. S.M. Kerwin, *Curr. Pharm. Des.*, 2000, **6**, 441.
37. J. Cuesta, M.A. Read and S. Neidle, *Mini. Rev. Med. Chem.*, 2003, **3**, 11.
38. T. Tauchi, K. Shin-Ya, G. Sashida, M. Sumi, A. Nakajima, T. Shimamoto, J.H. Ohyashiki and K. Ohyashiki, *Oncogene*, 2003, **22**, 5338.
39. D. Gomez, N. Aouali, A. Renaud, C. Douarre, K. Shin-Ya, J. Tazi, S. Martinez, C. Trentesaux, H. Morjani and J.F. Riou, *Cancer Res.*, 2003, **63**, 6149.
40. M.A. Shammas, R.J. Shmookler Reis, M. Akiyama, H. Koley, D. Chauhan, T. Hideshima, R.K. Goyal, L.H. Hurley, K.C. Anderson and N.C. Munshi, *Mol. Cancer Ther.*, 2003, **2**, 825.
41. S.M. Gowan, J.R. Harrison, L. Patterson, M. Valenti, M.A. Read, S. Neidle and L.R. Kelland, *Mol. Pharmacol.*, 2002, **61**, 1154.
42. C. Leonetti, S. Amodei, C. D'Angelo, A. Rizzo, B. Benassi, A. Antonelli, R. Elli, M. Stevens, M. D'Incalci, G. Zupi and A. Biroccio, *Mol. Pharmacol.*, 2004, **66**, 1138.
43. G. Pennarun, C. Granotier, L.R. Gauthier, D. Gomez, F. Hoffschir, E. Mandine, J.F. Riou, J.L. Mergny, P. Mailliet and F.D. Boussin, *Oncogene*, 2005.
44. E. Izbicka, D. Nishioka, V. Marcell, E. Raymond, K.K. Davidson, R.A. Lawrence, R.T. Wheelhouse, L.H. Hurley, R.S. Wu and D.D. Von Hoff, *Anticancer. Drug. Des.*, 1999, **14**, 355.
45. A. Rangan, O.Y. Fedoroff and L.H. Hurley, *J. Biol. Chem.*, 2001, **276**, 4640.
46. D. Gomez, T. Lemarteleur, L. Lacroix, P. Mailliet, J.L. Mergny and J.F. Riou, *Nucleic Acids Res.*, 2004, **32**, 371.
47. A.M. Burger, F. Dai, C.M. Schultes, A.P. Reszka, M.J. Moore, J.A. Double and S. Neidle, *Cancer Res.*, 2005, **65**, 1489.
48. I. Faraoni, M. Turriziani, G. Masci, L. De Vecchis, J.W. Shay, E. Bonmassar and G. Graziani, *Clin. Cancer Res.*, 1997, **3**, 579.
49. Y. Mo, Y. Gan, S. Song, J. Johnston, X. Xiao, M.G. Wientjes and J.L. Au, *Cancer Res.*, 2003, **63**, 579.
50. C. Douarre, D. Gomez, H. Morjani, J.M. Zahm, F.M. O'Donohue, L. Eddabra, P. Mailliet, J.F. Riou and C. Trentesaux, *Nucleic Acids Res.*, 2005, **33**, 2192.
51. L. Harrington and M.O. Robinson, *Oncogene*, 2002, **21**, 592.
52. C.M. Incles, C.M. Schultes, H. Kempski, H. Koehler, L.R. Kelland and S. Neidle, *Mol. Cancer Ther.*, 2004, **3**, 1201.
53. D. Gomez, N. Aouali, A. Londono-Vallejo, L. Lacroix, F. Megnin-Chanet, T. Lemarteleur, C. Douarre, K. Shin-ya, P. Mailliet, C. Trentesaux, H. Morjani, J.-L. Mergny and J.-F. Riou, *J. Biol. Chem.*, 2003, **278**, 50554.

54. M.Y. Kim, M. Gleason-Guzman, E. Izbicka, D. Nishioka and L.H. Hurley, *Cancer Res.*, 2003, **63**, 3247.

55. C.M. Incles, C.M. Schultes, L.R. Kelland and S. Neidle, *Mol. Pharmacol.*, 2003, **64**, 1101.

56. V.L. Makarov, Y. Hirose and J.P. Langmore, *Cell*, 1997, **88**, 657.

57. M.T. Hemann and C.W. Greider, *Nucleic Acids Res.*, 1999, **27**, 3964.

58. N.K. Jacob, K.E. Kirk and C.M. Price, *Mol. Cell*, 2003, **11**, 1021.

59. T. de Lange, *Genes. Dev.*, 2005, **19**, 2100.

60. S. Neidle and G. Parkinson, *Nat. Rev. Drug Discov.*, 2002, **1**, 383.

61. G. Cimino-Reale, E. Pascale, E. Battiloro, G. Starace, R. Verna and E. D'Ambrosio, *Nucleic Acids Res.*, 2001, **29**, E35.

62. D. Gomez, R. Paterski, T. Lemarteleur, K. Shin-Ya, J.L. Mergny and J.F. Riou, *J. Biol. Chem.*, 2004, **279**, 41487.

63. R. Court, L. Chapman, L. Fairall and D. Rhodes, *EMBO Rep.*, 2005, **6**, 39.

64. M. Lei, E.R. Podell and T.R. Cech, *Nat. Struct. Mol. Biol.*, 2004, **11**, 1223.

65. J.D. Griffith, L. Comeau, S. Rosenfield, R.M. Stansel, A. Bianchi, H. Moss and T. de Lange, *Cell*, 1999, **97**, 503.

66. R.M. Stansel, T. de Lange and J.D. Griffith, *Embo J.*, 2001, **20**, 5532.

67. D. Hockemeyer, A.J. Sfeir, J.W. Shay, W.E. Wright and T. de Lange, *Embo J.*, 2005.

68. X.D. Zhu, L. Niedernhofer, B. Kuster, M. Mann, J.H. Hoeijmakers and T. de Lange, *Mol. Cell*, 2003, **12**, 1489.

69. G.B. Celli and T. de Lange, *Nat. Cell. Biol.*, 2005, **7**, 712.

70. H. Takai, A. Smogorzewska and T. de Lange, *Curr. Biol.*, 2003, **13**, 1549.

71. S.H. Kim, C. Beausejour, A.R. Davalos, P. Kaminker, S.J. Heo and J. Campisi, *J. Biol. Chem.*, 2004.

72. F. d'Adda di Fagagna, P.M. Reaper, L. Clay-Farrace, H. Fiegler, P. Carr, T. Von Zglinicki, G. Saretzki, N.P. Carter and S.P. Jackson, *Nature*, 2003, **426**, 194.

73. K. Paeschke, T. Simonsson, J. Postberg, D. Rhodes and H.J. Lipps, *Nat. Struct. Mol. Biol.*, 2005, **12**, 847.

74. A.J. Zaug, E.R. Podell and T.R. Cech, *Proc. Natl. Acad. Sci. USA*, 2005, **102**, 10864.

75. Q. Yang, Y.L. Zheng and C.C. Harris, *Mol. Cell. Biol.*, 2005, **25**, 1070.

76. P.L. Opresko, P.A. Mason, E.R. Podell, M. Lei, I.D. Hickson, T.R. Cech and V.A. Bohr, *J. Biol. Chem.*, 2005, **280**, 32069.

77. M. Read, R.J. Harrison, B. Romagnoli, F.A. Tanious, S.H. Gowan, A.P. Reszka, W.D. Wilson, L.R. Kelland and S. Neidle, *Proc. Natl. Acad. Sci. USA*, 2001, **98**, 4844.

78. F. Rosu, V. Gabelica, K. Shin-ya and E. De Pauw, *Chem. Commun. (Camb.)*, 2003, **21**, 2702.

79. S. Rankin, A.P. Reszka, J. Huppert, M. Zloh, G.N. Parkinson, A.K. Todd, S. Ladame, S. Balasubramanian and S. Neidle, *J. Am. Chem. Soc.*, 2005, **127**, 10584.

80. A.K. Todd, M. Johnston and S. Neidle, *Nucleic Acids Res.*, 2005, **33**, 2901.

81. A. Siddiqui-Jain, C.L. Grand, D.J. Bearss and L.H. Hurley, *Proc. Natl. Acad Sci. USA*, 2002, **99**, 11593.
82. M.A. Shammas, R.J. Reis, C. Li, H. Koley, L.H. Hurley, K.C. Anderson and N.C. Munshi, *Clin. Cancer Res.*, 2004, **10**, 770.
83. J. Seenisamy, S. Bashyam, V. Gokhale, H. Vankayalapati, D. Sun, A. Siddiqui-Jain, N. Streiner, K. Shin-Ya, E. White, W.D. Wilson and L.H. Hurley, *J. Am. Chem. Soc.*, 2005, **127**, 2944.
84. C. Granotier, G. Pennarun, L. Riou, F. Hoffschir, L.R. Gauthier, A. De Cian, D. Gomez, E. Mandine, J.F. Riou, J.L. Mergny, P. Mailliet, B. Dutrillaux and F.D. Boussin, *Nucleic Acids Res.*, 2005, **33**, 4182.
85. H. Sun, J.K. Karow, I.D. Hickson and N. Maizels, *J. Biol. Chem.*, 1998, **273**, 27587.
86. P. Mohaghegh, J.K. Karow, R.M. Brosh, Jr., V.A. Bohr and I.D. Hickson, *Nucleic Acids Res.*, 2001, **29**, 2843.
87. M. Fry and L.A. Loeb, *J. Biol. Chem.*, 1999, **274**, 12797.
88. D.J. Stavropoulos, P.S. Bradshaw, X. Li, I. Pasic, K. Truong, M. Ikura, M. Ungrin and M.S. Meyn, *Hum. Mol. Genet.*, 2002, **11**, 3135.
89. L. Crabbe, R.E. Verdun, C.I. Haggblom and J. Karlseder, *Science*, 2004, **306**, 1951.
90. J.P. Vaughn, S.D. Creacy, E.D. Routh, C. Joyner-Butt, G.S. Jenkins, S. Pauli, Y. Nagamine and S.A. Akman, *J. Biol. Chem.*, 2005.
91. H. Ding, M. Schertzer, X. Wu, M. Gertsenstein, S. Selig, M. Kammori, R. Pourvali, S. Poon, I. Vulto, E. Chavez, P.P. Tam, A. Nagy and P.M. Lansdorp, *Cell*, 2004, **117**, 873.
92. I. Cheung, M. Schertzer, A. Rose and P.M. Lansdorp, *Nat. Genet.*, 2002, **31**, 405.
93. J.L. Li, R.J. Harrison, A.P. Reszka, R.M. Brosh, Jr., V.A. Bohr, S. Neidle and I.D. Hickson, *Biochemistry*, 2001, **40**, 15194.

CHAPTER 7

DNA Quadruplexes and Gene Regulation

THOMAS S. DEXHEIMER,[a] MICHAEL FRY[b] AND
LAURENCE H. HURLEY[a, c, d, e]

[a] College of Pharmacy, University of Arizona, Tucson, Arizona 85721
[b] Department of Biochemistry, Rappaport Faculty of Medicine, Technion –
Israel Institute of Technology, Haifa 31096, Israel
[c] BIO5 Institute for Collaborative Bioresearch, University of Arizona, Tucson,
Arizona 85721
[d] Arizona Cancer Center, 1515 N. Campbell Ave., Tucson, Arizona 85724
[e] Department of Chemistry, University of Arizona, Tucson, Arizona 85721

7.1 Transcription

In 1970 Francis Crick enunciated the fundamental dogma of molecular biology, which dictated the one-way flow of genetic information from DNA to RNA to protein.[1] This view implies that proteins, which produce the observable physical and biochemical characteristics of an individual (phenotype), are determined by an individual's genome or DNA sequence (genotype). In decoding DNA, information-carrying RNA intermediary molecules are copied from DNA in a process called transcription.[2] Transcription is a major determinant of gene expression that allows cells to proliferate, differentiate, and maintain proper homeostasis. This process serves as a major molecular "on/off" switch for the expression of genes.[3] Thus, a failure to generate the initial RNA transcript renders unnecessary all regulatory steps that follow, such as transcript processing and transport, and translation of the RNA transcript into protein. There are actually several distinct levels of gene transcription. A so-called "basal level," which typically refers to the constitutive or low-level expression of a given gene, may increase or decrease, attaining activated or repressed states, which are commonly referred to as "on" and "off" states, respectively.

Transcription involves synthesis, catalyzed by RNA polymerases, of an RNA molecule identical in sequence to one strand of the DNA duplex (termed the coding strand). There are three classes of eukaryotic RNA polymerases

(I, II, and III), each consisting of two large subunits and 12–15 smaller subunits. Of special importance is RNA polymerase II (RNAP II), which is responsible for the transcription of all protein-coding genes. A simplified conception of transcription is that DNA, in the presence of RNAP II and ribonucleoside triphosphates, is transcribed into RNA molecules. However, this minimal representation does not account for the intricacies in getting the transcription of a particular gene to take place only at particular times and in specific cell types. RNAP II alone is incapable of initiating *in vitro* transcription from DNA, because it lacks a subunit equivalent to the σ subunit of pro-karyotic RNA polymerases that functions in the recognition of gene-promoter regions. In its place, eukaryotes have a group of general transcription factors, which are independent of RNAP II, that assist in the recruitment and posi-tioning of RNAP II at the gene promoter and in the initiation and elongation of the transcriptional process.[4] However, gene promoters tend to be modular in architecture and can be very complex if they are differentially regulated in different tissue types, during tissue differentiation, or in response to endo-genous or exogenous cellular signals.

The estimated number of about 30,000 protein-encoding genes in the human genome[5] implies the existence of an equivalent number of gene promoters. How then is the transcriptional process regulated to allow genes to be transcribed at the right time and at the proper level in particular cell types? For example, precursors of mature erythrocytes must transcribe globin genes at a high rate, whereas cells of the pancreas are required to specifically express excreted digestive enzymes. Precisely timed cell-specific gene expression is attained mostly through the action of gene-specific transcription protein factors. These factors function as activators or repressors of gene expression by respectively increasing or decreasing the rate of transcription. *Trans*-acting activators bind *cis* DNA sequences to designated enhancer sites, whereas repressors bind to silencer sites. The interaction of these *trans*-acting factors with the DNA *cis* elements is a major contributor to the regulation of the transcriptional process.[6] A case in point of such cooperative action is displayed in genes whose transcription is increased in response to elevated temperature. These genes contain a common *cis*-acting regulatory sequence known as the heat shock element that is recognized and bound by a heat shock *trans*-acting transcription factor upon elevation of temperature. The binding of this protein to the heat shock DNA motif enhances transcription.[7] Moreover, transferring the heat shock element to a non-heat-inducible gene renders the recipient gene heat inducible.[8]

An additional cardinal regulatory mechanism of differential control of gene expression is chromatin structure. In eukaryotic cells, DNA is not "naked" but is packaged into nucleosomes, subunits of chromatin in which short tracts of DNA are wrapped around a core of histone proteins.[9] As a consequence, accessibility of RNAP II and regulatory transcription factors to DNA is limited. More often than not, in the presence of a stable, inaccessible chromatin structure (heterochromatin), transcription is repressed, whereas the formation of an open accessible chromatin structure (euchromatin) is transcribed.[10]

Transition between heterochromatin and euchromatin is mediated by chromatin-remodeling enzymes, which alter the folding, flexibility, and basic structure of chromatin. There are two classes of chromatin remodeling enzymes: those that covalently modify nucleosomal histone proteins through acetylation, phosphorylation, or methylation, and those that use the energy of ATP hydrolysis to disrupt histone–DNA interactions.[11] The emerging view is that chromatin remodeling complexes and histone acetyltransferase complexes cooperate with sequence-specific binding factors to enable productive interaction between the transcriptional machinery and the promoter of a given gene.

A chromatin modification that constitutes an additional level of regulation of gene transcription is DNA methylation by DNA methyltransferases. In the mammalian genome, this process consists of the covalent addition of a methyl group to the 5-position of the cytosine base predominantly when it is located 5' to a guanosine in a CpG dinucleotide.[12] Genomic DNA methylation patterns are non-randomly distributed. Rather, distinct regions, including most repetitive and transposable elements, are hypermethylated, whereas other regions, such as CpG islands in regulatory regions of house-keeping genes, are hypomethylated.[13] In general, hypermethylation of CpG islands adjacent to regulatory elements is associated with gene silencing, whereas hypomethylation in these sequences is linked to gene activation. The importance of methylation as a mechanism for gene silencing is underlined by the observation that one of the two X chromosomes in female mammals is inactivated as a result of methylation.[14] Further, aberrant gene silencing or activation is associated with disruption of normal DNA methylation patterns in tumors,[15,16] as well as in human disorders such as ATRX, Fragile X, and ICF syndromes.[17] Yet, while methylation can affect gene activity, it alone is insufficient to repress transcription.

Finally, as expounded in this review, torsional forces generated by positive or negative superhelical stress in DNA contribute to transcriptional regulation by modulating the structural transition of B-DNA into non-B-DNA structures.[18–21] Such conformational alterations can occur under specific conditions. In general, negative supercoiling takes place when histones are stripped from chromatin or subsequent to the progression of the transcription machinery along the template DNA, which entails generation of negative supercoiling behind the enzyme complex. Negative supercoiling destabilizes DNA and often results in local DNA unwinding at the proximal promoter region of both prokaryotic and eukaryotic genes. This conformational change is believed to enhance open promoter complex formation and to lead to transcriptional activation.[18–21] Such a structural transition within B-DNA was discovered by the appearance of nuclease hypersensitive elements (NHEs) and has been reported in a number of TATA-less mammalian genes.[22–26] NHEs are believed to represent regions that lack canonical nucleosomal structure that exposes the DNA, which then becomes accessible to nucleases. Most of the time-altered conformational alteration takes place along polypurine/polypyrimidine stretches, which tend to easily adopt non-B-DNA conformations such as melted DNA, hairpin structures, or slipped helices. Of special interest are tetrahelical structures of guanine-rich sequences (G-quadruplexes or tetraplexes) that are readily generated *in vitro* under physiological-like conditions. At the core of G-quadruplex DNA structures, which may possess parallel or antiparallel orientation, are two or more G-tetrads that require for their stabilization coordinate bonding of monovalent cations, most commonly Na^+ or K^+.[27,28] Direct evidence for the existence of G-quadruplexes *in vivo* is just beginning to emerge, and the ability of guanine-rich promoter sequences to form very stable G-quadruplex structures *in vitro* suggests that these DNA formations may play a role in transcriptional regulation. Recent studies of both crystal and solution structures of various G-quadruplexes revealed a high extent of diversity in their folding patterns.[27,28] Such structural variation implies that structural transitions in guanine-rich promoter DNA tracts may potentially play a role in regulating the transcription of specific genes. It has recently been shown that there are many hundreds of thousands of DNA sequences in the human genome that potentially could form G-quadruplex structures.[29] With the exception of telomeres, all of these sequences are in duplex regions of the genome. Thus, a major impediment to the formation of G-quadruplexes is the stability of double-stranded DNA. The requirement for the generation of negative superhelicity during the progression of the transcriptional machinery could severely limit the emergence of these structures from duplex DNA. Therefore, unwinding of DNA and the formation of single-stranded tracts during transcription or replication may provide the necessary conditions for the formation of G-quadruplexes. We will now take a closer look at some specific examples of how genes can be controlled by the formation of G-quadruplex structures within their specific regulatory regions.

7.2 G-Rich Clusters in the Promoter Regions of Mammalian Genes

7.2.1 Chicken β-Globin Gene

Studies of the chicken β-globin gene offered the earliest evidence for a relationship between DNA secondary-structure formation and transcription based on an observed correlation between changes in chromatin structure and gene activity. During chick erythrocyte development, chromatin structure in the vicinity of the β-globin gene becomes altered prior to gene expression. The timing of these changes suggests that they are associated with transcription activation rather than with the process of transcription itself.[30,31] Larsen and Weintraub detected these changes by the presence of preferential hypersensitivity to nuclease digestion in the 5′ flanking region of a transcriptionally active chicken β-globin gene.[32] This hypersensitivity was absent in cells whose β-globin gene was inactive.[33] Gene-activity-associated nuclease sensitivity was independently demonstrated by Wood and Felsenfeld with both nonspecific nucleases and restriction endonucleases.[34] Comparable nuclease hypersensitive sites were observed in intact cells, isolated nuclei, and supercoiled plasmids, while linear or relaxed plasmids were not susceptible to nuclease digestion.[35] Specifically, a 115-base-pair fragment within the nuclease hypersensitive region could be excised from nuclei by MspI digestion at a greater than 50% yield, of which one-third behaved as nucleosomal-free DNA.[36] In a supercoiled plasmid, this region was also sensitive to the chemical probe bromoacetaldehyde, which reacts with unpaired adenines and cytosines,[35] suggesting that this region existed in a partially denatured form. On the basis of these results it was proposed that the association of nuclease hypersensitivity with gene activation is related to the conversion of some fraction of the DNA sequence into an altered DNA secondary structure, which is dependent on superhelical strain and chromatin structure.

The nuclease hypersensitive region was mapped to the chicken β-globin gene promoter, wherein it extends roughly −60 to −260 base pairs 5′ to the start of mRNA transcription.[36] Examination of this 200-base-pair segment revealed a number of unusual features of nucleotide sequence that might affect the local conformation of the DNA. Most notably, this region is highly GC-rich (70%), including a stretch of 16 consecutive guanine residues (Table 1). It also exhibits short runs of alternating purines and pyrimidines, which are classical motifs of left-handed Z-DNA. Last, methylation sites within this region are in high abundance, present at approximately six times the average genomic frequency. Yet, subsequent *in vitro* investigation of this region using supercoiled plasmids revealed that there was no conversion of any measurable segment of DNA to left-handed Z-DNA. Moreover, cytosine methylation had no effect on the supercoiling properties or the DNA conformation of a plasmid containing this region.[37] Nevertheless, *in vitro* mapping at higher resolution of this 200-base-pair β-globin hypersensitive domain demonstrated that the stretch of consecutive guanine residues within this region was sensitive to bromoacetaldehyde and was cut preferentially by S1 nuclease.[37,38] *In vivo* high-resolution mapping revealed that this element was protected from nuclease attack in cells expressing the β-globin gene, whereas this region was accessible to nuclease attack in cells

Table 1 *G-quadruplex-forming sequences and their proximity to transcriptional start sites*

Gene	G-quadruplex-forming sequence (5′ to 3′)	Proximity to transcriptional start site
Chicken β-globin	GGGGGGGGGGGGGGGGGGCGGGTGGTGG	−195 to −170
Human sMtCK	CTGAGGAGGGGCTGGAGGGACCAC	−320 to −296
Mouse MCK	TCCGGAGGGGCAGGCTGAGGGCGGC	−1045 to −1021
Mouse α7 integrin	AAAGGTGGGGCGGCAGGGCGCAAGGCAAT	−215 to −186
Human c-*myc*	TGGGGAGGGTGGGGAGGGTGGGGAAGG	−142 to −115
Human *FMR1*	[(A/C)GG]$_n$	+135
Human insulin	[ACAGGGGTGTGGGGG]$_n$	−363

that did not express this gene.[39] The protection within this guanine tract was also reproducible *in vitro* using partially purified chicken erythrocyte nuclear factors. These results were contradictory to the previous studies in which widespread nuclease hypersensitivity within this 200-base-pair region was associated with chicken β-globin gene activity. Still, they attracted considerable attention to the oligoguanine tract in part because of its unusual physical properties and potential regulatory role in transcription.

Initial studies proposed that the guanine-rich region within the promoter of the adult chicken β-globin gene could adopt a supercoil-dependent triplex structure.[40] However, further analysis suggested a more complex DNA conformation within this region. For instance, this DNA stretch contains several runs of contiguous guanines, conforming to the general motif capable of assembling into an intramolecular G-quadruplex structure. Subsequent to sequence analysis, Woodford *et al.* provided the first indication for the formation of a G-quadruplex structure within the chicken β-globin gene promoter region.[41] Using a primer extension assay, they showed that a template containing the sequence $G_{16}CG(GGT)_3$, which corresponds to the guanine-rich region of interest, was necessary and sufficient to arrest DNA synthesis *in vitro* under physiological conditions. In addition, these authors demonstrated that the polymerase arrest site consisted of three successive blocks to DNA synthesis opposite the successive thymidine residues in the template, suggesting that this oligonucleotide tract forms three individual, yet potentially similar, DNA secondary structures. This result on its own could not establish the existence of a G-quadruplex structure, since the propensity to block DNA synthesis has been implicated in the formation of other structures, such as hairpins[42] and triplexes.[43,44] That these DNA secondary structures were, in fact, intramolecular G-quadruplexes was verified by Howell *et al.*[45] through an examination of the reaction conditions under which these polymerase arrest sites were generated. First, they demonstrated that the polymerase arrest sites were only observed using the single-stranded G-rich strand as a template. Second, they showed that the arrest sites were generated only in the presence of K^+ ions, which are required for the formation and stabilization of G-quadruplex structures. Third, the intrastrand nature of these structures was confirmed by demonstrating that generation of the arrest sites was independent of template concentration, indicating zero-order kinetics of the formation of unimolecular quadruplex DNA. Finally, specific guanine residues were substituted with 7-deaza-dGTP, which lacks a guanine N7 position and thus is incapable of forming the guanine–guanine Hoogsteen hydrogen bonds that are the hallmark of G-quadruplex structures. As a result, all three DNA synthesis arrest sites were eliminated. These results provided substantial support for the proposal that primer extension was arrested at three template domains that folded into intramolecular G-quadruplex structures.

Evaluation of the relative potencies of the three polymerase arrest sites revealed that the major molecular blocking structure was located at the 3′-most thymine residue, whereas the two other sites blocked polymerase progression more weakly. By employing the chemical probes dimethyl sulfate and OsO_4,

Howell *et al.* identified residues involved in tetrad formation as well as residues that were displaced from the core structure [Figure 1(A)].[45] These results were used to construct a model of the G-quadruplex structure as shown in Figure 1(B). This model was further corroborated by the introduction of specific mutations that affected the structural stability of the G-quadruplex, as assessed by the formation of blocks to DNA synthesis. Specifically, it was found that the structure formed differed from canonical G-quadruplex structures in some respects, including the presence of only one classical tetrad configuration, while the other tetrads were incomplete and contained non-guanine bases. The authors suggested that the hydrogen-bonding interactions between the loop bases and the bases within the flanking regions, which are illustrated in their model [Figure 1(B)], counteract the loss in G-quadruplex stability due to

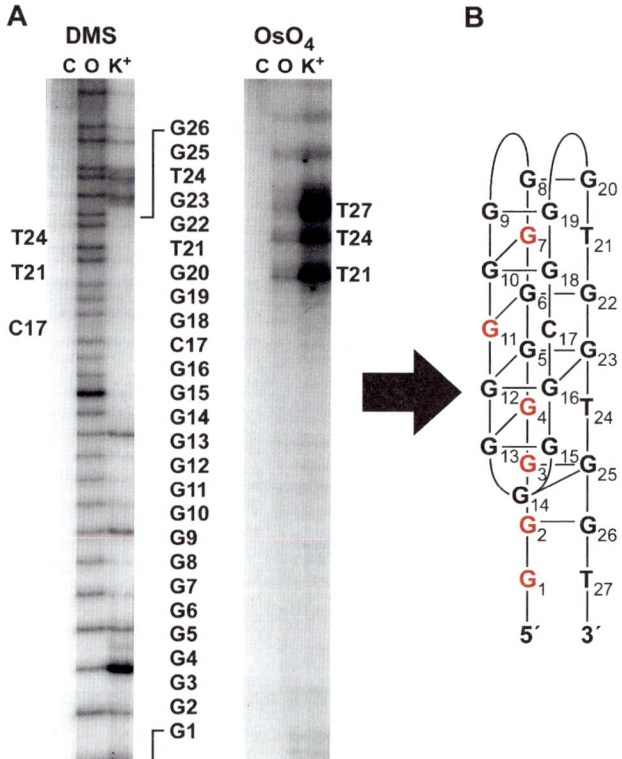

Figure 1 *(A) Chemical modification of the chicken β-globin G-quadruplex-forming sequence $G_{16}CG(GGT)_2GG$ with dimethyl sulfate and OsO_4 in the absence (O) and presence (K^+) of 40 mM KCl. The lanes labeled C represent control reactions to which no dimethyl sulfate or OsO_4 was added. (B) Proposed structural model of the "cinched" intramolecular G-quadruplex formed by the chicken β-globin sequence $G_{16}CG(GGT)_2GG$ in the presence of K^+. The G residues shown in red are those modified by dimethyl sulfate[45]*

    (Reprinted from reference 45 with permission from American Society for Biochemistry and Molecular Biology)

incomplete tetrad formation. In addition, the base pairing between bases 3′ and 5′ to the G-quadruplex-forming region creates a molecular "cinch," which further stabilizes this unique G-quadruplex structure. These findings, together with those that describe the triplex-forming ability of the same sequence, indicate the variety of complex structures that can be formed by the chicken β-globin promoter sequence.

Prior to the recognition of the formation of a DNA secondary structure within the chicken β-globin promoter, a poly(dG)-binding protein called BGP1 was discovered that was later found to bind specifically to the β-globin G-tract and to activate gene expression.[46] This protein is found only in chicken erythroid cells and is most abundant in the cells at a time of maximum expression of the β-globin gene. In addition, BGP1 expression becomes elevated at the same time, or shortly before, the changes in chromatin structure appear.[47] These characteristics are indicative of a relationship between the previously described *in vivo* nuclease protection pattern within the G-tract of an active β-globin gene and the binding of BGP1 to this same region. Conversely, when the β-globin gene is inactive, this region is sensitive to nuclease digestion and exhibits a non-B-DNA conformation, which suggests a relationship between gene silencing and the formation of a DNA secondary structure. Taken together, these data suggest a model in which BGP1 activates gene expression through its binding to the DNA and the induction of subsequent DNA structural transactions. In other words, the G-quadruplex-forming region within the chicken β-globin gene promoter could act as a conformational switch, which is subject to modulation by BGP1.

7.2.2 Muscle-Specific Genes

Myogenesis involves the coordinated transcriptional activation of multiple genes. The components of this regulation were elucidated by the identification and characterization of four myogenic regulatory factors that activate muscle tissue differentiation by specifically binding to regulatory elements of genes that encode muscle-specific proteins, thus stimulating their transcription.[48] These transcription factors belong to a subgroup within the superfamily of basic helix-loop-helix proteins.[49] Initially, these muscle cell–specific transcription factors were reported to bind specifically to the major groove of the double-stranded sequence motif CANNTG, or E-box, that is present in several copies within promoter and enhancer regions of muscle-specific genes.[50,51] Further investigation by Santoro *et al.* revealed that a major member of the myogenic transcription factors, MyoD, could also bind to single-stranded DNA probes.[52] Specifically, MyoD bound in a sequence-specific manner to the DNA single-strand d(GGGGGTT**GTGGAC**GACGGACT), an E-box-containing (bold) portion of the mouse muscle creatine kinase (MCK) non-coding enhancer strand. By contrast, MyoD did not form a complex with the complementary coding strand of this motif. The sequence-specific binding of MyoD to the guanine-rich DNA tract suggested potential recognition by this protein of a specific DNA conformation whose nature remained unidentified in this early work.

Serendipitously, while assaying the interactions of basic helix-loop-helix proteins with the single-stranded MCK sequence, Walsh and Gualberto identified a slower mobility band in the absence of protein that was made up of less than 2% of the total DNA.[53] Through further investigation these authors showed that MyoD bound preferentially to the slowly migrating DNA species. Methylation protection analysis revealed that the 3'-end of the MCK sequence at each of the five consecutive guanines (underlined in the sequence above) remained protected in both the free DNA and the MyoD–DNA complex. These results led the authors to propose that the slower mobility DNA form represented an intermolecular G-quadruplex structure. This hypothesis gained support from the observation that mutations of residues included in the cluster of protected guanines resulted in loss of the slower mobility band. An affinity of MyoD for tetrahelical DNA was also demonstrated by its binding to a G-quadruplex form of the *Tetrahymena* telomeric DNA. Last, measurement of dissociation constants indicated that MyoD had a significantly higher affinity for the G-quadruplex structure of the E-box sequence than for its double-stranded form.[53] Together, these data suggested that through their high-affinity binding of MyoD, DNA G-quadruplex structures could possibly play a role in the transcriptional regulation of muscle-specific genes.

In extending the described studies, Yafe *et al.* searched regulatory regions of other muscle-specific genes for sequences with a propensity to form altered DNA structures that could potentially interact with MyoD.[54] Significantly, these authors detected a disproportionately high frequency of clusters of contiguous guanine residues in promoter and enhancer regions of multiple muscle-specific genes. In addition to the previously studied mouse MCK enhancer region, sequences derived from promoter regions of both human mitochondrial creatine kinase (sMtCK) and mouse $\alpha 7$ integrin genes (Table 1) were investigated for their ability to adopt an altered DNA conformation. Using non-denaturing gel electrophoresis, dimethyl sulfate footprinting, and CD spectroscopy, it was demonstrated that all three single-stranded guanine-rich sequences had the ability to form bimolecular G-quadruplex structures. Additionally, the sMtCK and the $\alpha 7$ integrin sequences were capable of adopting hairpin and intramolecular G-quadruplex structures, respectively.[54] In analogy to previous studies with the guanine-rich MCK enhancer sequence, MyoD was shown to bind to all three bimolecular G-quadruplex structures at high affinity (K_d values of ~ 2.5–5.0 nM), whereas it did not detectably bind to the double-stranded forms of the sMtCK and $\alpha 7$ integrin sequences or the hairpin or unimolecular DNA structures of their guanine-rich strand.[55] Further, the affinity of MyoD for the biomolecular G-quadruplex structures of the sMtCK and $\alpha 7$ integrin sequences, as reflected by measured K_d values, was higher by ~ 20–40-fold than its affinity for the MyoD E-box duplex DNA target. As noted by Etzioni *et al.*,[55] the observed bimolecular G-quadruplex structures cannot be formed *in vivo* because each sequence is represented by a single copy in genomic DNA. It was thus proposed that bimolecular

G-quadruplex structures could potentially be generated *in vivo* by the association of two adjacent hairpin structures of two neighboring G-rich sequences within single-stranded DNA. This model was supported to some degree by the demonstration that two representative G-rich oligonucleotides from the mouse α7 integrin promoter, which were located 85 base pairs apart, adopted a heterodimeric G-quadruplex structure.[55]

By virtue of its helix-loop-helix motif, MyoD is able to form homodimers, as well as heterodimers, with other ubiquitously expressed helix-loop-helix proteins, such as E12 and E47. MyoD–E-protein heterodimers bind to E-box DNA more tightly than MyoD homodimers and act as stronger activators of transcription.[56,57] MyoD–E47 heterodimers were found to bind to bimolecular G-quadruplex structures more weakly than MyoD homodimers.[55] Specifically, MyoD heterodimers bound more tightly to E-box DNA than to bimolecular G-quadruplex structures of promoter sequences of the sMtCK and α7 integrin, as reflected by 7–19-fold lower K_d values of the heterodimer–E-box complexes. Etzioni *et al.*[55] hypothesized that the preferential binding of the relatively transcriptionally inactive MyoD homodimers to bimolecular G-quadruplexes may repress untimely activation of muscle-specific genes. Thus, properly timed activation of myogenesis takes place only upon formation of heterodimers that lose their affinity for the G-quadruplex DNA domains and gain increased binding affinity for E-box DNA. A model for switched DNA targets as a mechanism for the regulation of timed myogenic gene activation is shown in Figure 2.

7.2.3 Human c-myc Gene

Perhaps the most well documented gene with the means to regulate transcription through the formation of non-B-DNA structures is the human *c-myc* oncogene. The protein product of the *c-myc* oncogene is a critical transcription factor that regulates a variety of genes associated with cellular proliferation, differentiation, and apoptosis.[58,59] Normally, *c-myc* gene expression is tightly regulated, yet it is well-established that alterations in this regulation are associated with a significant number of human cancers.[58,60] In addition, it has been shown that the *c-myc* oncoprotein has an extremely short half-life relative to other cellular proteins,[61] and as a result *c-myc* transcriptional control has emerged as an attractive target for anticancer therapeutic strategies. The oncogenic properties of *c-myc* can arise from a variety of mechanisms, such as chromosomal translocation[62] and gene amplification,[63] or through simple upregulation of transcription. However, irrespective of how *c-myc* attains its oncogenic properties, the mechanisms behind the altered transcriptional regulation of *c-myc* expression are complex and involve multiple promoters and transcriptional start sites.

The first step to *c-myc* transcriptional activation is change of chromatin structure, which is required to allow access of the transcriptional machinery to the promoter.[64] Several NHEs have been shown to play important roles in this process. Notably, one of these, the NHE III$_1$, has been the focus of considerable research over the past two decades[64–67] because of its major role in controlling

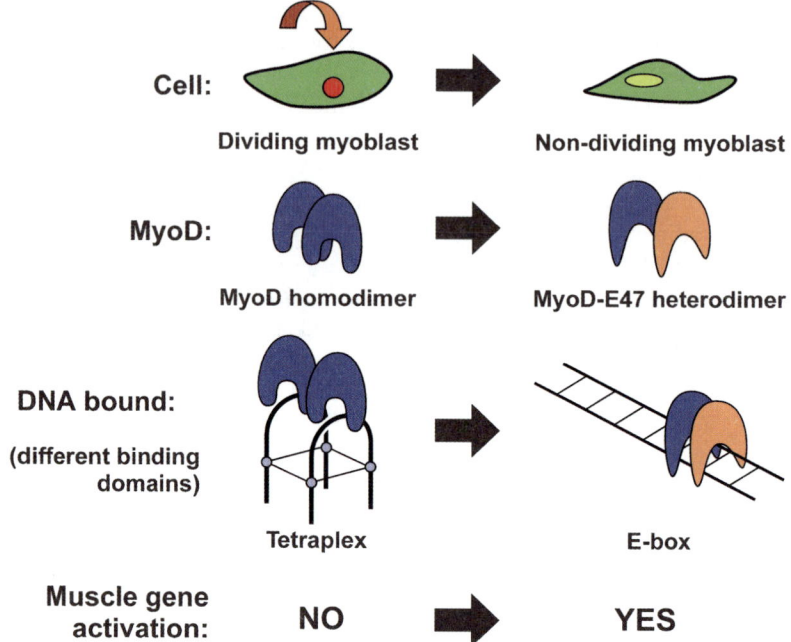

Cell:

Dividing myoblast   Non-dividing myoblast

MyoD:

MyoD homodimer   MyoD-E47 heterodimer

DNA bound:

(different binding domains)

Tetraplex   E-box

Muscle gene activation: NO   YES

Figure 2   *A model for switched DNA targets as a mechanism for the regulation of timed myogenic gene activation (see text for details)*

about 75–85% of the total *c-myc* transcription.[68,69] NHE III$_1$ is a 27-base-pair sequence, located approximately 100 base pairs upstream from the P1 promoter, that is cytosine-rich on the coding strand and guanine-rich on the noncoding strand. Further characterization of this region *in vitro* revealed that it was capable of engaging in a slow equilibrium between a normal duplex helix structure and both unwound and non-B-form regions, which are sensitive to S1 nuclease.[70] The dynamic character of this region resulted in the first attempts to inhibit the activation of *c-myc* transcription through the use of a G-rich oligonucleotide that was targeted to the coding strand of the NHE III$_1$. It was demonstrated that this oligonucleotide could reduce *c-myc* transcription in HeLa cell extracts[71] as well as in live HeLa cells.[72] However, the proposed mechanism of transcription-factor binding inhibition through the formation of an NHE III$_1$ triplex structure could not be established under physiological conditions. Additional *in vitro* analysis suggested that the inability to form triplex structures *in vivo* was due to a competing equilibrium wherein the guanine-rich oligonucleotides formed aggregates involving G-quadruplex structures at physiological K$^+$ concentrations.[73]

As indicated previously, an intramolecular G-quadruplex structure can be formed in a DNA sequence that contains four runs of at least two or more contiguous guanines separated by one or more bases. Specifically, the noncoding guanine-rich strand within the *c-myc* NHE III$_1$ region consists of five runs of three or four contiguous guanines separated by a single adenine or thymine (Table 1) and, therefore, theoretically can form five possible G-quadruplex structures, based on the number of different combinations of four G-tracts. Simonsson *et al.* proposed the first intramolecular G-quadruplex structure within the *c-myc* NHE III$_1$ region.[74] They suggested a model of an antiparallel-stranded structure involving three G-tetrads formed by the two 5'- and 3'-end G-tracts (1–2,4–5; see Figure 3) consisting of two lateral loops and a central diagonal loop. However, further examination by one of our laboratories revealed that the *c-myc* NHE III$_1$ could form at least two distinct intramolecular G-quadruplex structures, which were in equilibrium in K$^+$-containing solution.[75] Using chemical probing, we proposed two models of intramolecular G-quadruplexes, in which both structures used a unique set of four G-tracts. The proposed predominant structure was an antiparallel-stranded G-quadruplex, which was formed by the four 3'-end G-tracts (1–4; see Figure 3) and involved three G-tetrads and three lateral loops, while the minor species was similar to the structure proposed by Simonsson *et al.* These models were based solely upon chemical probing, which can only identify the guanines that are involved in G-tetrad formation and not the orientation of the strands within a given G-quadruplex structure. Therefore, further investigation was required to define the structural complexity of the G-quadruplexes formed within the *c-myc* NHE III$_1$ region.

Comparative CD revealed that the spectra of the HIV aptamer T30695 and a 27-base-pair oligonucleotide corresponding to the G-rich strand of the *c-myc* NHE III$_1$ (Pu27-mer) were relatively comparable, while the thrombin-binding aptamer resulted in a different CD signature.[76] Specifically, the thrombin-binding aptamer had an absorbance maximum at 295 nm, which is in agreement with an antiparallel structure, while the T30695 and the *c-myc* Pu27-mer exhibited maxima at 262 nm, which is consistent with a parallel structure.[77] The tentative conclusion from the CD spectra examination was the existence of parallel-type G-quadruplex structures within the *c-myc* NHE III$_1$ region. Next, NMR studies by Phan *et al.*[78] were employed to define the folding of the G-quadruplex structures within the wild-type *c-myc* Pu27-mer sequence. Initial spectra showed broad envelopes of proton signals, which indicated the presence of multiple G-quadruplex forms. However, a truncated sequence using the four 3'-end G-tracts (1–4) gave a well-resolved spectrum from which a parallel folding pattern was determined.[78]

For a given set of G-tracts an additional level of complexity can occur, which is derived from a particular set of four G-tracts containing disproportionate numbers of guanine bases. For example, the intramolecular G-quadruplex structure formed by the four 3'-end G-tracts (1–4) is limited to having only three G-tetrads, due to the occurrence of two G-tracts containing only three guanine bases. The remaining two G-tracts containing four guanine bases can lead to redundancy and four possible loop isomers, based on the principle that the guanines located at the 3'- or 5'-ends of these redundant G-tracts could assume either loop or tetrad positions. Therefore, to obtain defined loop isomers, dual G-to-T mutations within these two redundant G-tracts are

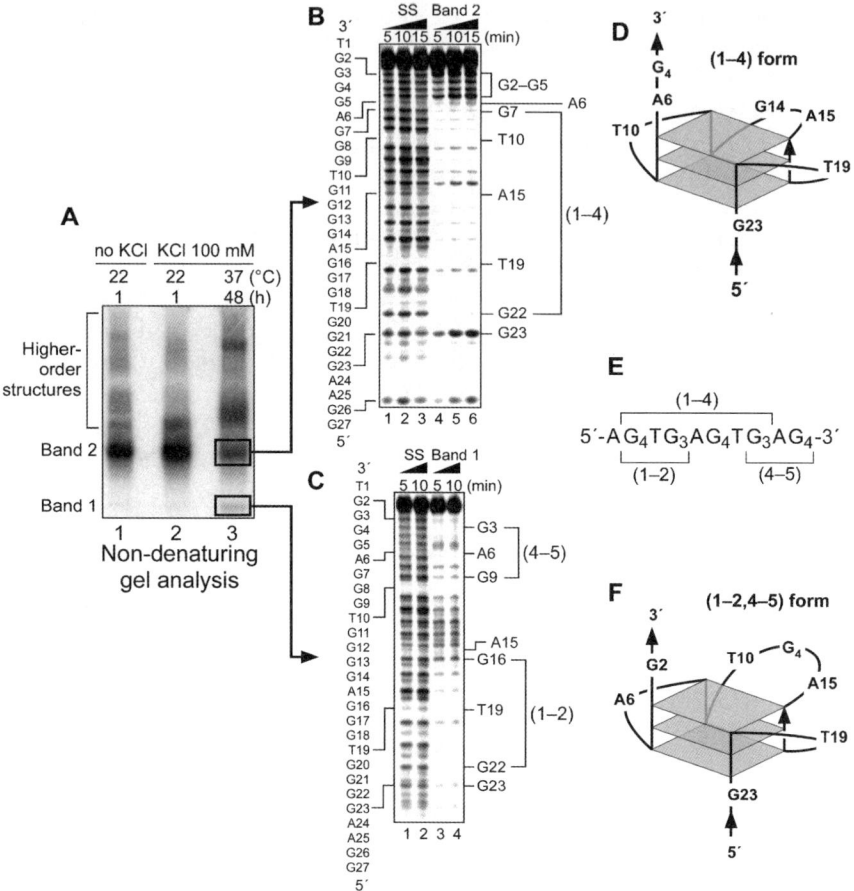

Figure 3 *(A) Nondenaturing gel analysis of Pu27. (B) DMS footprinting (lanes 4–6) of band 2 in (A). Controls (lanes 1–3) show the results of DMS treatment of the denatured Pu27. The Pu27 base sequence is shown to the left of the left panel. (C) As in (B), but for band 2 in (A). (D) NMR-determined folding pattern of the G-quadruplex from band 1. (E) Sequence of the NHE III₁, showing the numbering of guanine runs in B–D and F. The numbers 1–5 represent the individual runs of guanines, numbered 5′ to 3′ in the Pu27. (F) NMR-determined folding pattern of the G-quadruplex from band 1*[75,78,79]

required. This led to an NMR spectrum from which a structure of a single-loop isomer was attained for the four 3′-end G-tracts (1–4) of the truncated Pu27-mer (Figure 4).[79] Last, a parallel-type folding pattern for the previously described minor species was obtained through four specific G-to-T mutations to the central G-tract of the full-length Pu27-mer.[78] Notably, both of these structures were consistent with the previously reported chemical probing data. Overall, these results provided convincing evidence for the presence of two parallel-stranded G-quadruplex forms within the *c-myc* NHE III₁ region.

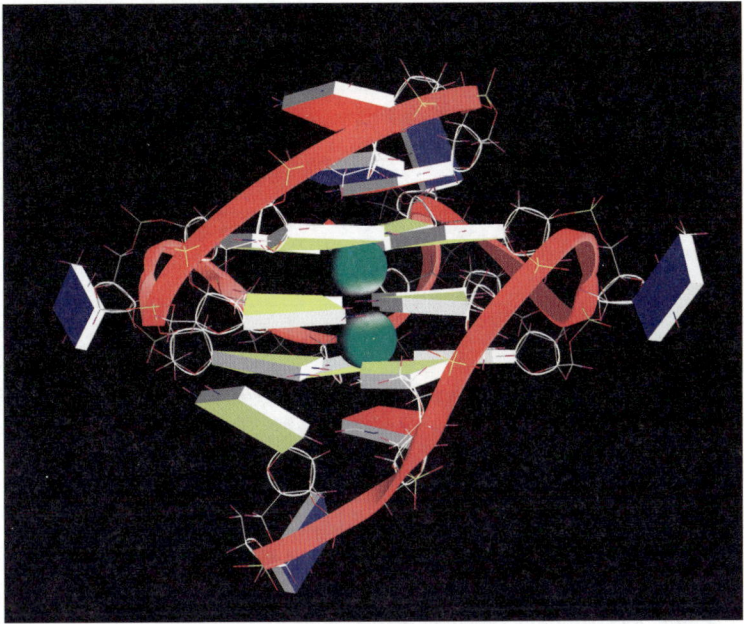

Figure 4 *A representative model of the NMR-refined MYC22 parallel-stranded G-quadruplex structure. Dual G-to-T mutations at positions 14 and 23 were used to attain an NMR structure*[78]

Specifically, the structure formed by the four 3'-end G-tracts (1–4), which is in dynamic equilibrium between different loop isomers, is the kinetically favored structure, while the thermodynamically favored structure is that formed by the two 3'- and 5'-end G-tracts (1–2,4–5) containing a large six-nucleotide central loop (Figure 3).

Prior to structural determination, *in vivo* studies by Michelotti *et al.* suggested that the *c-myc* NHE III₁ region assumes at least three different states, two that are transcriptionally active forms and a third that has a topologically distinct open but repressed form.[80] Consequently, the observation that G-quadruplex structures were readily generated by the single-stranded guanine-rich strand within the *c-myc* NHE III₁ region led to the proposal that these structures could assume the role of a silencer element to repress *c-myc* transcription. In order to test this hypothesis, potential destabilizing mutations were introduced into the G-quadruplex-forming region and evaluated for both promoter activity and their ability to form G-quadruplex structures.[75] It was found that two specific G-to-A single-base mutations, which destabilized the kinetically favored G-quadruplex structure, resulted in a 3-fold increase in *c-myc* promoter activity. To further link the repressing properties and the ability to form a G-quadruplex structure within the NHE III₁ region of the *c-myc* promoter, the Hurley lab demonstrated that the G-quadruplex-interactive agent TMPyP4 was able to decrease *c-myc* expression both at the mRNA

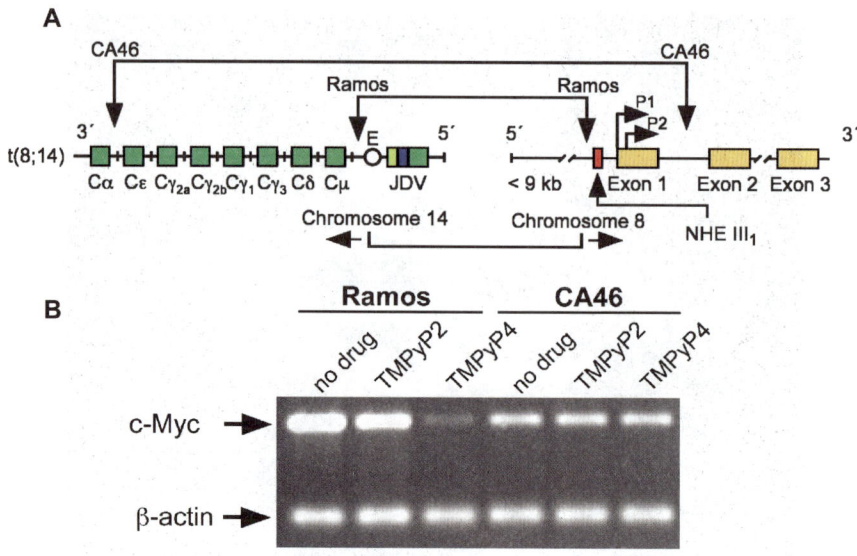

Figure 5   *(A) Diagram of the rearrangements involved in the Ramos and CA46 Burkitt's lymphoma cell lines. Vertical arrows indicate the breakage and rejoining points between chromosomes 14 and 8 for each translocation. (B) RT-PCR for c-myc and β-actin in Ramos (lanes 1–3) and CA46 (lanes 4–6) cell lines after no treatment (lanes 1 and 4) and treatment with 100 μM TMPyP2 (lanes 2 and 5) and TMPyP4 (lanes 3 and 6) for 48 h. The experiment was repeated, with comparable results[75]*

and protein levels, as well as lower the level of several *c-myc*-regulated genes.[81] By contrast, TMPyP2, a structural isomer of TMPyP4, which lacks the ability to interact with the silencer element,[82] had a much reduced effect on *c-myc* transcription. Last, the Hurley lab assessed the requirement for the NHE III$_1$ region for the down regulation of *c-myc* by TMPyP4 through the use of the two Burkitt's lymphoma cell lines Ramos and CA46, which respectively have retained or lost the NHE III$_1$ region because of different translocation break points.[75] As predicted, when the NHE III$_1$ was deleted, as in the CA46 cell line, TMPyP4 had no effect on *c-myc* expression, whereas in the Ramos cell line, in which the NHE III$_1$ was present, TMPyP4 lowered *c-myc* transcriptional activation (Figure 5). Taken together, these results provided convincing evidence that specific G-quadruplex structures within the NHE III$_1$ of the *c-myc* promoter represent the silenced state of the gene.

The discovery of both DNA secondary-structure formation within the NHE III$_1$ region of the *c-myc* promoter and proteins that bind specifically to the cytosine-rich strand, guanine-rich strand, or the duplex of the same region has led to a model of how the NHE III$_1$ controls *c-myc* expression (Figure 6). The model suggests the presence of three different DNA structural populations within the NHE III$_1$: two that cause activation and one that results in repression of *c-myc* transcription.[80] First, when the NHE III$_1$ assumes a normal

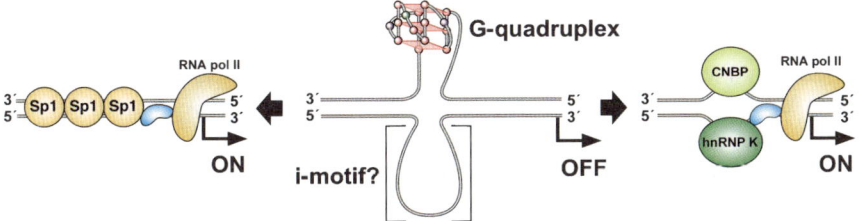

Figure 6 *Model of the three different DNA structural populations within the c-myc NHE III₁. Two cause activation through the binding of transcription factors and one causes repression through the formation of DNA secondary structures (i.e., G-quadruplex and i-motif)*[75]

duplex B-form conformation, it contains several binding sites for Sp1, the well-known zinc finger transcription factor, which has been subsequently demonstrated to bind to the NHE III₁ and promote *c-myc* transcription.[83] Second, the NHE III₁ is also capable of forming a denatured, or open, form, which is involved in the activation of *c-myc* transcription due to the recognition and co-regulation by two single-stranded binding proteins. Specifically, cellular nucleic acid binding protein (CNBP) binds to the guanine-rich strand of the NHE III₁ opposite to heterogeneous ribonucleoprotein K (hnRNP K), which binds to the cytosine-rich strand.[84,85] Last, the G-quadruplex structures and the possibly equivalent i-motif structures formed in the complementary strand represent the third and final structural form within the NHE III₁, leading to silencing of *c-myc* expression. Overall, both the folded and unfolded form of the NHE III₁ are significant components of *c-myc* transcriptional control, which could be orchestrated by the presentation of different molecular topologies as transcription-factor-binding sites.

7.3 DNA Tandem Repeat Sequences

Tandemly repetitive DNA sequences are found ubiquitously in the genomes of nearly all eukaryotic species.[86] Approximately 25% of the human genome consists of repetitive DNA.[86,87] These repetitive elements differ by their position in the genome, sequence, size, number of copies, and content or lack thereof of coding regions. These sequences were originally referred to as "junk DNA" because it was believed they had no biological functions.[88] However, accumulating evidence now indicates that at least some of these repeat sequences are functionally significant within the genome. Depending on the repeat sequence composition, some guanine-rich repeat sequences are archetypical candidates for the formation of G-quadruplex structures. Several studies have shown that the biological function of particular DNA repeat sequences may depend on their capacity to form such structures.

In general, these DNA repeat sequences can be classified into two categories based on their locations within the genome.[89] The first class includes those DNA repeats that possess a biological function but are not associated with genes. For instance, a hairpin structure formed by the human centromere pentanucleotide repeat d(AATGG) has been proposed to provide specific recognition sites for the kinetochore during mitosis.[90] Also, the hexanucleotide repeat d(TTAGGG)$_n$ at the 3'-end DNA overhang of the human telomere readily folds into G-quadruplex structures, which are believed to be involved in the maintenance of chromosomal integrity.[91] The second category of DNA repeat sequences consists of sequences that are located in regulatory regions of genes. These repeats are generally referred to as microsatellites or minisatellites, depending on the repeat unit length. Recently there has been increased interest in sequences of these types because of their propensity to expand, leading to abnormal expression of the associated gene. Coupled alterations in repeat length and gene expression are associated with more than 40 human diseases.[92,93] Two well-studied examples, which will be discussed in more detail, are the expansion of the trinucleotide d(CGG)$_n$ repeat sequence located at the transcribed but untranslated 5'-end of the *FMR1* (fragile X mental retardation 1) gene and the insulin-linked polymorphic region (ILPR) consensus repeat sequence d(ACAGGGGTGTGGGG), located in the 5'-flanking region of the human insulin gene (see Table 1).

7.3.1 Human *FMR1* Gene

Fragile X syndrome is the most common form of inherited mental-retardation disorder.[94] It is associated with dynamic expansion of a d(CGG) trinucleotide repeat located in the 5'-untranslated region of the first exon of the *FMR1* gene.[95–97] The number of d(CGG) repeats in normal alleles is 2 to 52 repeats, whereas asymptomatic or minimally affected bearers of a premutation have 55 to 200 repeats, and affected individuals with a full mutation have >200 to 2000 repeats.[98] A full mutation of the d(CGG) repeat element is accompanied by hypermethylation of adjacent CpG islands in the promoter and in the repeat region itself and by histone deacetylation, leading to the suppression of *FMR1* transcription.[99,100] Initial indication of an altered chromatin structure at the 5'-end of the *FMR1* gene was derived from the loss of acetylation of histones H3 and H4[101] and the gain of nuclease sensitivity[102] in fragile X patient cell lines containing the d(CGG) repeat expansion. Indeed, treatment of fragile X cells with inhibitors of histone deacetylase and demethylation of the DNA by 5-aza deoxycytidine led to partial restoration of *FMR1* mRNA synthesis, although its protein product (FMRP) remained minimal.[103] However, these results suggested that full reactivation of *FMR1* expression and FMRP synthesis requires the removal of at least one additional inhibitory level of regulation.

The potential for the formation of unusual DNA structures in the *FMR1* trinucleotide repeat has been recognized and suggested as a contributing factor in the expansion of fragile X syndrome and, accordingly, the silencing of the *FMR1* gene. Several lines of evidence reveal that d(CGG)$_n$ oligomers can fold into unimolecular hairpins.[104–108] In addition, d(CGG)$_n$ oligomers have been shown to assemble into bimolecular and tetramolecular G-quadruplex structures.[109–111] These types of DNA secondary structures were found to block the progression of DNA polymerase *in vitro*,[109,112] raising the prospect that they may trigger polymerase slippage or displacement synthesis of an Okazaki fragment, resulting in expansion of the trinucleotide repeat sequence. Subsequent to expansion, the stabilization of these unusual DNA structures could facilitate the impediment of *FMR1* transcription, the activation of hypermethylation, and the clinical phenotype of fragile X syndrome.

Genotypically, the fragile X trinucleotide repeating unit is d(CGG), yet the presence of stable and unstable alleles of similar sizes led to the proposal that a feature other than repeat length may be involved in stability and repeat expansion. Stable alleles were found to have d(AGG) triplets at an average periodicity of 10 repeats, whereas interspersed d(AGG) triplets have been shown to be partially or completely missing in individuals with premutation and are completely absent in patients with full mutation.[113,114] Further analysis suggested that a threshold of approximately 33 uninterrupted d(CGG) repeats denotes the start of instability.[115] In addition, d(CGG) repeats containing interspersed d(AGG) triplets reduced the formation of slipped-strand DNA structures compared to uninterrupted d(CGG) repeats,[116] suggesting the existence of pure d(CGG) repeats as a precursor to repeat expansion. These results demonstrate dependence on both the repeat length and its monotony for repeat expansion and instability. This additional dependency on the composition of the trinucleotide repeat sequence initiated the investigation of the affect of interspersed d(AGG) triplets on the formation and stability of $d(CGG)_n$ quadruplex structures. Usdin reported that $d(CGG)_{20}$ and $d(AGG)_{20}$ oligomers could adopt intramolecular G-quadruplex structures with comparable stability; in contrast, the $d(CGG)_{20}$ oligomer formed much more stable hairpin structures than the $d(AGG)_{20}$ oligmer.[117] Additional studies showed that there was a reduction in the formation and the thermal stability of bimolecular G-quadruplex structures utilizing oligomers containing interspersed d(AGG) within $d(CGG)_{18}$ at intervals greater or equal to the average physiological frequency compared to an uninterrupted $d(CGG)_{18}$ oligmer.[118] Therefore, due to their G-quadruplex destabilizing effect, the interfering d(AGG) triplets could potentially prevent repeat expansion and the ensuing transcriptional repression of *FMR1*.

The final event in the transmission of fragile X syndrome is the abnormal methylation of the *FMR1* promoter and correlated gene inactivation.[119] The rate of methylation is determined by the initial substrate–enzyme recognition. DNA methyltransferases are remarkably better at recognizing and methylating DNA sites undergoing local conformational changes, such as hairpins or G-quadruplex structures, relative to B-form duplex DNA.[120,121] Methylation at the 5 position of cytosine has been shown to stabilize G-quadruplex structures by producing a "flipped out" cytosine, unstacked from its normal position in B-DNA.[122] Specifically, methylation of cytosine residues in a short $d(^mCGG)_n$ tract has been shown to stabilize the formation of a tetramolecular G-quadruplex structure.[111] Therefore, if formed, these DNA secondary structures could provide a greater number of high-affinity methylation sites for methyltransferases, resulting in hypermethylation of the fragile X repeat and subsequent suppression of the

FMR1 gene. By this token, instability or complete absence of secondary structures of the shorter d(CGG)$_n$ tracts in normal or premutation alleles prohibit methylation.

Supplementary support for the existence *in vivo* of G-quadruplex structures of the d(CGG)$_n$ repeat sequence was provided by the discovery of proteins with the ability to stabilize or destabilize G-quadruplex forms of this repeat sequence. Specifically, human Werner syndrome DNA helicase[123] and two hnRNP murine telomeric DNA binding proteins[124,125] have been shown to unwind d(CGG)$_n$ bimolecular G-quadruplex structures and potentially alleviate the blocking of replication and/or transcription of the *FMR1* gene. In addition, the G-quadruplex-interactive agent TMPyP4, which had previously been shown to stabilize G-quadruplex structures of telomeric DNA, has been shown to conversely destabilize *in vitro* bimolecular G-quadruplex structures of d(CGG)$_n$.[126] In contrast, the human Ku antigen was reported to bind to and stabilize bimolecular G-quadruplex structures of d(CGG)$_n$ and to protect them from digestion and unwinding by nucleases and destabilizing proteins, respectively.[127] The discovery of proteins that stabilize or destabilize G-quadruplex structures of d(CGG)$_n$ adds to the complexity in the regulation of the expression of the *FMR1* gene. Overall, controversy still remains whether hairpins or G-quadruplexes constitute the predominant secondary structure of the d(CGG)$_n$ sequence. However, there is an agreement that secondary structures formed by this repeat sequence are at the root of its expansion and the ensuing silencing of *FMR1* transcription.

7.3.2   Human Insulin Gene

Insulin-dependent diabetes mellitus (IDDM) is associated with a deficiency in insulin production or ineffective insulin. Susceptibility to IDDM is complex, involving both genetic and environmental factors. Intriguingly, one of the genetic components implicated in IDDM susceptibility is the insulin minisatellite located on chromosome 11p15, 363 base pairs upstream from the human insulin gene transcription start site.[128] The location of the ILPR element at the proximal promoter region of the insulin gene suggested that it functions as a transcriptional regulator involved in the regulation of normal insulin production and its deterioration in IDDM. This element is highly polymorphic in human populations, displaying variations in both length and sequence.[129] Human ILPRs are grouped into three classes according to length: class I ($\sim$40 repeats), class II ($\sim$85 repeats), and class III ($\sim$150 repeats).[130] Moreover, at least eleven different minor sequence variants related to the ILPR consensus sequence d(ACAGGGGTGTGGGG) have been reported and termed "a" through "k."[131] The genomic location of the ILPR element as well as its guanine-rich sequence raised the possibility that the transcriptional activity of the insulin gene may be modulated by the quaternary DNA topology of the ILPR.

Preliminary evidence of an altered DNA structure within the ILPR element was first discovered by Hammond-Kosack *et al.* through the use of supercoiled plasmids containing a 5-kb genomic insulin insert and employment of probes to query DNA structure.[132] The chemical probe bromoacetaldehyde and the enzymatic probe S1 nuclease produced multiple hypersensitive sites within the insulin fragment, with the major site corresponding to the ILPR. It is noteworthy that the generation of these hypersensitive sites depended on torsional stress in the DNA such that no nuclease reactivity was observed in linearized plasmids. Further evidence for an altered DNA structure at this region was provided by electron microscopic homoduplex analysis. Results revealed looping-out of the two strands in 30% of the sequences tested following denaturation and renaturation of ILPR-containing DNA fragments. Specifically, one strand within this looped-out region was consistently 50% the length of the other strand, suggesting that one strand adopted a compact structure [Figure 7(A)]. In a more physiological system, in which the DNA was assembled into chromatin *in vitro* using histones and a *Xenopus laevis* unfertilized egg extract, a single P1 nuclease hypersensitive site was created within the ILPR, suggesting that nucleosomal structure formation was inhibited in the ILPR region, supposedly because of the formation of an unusual DNA structure.[133]

In view of the existence of two contiguous runs of four guanines per repeat and the nature of the repeat length of the ILPR consensus sequence, it is anticipated that the guanine-rich strand within the ILPR has the capacity to form multiple G-quadruplex structures. Premature pausing of DNA polymerase upstream to every guanine tract and resistance of the guanine residues to methylation by dimethyl sulfate are in line with the formation of intramolecularly folded G-quadruplex structures within the G-rich strand of the ILPR.[132] NMR-based modeling of two consecutive repeats of the ILPR consensus motif d(ACAGGGGTGTGGGG) confirmed the formation of an antiparallel intramolecular G-quadruplex structure containing four stacked G-tetrads and three lateral loops, consisting of three residues each [Figure 7(B)].[134] NMR analysis, CD spectroscopy, and nondenaturing gel electrophoresis demonstrated that the intramolecular structure is very stable, with a melting temperature of 85°C. This DNA species required a minimum number of two ILPR repeats to produce an individual folded structure.[134] Conversely, in an *in vitro* replication assay in the presence of the complementary cytosine-rich strand, replication was completely arrested using a unit number of ≥ 6, whereas two repeats produced only a minor arrest. This result suggested that individual G-quadruplex structures assembled into higher-order structures or intermolecular G-quadruplex structures.[89] Such structures had previously been shown to exist in two- and four-stranded forms of a single ILPR repeat unit.[135] With a model NMR structure available of the ILPR G-quadruplex structure, the analysis of key interactions that stabilize or destabilize the structure became feasible. Thus, the relation between the stability of the ILPR G-quadruplex structure *in vitro* and its effect on transcriptional activity *in vivo* was investigated.

Initial studies to examine the ILPR element as a transcriptional regulatory region in the human insulin gene indicated that it affected only minimally the total transcriptional activity of the 5'-flanking region and that other sequences

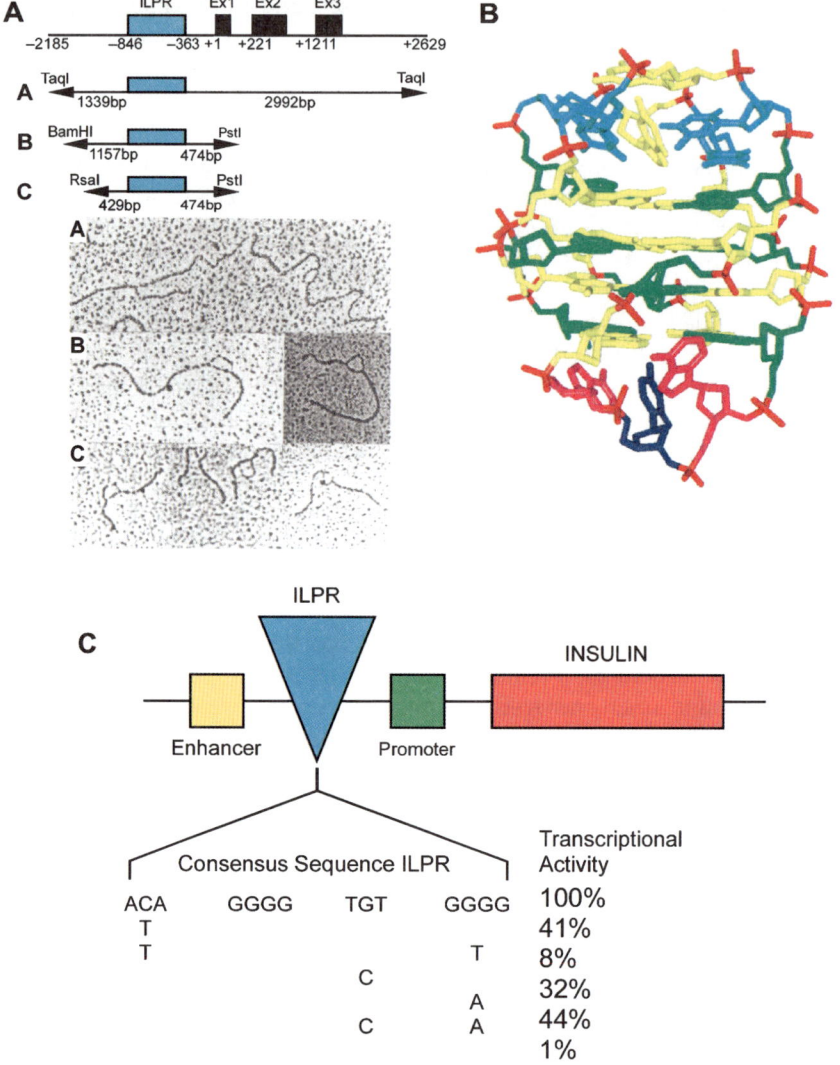

Figure 7 *(A) Electron microscopic homoduplex analysis of DNA fragments containing the ILPR. A, B, and C refer to the three different fragments used.[132] (B) NMR-based model of an intramolecular G-quadruplex formed by two consecutive repeats of the ILPR consensus motif d(ACAGGGGTGTGGGG) (guanines are shown in green).[134] (C) Transcriptional activity of individual ILPR polymorphic repeats[134,137]*

(A reprinted from reference 132, with permission from Oxford University Press, B and C reprinted from reference 134, with permission from Elsevier)

played a much larger role.[136] However, these studies were carried out using the only human insulin gene available at the time, which was an allele containing a class I ILPR sequence. Upon the discovery of expanded ILPR repeats, studies carried out by Kennedy *et al.* revealed that a class III ILPR allele exhibited more than 2-fold the transcriptional activity of a class I ILPR allele.[137] These results were reproducible and were amplified using heterologous promoters. In addition, co-transfection of ILPR constructs with the transcription factor purine-rich element binding protein A (Pur-1), which has been shown to bind to ILPR at a high affinity,[138] resulted in greater activation of transcription. All the described experiments were carried out using the ILPR consensus motif, yet sequence heterogeneity can also occur between individuals. At least eleven different variants containing single, double, or even triple nucleotide changes have been shown to exist. Kennedy *et al.* discovered that these ILPR variants were bound by Pur-1 with different affinities.[137] This could potentially lead to a differential effect on transcription. Similar to the ILPR length heterogeneity studies, these authors performed co-transfection experiments to examine if transcriptional activation by the different ILPR variants was consistent with the respective *in vitro*-measured Pur-1 binding affinities. For most of the repeat variants, it was found that *in vivo* transcriptional activity was positively correlated with the tightness of Pur-1 binding *in vitro*. However, such positive correlation did not exist at all for some repeat variants, such as "h" and "k." It was postulated, therefore, that other factors, such as the propensity of ILPR variants to form G-quadruplex structures, might contribute to the overall transcriptional activity of the ILPR.

The ability of the ILPR to form an inter- or intramolecular G-quadruplex structure is dependent on the repeat-sequence composition. Therefore, the nucleotide changes observed within some of the ILPR repeat variants may affect their ability to adopt stable DNA secondary structures. This proposition was tested directly and indirectly using native gel electrophoresis and NMR analysis, respectively. Interestingly, repeat variants that were predicted *in silico* and demonstrated *in vitro* to destabilize the G-quadruplex structure were shown to be associated with the level of transcriptional activity *in vivo* [Figure 7(C)].[134,139] For example, any repeat variant containing an interruption within the G-tracts of the ILPR sequence resulted in both destabilization of the G-quadruplex structure and a decrease in transcription. In addition, NMR studies revealed that the guanine residue in the TGT loop contains loop–loop and loop–tetrad interactions, and in fact, mutations within this loop also decreased the transcriptional activity.[134] Consequently, introduction of counteracting mutations into low transcriptional ILPR repeat variants, which restored the ability of the repeats to adopt a G-quadruplex structure and bind to Pur-1, resulted in the regeneration of high transcriptional activity of the mutated ILPR repeats.[139]

In contrast to the negative regulatory function of G-quadruplex formation in the repeat element of the *FMR1* gene, inter- or intramolecular G-quadruplexes in the ILPR element serve as positive *cis*-acting transcriptional regulators of the human insulin gene through recognition by Pur-1. The length and sequence heterogeneity of the ILPR contribute to the overall transcriptional activity based on the quantity (number) and quality (stability) of the G-quadruplexes formed. In addition, these studies have provided a specific model of the interplay between chromatin structure and torsional stress, as well as the G-quadruplex-forming ability and the binding of transcription factors in the regulation of gene expression, although it is still not clear if there is a relationship between transcriptional activity and genetic susceptibility to IDDM. These studies demonstrated that both inter- and intramolecular G-quadruplex formation in the ILPR can influence transcriptional activity of the human insulin gene and thus may potentially contribute to that portion of diabetes susceptibility attributed to the ILPR minisatellite locus.

7.4 Summary

A bioinformatics search of the human genome for the simplest G-quadruplex-forming motif exposed a vast number of sequences that can potentially fold into G-quadruplex structures.[140] However, no similar genome-wide search has yet been conducted for G-quadruplex-forming motifs located in gene promoter, enhancer, and silencer regions. Nevertheless, in addition to the examples described here in detail, the existence of polypurine/polypyrimidine tracts has been identified in several other mammalian gene-promoter regions, such as androgen receptors,[141] Hmga2,[22] c-Scr,[26] TGF-β,[25] c-Ki-Ras,[23] PDGF-A-chain,[142] VEGF,[143] Hif-1α,[144] c-kit,[145] and bcl-2.[146] For most of these genes, however, the formation of a G-quadruplex structure and its function in gene transcription have not been investigated. In general, G-quadruplexes formed in promoter regions of genes such as *c-myc* and ILPR have very high melting points. This suggests that once they form, these structures will prevail unless there are specific helicases or G-quadruplex-destabilizing proteins that convert them back to single- and double-stranded structures. A second consequence of the formation of tetrahelical structures by guanine-rich DNA strands is that the cytosine-rich strands should loop out, a phenomenon observed with the ILPR sequence [Figure 7(B)]. If these cytosine-rich strands are able to form DNA secondary structures, such as i-motifs, then they could also act as transcriptional modulating elements. Overall, the insight gained from the fundamental properties of the diverse molecular topologies of DNA in comparison to duplex DNA has made them appealing potential contributors to the selective modulation of gene transcription and possible targets for drug design.

References

1. F. Crick, *Nature*, 1970, **227**, 561.
2. D. Latchman, *Eukaryotic Transcription Factors*, fourth edition, Elsevier Academic Press, London, 2004, pp 1–22.
3. P.J. Mitchell and R. Tjian, *Science*, 1989, **245**, 371.
4. B. Lewin, *Gene VII*, Oxford University Press and Cell Press, New York, 2000, pp 233–271.
5. J.C. Venter *et al*, *Science*, 2001, **291**, 1304.
6. R.H. Garrett and C.M. Grisham, *Biochemistry*, second edition, Saunders College Publishing, Fort Worth, 1999, pp 1014–1046.
7. E.H. Davidson, H.T. Jacobs and R.J. Britten, *Nature*, 1983, **301**, 468.
8. H.R. Pelham, *Cell*, 1982, **30**, 517.

9. P. Ridgway and G. Almouzni, *J. Cell. Sci.*, 2001, **114**, 2711.

10. G.J. Narlikar, H.Y. Fan and R.E. Kingston, *Cell*, 2002, **108**, 475.

11. C.J. Fry and C.L. Peterson, *Curr. Biol.*, 2001, **11**, R185.

12. A. Bird, *Genes Dev.*, 2002, **16**, 6.

13. P.A. Jones and S.B. Baylin, *Nat. Rev. Genet.*, 2002, **3**, 415.

14. E. Heard, P. Clerc and P. Avner, *Annu. Rev. Genet.*, 1997, **31**, 571.

15. S.B. Baylin and J.G. Herman, *Trends Genet.*, 2000, **16**, 168.

16. P.A. Jones and P.W. Laird, *Nat. Genet.*, 1999, **21**, 163.

17. K.D. Robertson, *Oncogene*, 2002, **21**, 5361.

18. C.P. Liang and W.T. Garrard, *Mol. Cell. Biol.*, 1997, **17**, 2825.

19. M.W. Leonard and R.K. Patient, *Mol. Cell. Biol.*, 1991, **11**, 6128.

20. G. Pan and J. Greenblatt, *J. Biol. Chem.*, 1994, **269**, 30101.

21. F.C. Holstege, D. Tantin, M. Carey, P.C. van der Vliet and H.T. Timmers, *EMBO J.*, 1995, **14**, 810.

22. A. Rustighi, M.A. Tessari, F. Vascotto, R. Sgarra, V. Giancotti and G. Manfioletti, *Biochemistry*, 2002, **41**, 1229.

23. D.G. Pestov, A. Dayn, E. Siyanova, D.L. George and S.M. Mirkin, *Nucleic Acids Res.*, 1991, **19**, 6527.

24. A.C. Johnson, Y. Jinno and G.T. Merlino, *Mol. Cell. Biol.*, 1988, **8**, 4174.

25. R. Lafyatis, F. Denhez, T. Williams, M. Sporn and A. Roberts, *Nucleic Acids Res.*, 1991, **19**, 6419.

26. S. Ritchie, F.M. Boyd, J. Wong and K. Bonham, *J. Biol. Chem.*, 2000, **275**, 847.

27. T. Simonsson, *Biol. Chem.*, 2001, **382**, 621.

28. M.A. Keniry, *Biopolymers*, 2000, **56**, 123.

29. J.L. Huppert and S. Balasubramanian, *Nucleic Acids Res.*, 2005, **33**, 2908.

30. H. Weintraub, H. Beug, M. Groudine and T. Graf, *Cell*, 1982, **28**, 931.

31. H. Weintraub, *Nucleic Acids Res.*, 1979, **7**, 781.

32. A. Larsen and H. Weintraub, *Cell*, 1982, **29**, 609.

33. B.M. Emerson and G. Felsenfeld, *Proc. Natl. Acad. Sci. USA*, 1984, **81**, 95.

34. W.I. Wood and G. Felsenfeld, *J. Biol. Chem.*, 1982, **257**, 7730.

35. T. Kohwi-Shigematsu, R. Gelinas and H. Weintraub, *Proc. Natl. Acad. Sci. USA*, 1983, **80**, 4389.

36. J.D. McGhee, W.I. Wood, M. Dolan, J.D. Engel and G. Felsenfeld, *Cell*, 1981, **27**, 45.

37. J.M. Nickol and G. Felsenfeld, *Cell*, 1983, **35**, 467.

38. E. Schon, T. Evans, J. Welsh and A. Efstratiadis, *Cell*, 1983, **35**, 837.

39. P.D. Jackson and G. Felsenfeld, *Proc. Natl. Acad. Sci. USA*, 1985, **82**, 2296.

40. Y. Kohwi, *Nucleic Acids Res.*, 1989, **17**, 4493.

41. K.J. Woodford, R.M. Howell and K. Usdin, *J. Biol. Chem.*, 1994, **269**, 27029.

42. D.T. Weaver and M.L. DePamphilis, *J. Mol. Biol.*, 1984, **180**, 961.

43. N. Baran, A. Lapidot and H. Manor, *Proc. Natl. Acad. Sci. USA*, 1991, **88**, 507.

44. C. Giovannangeli, N.T. Thuong and C. Helene, *Proc. Natl. Acad. Sci. USA*, 1993, **90**, 10013.
45. R.M. Howell, K.J. Woodford, M.N. Weitzmann and K. Usdin, *J. Biol. Chem.*, 1996, **271**, 5208.
46. S.P. Clark, C.D. Lewis and G. Felsenfeld, *Nucleic Acids Res.*, 1990 **18**, 5119.
47. C.D. Lewis, S.P. Clark, G. Felsenfeld and H. Gould, *Genes Dev.*, 1988 **2**, 863.
48. C.A. Berkes and S.J. Tapscott, *Semin. Cell. Dev. Biol.*, 2005, **16**, 585.
49. L.A. Sabourin and M.A. Rudnicki, *Clin. Genet.*, 2000, **57**, 16.
50. A.B. Lassar, J.N. Buskin, D. Lockshon, R.L. Davis, S. Apone, S.D. Hauschka and H. Weintraub, *Cell*, 1989, **58**, 823.
51. T.K. Blackwell and H. Weintraub, *Science*, 1990, **250**, 1104.
52. I.M. Santoro, T.M. Yi and K. Walsh, *Mol. Cell. Biol.*, 1991, **11**, 1944.
53. K. Walsh and A. Gualberto, *J. Biol. Chem.*, 1992, **267**, 13714.
54. A. Yafe, S. Etzioni, P. Weisman-Shomer and M. Fry, *Nucleic Acids Res.*, 2005, **33**, 2887.
55. S. Etzioni, A. Yafe, S. Khateb, P. Weisman-Shomer, E. Bengal and M. Fry, *J. Biol. Chem.*, 2005, **280**, 26805.
56. A.B. Lassar, R.L. Davis, W.E. Wright, T. Kadesch, C. Murre, A. Voronova, D. Baltimore and H. Weintraub, *Cell*, 1991, **66**, 305.
57. C. Murre, P.S. McCaw, H. Vaessin, M. Caudy, L.Y. Jan, Y.N. Jan, C.V. Cabrera, J.N. Buskin, S.D. Hauschka, A.B. Lassar, H. Weintraub and D. Baltimore, *Cell*, 1989, **58**, 537.
58. K.B. Marcu, S.A. Bossone and A.J. Patel, *Annu. Rev. Biochem.*, 1992 **61**, 809.
59. S. Pelengaris, B. Rudolph and T. Littlewood, *Curr. Opin. Genet. Dev.*, 2000, **10**, 100.
60. C.A. Spencer and M. Groudine, *Adv. Cancer. Res.*, 1991, **56**, 1.
61. B. Luscher and R.N. Eisenman, *Princess Takamatsu Symp.*, 1986, **17**, 291.
62. P.D. Fahrlander, J. Sumegi, J.Q. Yang, F. Wiener, K.B. Marcu and G. Klein, *Proc. Natl. Acad. Sci. USA*, 1985, **82**, 3746.
63. K. Alitalo, M. Schwab, C.C. Lin, H.E. Varmus and J.M. Bishop, *Proc. Natl. Acad. Sci. USA*, 1983, **80**, 1707.
64. U. Siebenlist, L. Hennighausen, J. Battey and P. Leder, *Cell*, 1984 **37**, 381.
65. E.H. Postel, S.E. Mango and S.J. Flint, *Mol. Cell. Biol.*, 1989, **9**, 5123.
66. A.B. Firulli, D.C. Maibenco and A.J. Kinniburgh, *Biochem. Biophys. Res. Commun.*, 1992, **185**, 264.
67. O. Sakatsume, H. Tsutsui, Y. Wang, H. Gao, X. Tang, T. Yamauchi, T. Murata, K. Itakura and K.K. Yokoyama, *J. Biol. Chem.*, 1996, **271**, 31322.
68. S.J. Berberich and E.H. Postel, *Oncogene*, 1995, **10**, 2343.
69. T.L. Davis, A.B. Firulli and A.J. Kinniburgh, *Proc. Natl. Acad. Sci. USA*, 1989, **86**, 9682.
70. T.C. Boles and M.E. Hogan, *Biochemistry*, 1987, **26**, 367.

71. M. Cooney, G. Czernuszewicz, E.H. Postel, S.J. Flint and M.E. Hogan, *Science*, 1988, **241**, 456.
72. E.H. Postel, S.J. Flint, D.J. Kessler and M.E. Hogan, *Proc. Natl. Acad. Sci. USA*, 1991, **88**, 8227.
73. W.M. Olivas and L.J. Maher 3rd, *Biochemistry*, 1995, **34**, 278.
74. T. Simonsson, P. Pecinka and M. Kubista, *Nucleic Acids Res.*, 1998 **26**, 1167.
75. A. Siddiqui-Jain, C.L. Grand, D.J. Bearss and L.H. Hurley, *Proc. Natl. Acad. Sci. USA*, 2002, **99**, 11593.
76. J. Seenisamy, E.M. Rezler, T.J. Powell, D. Tye, V. Gokhale, C.S. Joshi, A. Siddiqui-Jain and L.H. Hurley, *J. Am. Chem. Soc.*, 2004, **126**, 8702.
77. V. Dapic, V. Abdomerovic, R. Marrington, J. Peberdy, A. Rodger, J.O. Trent and P.J. Bates, *Nucleic Acids Res.*, 2003, **31**, 2097.
78. A.T. Phan, Y.S. Modi and D.J. Patel, *J. Am. Chem. Soc.*, 2004, **126**, 8710.
79. A. Ambrus, D. Chen, J. Dai, R.A. Jones and D. Yang, *Biochemistry*, 2005, **44**, 2048.
80. G.A. Michelotti, E.F. Michelotti, A. Pullner, R.C. Duncan, D. Eick and D. Levens, *Mol. Cell. Biol.*, 1996, **16**, 2656.
81. C.L. Grand, H. Han, R.M. Muñoz, S. Weitman, D.D. Von Hoff, L.H. Hurley and D.J. Bearss, *Mol. Cancer Ther.*, 2002, **1**, 565.
82. H. Han, D.R. Langley, A. Rangan and L.H. Hurley, *J. Am. Chem. Soc.*, 2001, **123**, 8902.
83. E. DesJardins and N. Hay, *Mol. Cell. Biol.*, 1993, **13**, 5710.
84. E.F. Michelotti, T. Tomonaga, H. Krutzsch and D. Levens, *J. Biol. Chem.*, 1995, **270**, 9494.
85. M. Takimoto, T. Tomonaga, M. Matunis, M. Avigan, H. Krutzsch, G. Dreyfuss and D. Levens, *J. Biol. Chem.*, 1993, **268**, 18249.
86. R.K. Moyzis, D.C. Torney, J. Meyne, J.M. Buckingham, J.R. Wu, C. Burks, K.M. Sirotkin and W.B. Goad, *Genomics*, 1989, **4**, 273.
87. R.L. Stallings, D.C. Torney, C.E. Hildebrand, J.L. Longmire, L.L. Deaven, J.H. Jett, N.A. Doggett and R.K. Moyzis, *Proc. Natl. Acad. Sci. USA*, 1990, **87**, 6218.
88. L.E. Orgel, F.H. Crick and C. Sapienza, *Nature*, 1980, **288**, 645.
89. P. Catasti, X. Chen, S.V. Mariappan, E.M. Bradbury and G. Gupta, *Genetica*, 1999, **106**, 15.
90. P. Catasti, G. Gupta, A.E. Garcia, R. Ratliff, L. Hong, P. Yau, R.K. Moyzis and E.M. Bradbury, *Biochemistry*, 1994, **33**, 3819.
91. W.E. Wright, V.M. Tesmer, K.E. Huffman, S.D. Levene and J.W. Shay, *Genes Dev.*, 1997, **11**, 2801.
92. I.V. Kovtun, G. Goellner and C.T. McMurray, *Biochem. Cell. Biol.*, 2001, **79**, 325.
93. R.R. Sinden and R.D. Wells, *Curr. Opin. Biotechnol.*, 1992, **3**, 612.
94. R.L. Nussbaum and D.H. Ledbetter, *Annu. Rev. Genet.*, 1986, **20**, 109.
95. I. Oberle, F. Rousseau, D. Heitz, C. Kretz, D. Devys, A. Hanauer, J. Boue, M.F. Bertheas and J.L. Mandel, *Science*, 1991, **252**, 1097.

96. A.J. Verkerk, M. Pieretti, J.S. Sutcliffe, Y.H. Fu, D.P. Kuhl, A. Pizzuti, O. Reiner, S. Richards, M.F. Victoria, F.P. Zhang, B.E. Eussen, G.-J.B. van Ommen, L.A.J. Blanden, G.J. Riggins, J.L. Chastain, C.B. Kunst, H. Galjaard, C.T. Caskey, D.L. Nelson, B.A. Oostra and S.T. Warran, *Cell*, 1991, **65**, 905.

97. S. Yu, M. Pritchard, E. Kremer, M. Lynch, J. Nancarrow, E. Baker, K. Holman, J.C. Mulley, S.T. Warren, D. Schlessinger, G.R. Sutherland and R.I. Richards, *Science*, 1991, **252**, 1179.

98. Y.H. Fu, D.P. Kuhl, A. Pizzuti, M. Pieretti, J.S. Sutcliffe, S. Richards, A.J. Verkerk, J.J. Holden, R.G. Fenwick Jr, S.T. Warren, B.A. Oostra, D.L. Nelson and C.T. Caskey, *Cell*, 1991, **67**, 1047.

99. A. El-Osta, *Biochem. Biophys. Res. Commun.*, 2002, **295**, 575.

100. R.S. Hansen, S.M. Gartler, C.R. Scott, S.H. Chen and C.D. Laird, *Hum. Mol. Genet.*, 1992, **1**, 571.

101. B. Coffee, F. Zhang, S.T. Warren and D. Reines, *Nat. Genet.*, 1999, **22**, 98.

102. D.E. Eberhart and S.T. Warren, *Somat. Cell. Mol. Genet.*, 1996, **22**, 435.

103. P. Chiurazzi and G. Neri, *Brain Res. Bull.*, 2001, **56**, 383.

104. Y. Nadel, P. Weisman-Shomer and M. Fry, *J. Biol. Chem.*, 1995, **270**, 28970.

105. M. Mitas, A. Yu, J. Dill and I.S. Haworth, *Biochemistry*, 1995, **34**, 12803.

106. A.M. Gacy, G. Goellner, N. Juranic, S. Macura and C.T. McMurray, *Cell*, 1995, **81**, 533.

107. X. Chen, S.V. Mariappan, P. Catasti, R. Ratliff, R.K. Moyzis, A. Laayoun, S.S. Smith, E.M. Bradbury and G. Gupta, *Proc. Natl. Acad. Sci. USA*, 1995, **92**, 5199.

108. S.V. Mariappan, P. Catasti, X. Chen, R. Ratliff, R.K. Moyzis, E.M. Bradbury and G. Gupta, *Nucleic Acids Res.*, 1996, **24**, 784.

109. K. Usdin and K.J. Woodford, *Nucleic Acids Res.*, 1995, **23**, 4202.

110. A. Kettani, R.A. Kumar and D.J. Patel, *J. Mol. Biol.*, 1995, **254**, 638.

111. M. Fry and L.A. Loeb, *Proc. Natl. Acad. Sci. USA*, 1994, **91**, 4950.

112. S. Kang, K. Ohshima, M. Shimizu, S. Amirhaeri and R.D. Wells, *J. Biol. Chem.*, 1995, **270**, 27014.

113. C.B. Kunst and S.T. Warren, *Cell*, 1994, **77**, 853.

114. M.C. Hirst, P.K. Grewal and K.E. Davies, *Hum. Mol. Genet.*, 1994, **3**, 1553.

115. E.E. Eichler, J.J. Holden, B.W. Popovich, A.L. Reiss, K. Snow, S.N. Thibodeau, C.S. Richards, P.A. Ward and D.L. Nelson, *Nat. Genet.*, 1994, **8**, 88.

116. C.E. Pearson, E.E. Eichler, D. Lorenzetti, S.F. Kramer, H.Y. Zoghbi, D.L. Nelson and R.R. Sinden, *Biochemistry*, 1998, **37**, 2701.

117. K. Usdin, *Nucleic Acids Res.*, 1998, **26**, 4078.

118. P. Weisman-Shomer, E. Cohen and M. Fry, *Nucleic Acids Res.*, 2000, **28**, 1535.

119. J.S. Sutcliffe, D.L. Nelson, F. Zhang, M. Pieretti, C.T. Caskey, D. Saxe and S.T. Warren, *Hum. Mol. Genet.*, 1992, **1**, 397.

120. S.S. Smith, J.L. Kan, D.J. Baker, B.E. Kaplan and P. Dembek, *J. Mol. Biol.*, 1991, **217**, 39.

121. S.S. Smith, A. Laayoun, R.G. Lingeman, D.J. Baker and J. Riley, *J. Mol. Biol.*, 1994, **243**, 143.

122. R.J. Roberts, *Cell*, 1995, **82**, 9.

123. M. Fry and L.A. Loeb, *J. Biol. Chem.*, 1999, **274**, 12797.

124. S. Khateb, P. Weisman-Shomer, I. Hershco, L.A. Loeb and M. Fry, *Nucleic Acids Res.*, 2004, **32**, 4145.

125. P. Weisman-Shomer, Y. Naot and M. Fry, *J. Biol. Chem.*, 2000, **275**, 2231.

126. P. Weisman-Shomer, E. Cohen, I. Hershco, S. Khateb, O. Wolfovitz-Barchad, L.H. Hurley and M. Fry, *Nucleic Acids Res.*, 2003, **31**, 3963.

127. L. Uliel, P. Weisman-Shomer, H. Oren-Jazan, T. Newcomb, L.A. Loeb and M. Fry, *J. Biol. Chem.*, 2000, **275**, 33134.

128. G.I. Bell, S. Horita and J.H. Karam, *Diabetes*, 1984, **33**, 176.

129. G.I. Bell, J.H. Karam and W.J. Rutter, *Prog. Clin. Biol. Res.*, 1982, **103 Pt A**, 57.

130. G.I. Bell, J.H. Karam and W.J. Rutter, *Proc. Natl. Acad. Sci. USA*, 1981, **78**, 5759.

131. P. Rotwein, S. Yokoyama, D.K. Didier and J.M. Chirgwin, *Am. J. Hum. Genet.*, 1986, **39**, 291.

132. M.C. Hammond-Kosack, B. Dobrinski, R. Lurz, K. Docherty and M.W. Kilpatrick, *Nucleic Acids Res.*, 1992, **20**, 231.

133. M.C. Hammond-Kosack, M.W. Kilpatrick and K. Docherty, *J. Mol. Endocrinol.*, 1993, **10**, 121.

134. P. Catasti, X. Chen, R.K. Moyzis, E.M. Bradbury and G. Gupta, *J. Mol. Biol.*, 1996, **264**, 534.

135. M.C. Hammond-Kosack and K. Docherty, *FEBS Lett.*, 1992, **301**, 79.

136. M.D. Walker, T. Edlund, A.M. Boulet and W.J. Rutter, *Nature*, 1983, **306**, 557.

137. G.C. Kennedy, M.S. German and W.J. Rutter, *Nat. Genet.*, 1995, **9**, 293.

138. G.C. Kennedy and W.J. Rutter, *Proc. Natl. Acad. Sci. USA*, 1992, **89**, 11498.

139. A. Lew, W.J. Rutter and G.C. Kennedy, *Proc. Natl. Acad. Sci. USA*, 2000, **97**, 12508.

140. A.K. Todd, M. Johnston and S. Neidle, *Nucleic Acids Res.*, 2005, **33**, 2901.

141. S. Chen, P.C. Supakar, R.L. Vellanoweth, C.S. Song, B. Chatterjee and A.K. Roy, *Mol. Endocrinol.*, 1997, **11**, 3.

142. Z. Wang, X.H. Lin, Q.Q. Qiu and T.F. Deuel, *J. Biol. Chem.*, 1992, **267**, 17022.

143. D. Sun, K. Guo, J.J. Rusche and L.H. Hurley, *Nucleic Acids Res.*, 2005, **33**, 6070.

144. R. DeArmond, S. Wood, D. Sun, L.H. Hurley and S.W. Ebbinghaus, *Biochemistry*, 2005, **44**, 16341.

145. S. Rankin, A.P. Reszka, J. Huppert, M. Zloh, G.N. Parkinson, A.K. Todd, S. Ladame, S. Balasubramanian and S. Neidle, *J. Am. Chem. Soc.*, 2005, **127**, 10584.

146. J. Dai, T.S. Dexheimer, D. Chen, M. Carver, A. Ambrus, R.A. Jones and D. Yang, *J. Am. Chem. Soc.*, 2006, **128**, 1096.

Quadruplexes in the Genome

JULIAN HUPPERT

Cambridge, UK

8.1 Introduction

G-quadruplexes are a significant and highly varied structure,[1,2] as has been discussed elsewhere in this book. As well as being of interest purely as a structure (especially in the nano-technology field), there has been interest in their physiological roles as a natural part of the genome. There has been particular focus on their role in telomeric regions[3,4] and in controlling regulation of the gene *c-myc*,[5–7] both of which are also given in detail in this book.

Over the years, a number of other quadruplex-forming regions have also received attention. The first was in 1988 when Sen and Gilbert[8] suggested that there was quadruplex formation in meiosis, and there is now an ever-expanding number of proposed physiological roles for quadruplexes. The rapid growth in the study of physiologically relevant quadruplexes has been particular fuelled by the availability of relatively complete genomic datasets[9] and bioinformatic tools. These allow rapid discovery of sequences capable of forming quadruplexes (putative quadruplex sequences, or PQS) in interesting genomic locations.

In this chapter, I will first consider exactly how one defines a PQS, and then outline some of the specific quadruplex-forming sequences that have been identified. I will highlight the proteins that have been shown to interact specifically with quadruplexes. I will then discuss the physiological roles that PQS in general could play. Lastly, I will discuss the genome-wide studies that have been performed to date. The chapter then ends with some proposed future directions for this field.

8.2 What is a Quadruplex?

In order to identify and investigate novel quadruplex-forming sequences, it is necessary to develop an understanding of which sequences will form a quadruplex. This involves the development of an algorithm which will predict secondary structure in a binary sense (quadruplex/non-quadruplex) from the primary sequence, in a manner analogous to that possible for protein sequences.[10] Ideally, further information would allow more detailed prediction of the full tertiary structure – predictions as to exactly what form the sequence could fold – and possibly estimated thermodynamic predictions of stability, analogous to the models already existing for duplex DNA.

So far, there have been three attempts to develop such a predictive algorithm.[11–13] Each of these use slightly different rules to identify PQS.

Todd and Neidle[11] defined a PQS sequence as having the form $G_{3-5}N_{1-7}G_{3-5}$ $N_{1-7}G_{3-5}N_{1-7}G_{3-5}$. This is similar to the "Folding Rule" of Huppert and Balasubramanian (implemented in the program *quadparser*), which was $G_{3+}N_{1-7}G_{3+}N_{1-7}G_{3+}N_{1-7}G_{3+}$.[12] These definitions differ essentially only for continuous runs of guanine in excess of 5 bases, which are counted in the latter case but not the former.

In contrast, d'Antonio and Bagga[13] used a more complex procedure, in which sequences were originally fitted to a structure of $G_{2+}N_{0+}G_{2+}N_{0+}G_{2+}N_{0+}G_{2+}$, in which runs of two or more guanines were allowed. The gaps were allowed to be of arbitrary length, but subject to the constraint that no more than one gap could be of zero bases in length, and the total length of the sequence could be no longer than 25 bases. Interestingly, this would exclude known quadruplex-forming sequences such as the *Oxytricha* telomeric repeat sequence d(G_4T_4 $G_4T_4G_4T_4G_4$), which is 28 bases in length. In addition, d'Antonio and Bagga used a scoring process, based on the evenness of the gap lengths between the G-tracts, and the length of the G-tracts. No biophysical evidence was advanced to support this scoring process.

A rationalisation for their approach is provided in Huppert and Balasubramanian,[12] who consider four principle parameters of importance. These are (a) the strand stoichiometry, (b) the number of stacked tetrads in the quadruplex core, (c) the presence of mutations or deletions, and (d) the length and composition of loops. Some of these parameters may be resolved by consideration of the structures and previously known results, while others are resolved by experimentation. The following descriptions and commentary are also valid for the approach of Todd and Neidle.[11]

8.2.1 Strand Stoichiometry

Quadruplexes can be uni-, bi- or tetramolecular. Since under physiological conditions the strand concentration of DNA is relatively low (order nM), except in rare exceptions such as *Stylonychia lemnae* macronuclei,[14] interstrand quadruplexes will be strongly disfavoured. Therefore, only sequences that can from intramolecular quadruplexes were considered.

8.2.2 Number of Tetrads

G-quadruplex structures can in principle form from any number of G-tetrad stacks. In general, the stability increases with increasing numbers of stacks. Single G-tetrads have only been reported in highly concentrated guanine solutions at mM concentrations,[15] which are unlikely to be physiologically relevant. There are a few examples of double-stack quadruplexes, such as the thrombin-binding aptamer (TBA)[16,17] and the sequences identified as responsible for the fragile X syndrome.[18] However, because these are in general less stable with regard either to single stranded forms or duplex formation,[19] only sequences capable of forming three or more G-tetrad stacks were considered.

8.2.3 Discontinuities in G-Tracts

Does a quadruplex have to be made up of perfect guanine tetrads, or can it tolerate discontinuities in the guanine bases? A few studies have recently been published identifying tetrads not comprised purely of guanine,[20–24] but most of these are artificially designed sequences, where a mixed tetrad is stabilised by flanking G-tetrads. In addition, there have recently been structures identified as having 'slipped' structures,[25] where two quadruplexes slide against each other, or intrastrand leaps,[26] in which a particular stack of guanines comes from more than one consecutive series of bases, but these are also artificial structures. An unpublished study exploring the effects of guanine replacements and deletions on a variant of the human telomeric sequence $d(GGTTAG)_n$ showed that variations result in sequences with significantly lower stability and for that reason only sequences with no discontinuities in the G-tracts were considered.

8.2.4 Loop Length and Composition

An intramolecular quadruplex must have three loops to link the tetrads together, and they play a large role in determining both the stability and folding pattern of the quadruplex. Hazel *et al.*[27] conducted a study of sequences with varying loop lengths, and showed that quadruplex formation could occur with sequences containing up to 7 bases, with decreasing stability as the number of bases increased. Zero-base loops, although having been demonstrated in one instance, were neglected.[28]

No significant efforts have been made to address the question of predicting tertiary structure from primary sequences, although some predictions may be made based on the study performed by Hazel *et al.*[27] They demonstrated using both molecular modelling and biophysics that very short loops (especially single-base loops) tend to produce parallel quadruplexes, whereas longer loops favour anti-parallel quadruplexes. This is demonstrated in the graph in Figure 1, showing circular dichroism traces for the family of sequences $d[(TG_3T_xG_3T_xG_3T_xG_3T), 1 \leq x \leq 7]$.

The problem of estimating thermodynamic stability from primary sequences has not received any attention to date, though such a question must involve taking into account the observations that longer loops have lower stabilities, longer G-tracts are more stable, and that the presence of multiple potential structures leads to greater overall stability.

8.3 Characterised Quadruplex-Forming Sequences

Aside from telomeric sequences and *c-myc*, discussed elsewhere in this book, other known quadruplex-forming sequences include the fragile X syndrome repeat $d(CGG)_n$ and[18,29,30] the Cystatin B promoter,[31] which has a region with sequence $(CGCG_4CG_4)_4$ and is involved in epilepsy. G-rich strands of the

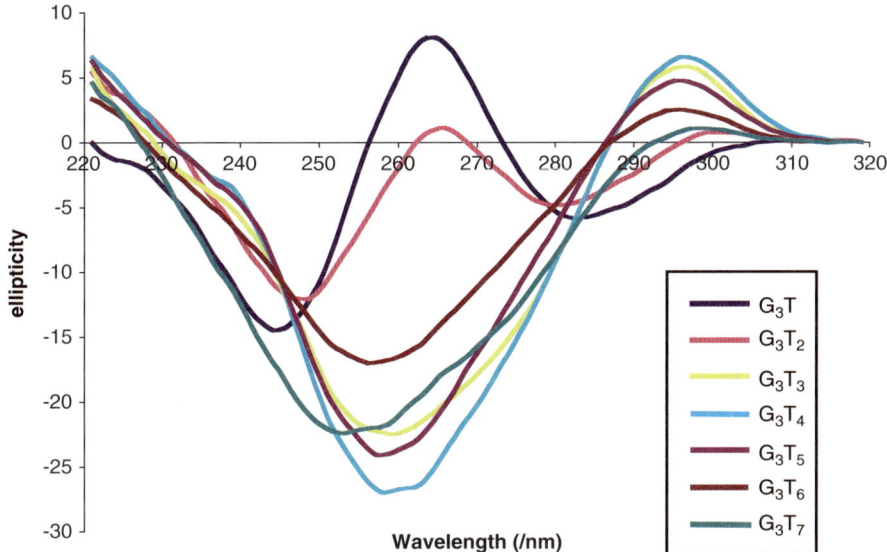

Figure 1   *CD spectra for sequences with loop lengths varying. Samples with loops of length 1 or 2 exhibit a peak at 265 nm characteristic of parallel folds (length 2 possibly being polymorphic); the other sequences have a peak at 295 nm, characteristic of anti-parallel folding*

human insulin gene can form quadruplexes,[32] as can the mouse *Ms6-hm* hypervariable satellite repeat,[33] with sequence (CAGGG)$_n$. It has recently been proposed that the promoter regions of the RET protooncogene[34] and Ki-ras[35] can each form a quadruplex. G-rich RNA can also fold into quadruplex structures, for example the insulin-like growth factor II (IGF II) mRNA.[36]

8.4   Proteins Interact with Quadruplexes

A variety of natural proteins have been found to specifically bind quadruplexes. In addition, others have been designed artificially to bind quadruplexes,[14,37,38] although these will not be discussed further here.

Naturally occurring quadruplex-binding proteins include the helicases implicated in Bloom's[39] and Werner's[29] syndromes, which have been shown to specifically unwind quadruplexes and to be 'inhibited' by ligands which bind quadruplexes.[40] The *Saccharomyces cerevisiae* protein RAP1[41,42] and the β-subunit of the *Oxytricha nova* telomere binding protein[43] have both been shown to promote quadruplex formation and bind the resulting quadruplexes. The rat hepatocyte protein qTPB42[44] acts to protect quadruplexes from heat denaturation and nuclease digestion. Quadruplexes also interact with various natural enzymes and have been shown to stimulate the activity of a DNA polymerase[45] and inhibit telomerase.[46] The fact that these proteins selectively recognize this non-standard quadruplex motif presumably means that it is capable of forming naturally *in vivo*.

This selective recognition of quadruplexes is perhaps best exemplified by considering Bloom's syndrome, a rare condition characterised by genomic instability and a high incidence of cancer. It is caused by defects in the *BLM* gene, which encodes a 1417-amino acid helicase. This helicase was shown by Sun *et al.*[39] to have an ATP-dependent unwinding activity on quadruplex substrates to yield single strands. It can also unwind duplex DNA, but preferentially acts on quadruplex DNA, as determined by competition experiments and the fact that less of the helicase is required to unwind quadruplex than duplex. Hence it is believed that the *BLM* helicase is required to remove quadruplexes that form prior to recombination and replication. This may explain why it cannot be substituted by other helicases, and hence why deficiency causes such severe effects. It also implies that quadruplex formation does occur *in vivo*, or else there would be no reason for such a helicase to exist, and its absence would have no significant effects.

8.5 Possible Functions of Quadruplexes in the Genome

What roles could be played by sequences in the genome capable of forming quadruplexes? One role, currently believed to be the function of a quadruplex formed in the promoter of the oncogene *c-myc*[5–7] and studied by Hurley and co-workers (d(T GGGG A GGG T GGGG A GGG T GGGG AAGG)), is to act as a downregulator of gene transcription. This function could possibly be achieved by acting as a steric block to transcription initiation, either directly or as a ligand to which a protein can bind and block transcription (Figure 2).

Another possible example of this form of control may be found in the *COL1A1* gene, which is connected to osteoporosis. This gene encodes one of the protein chains of collagen, and when over-expressed results in lowered bone mineral density and increased risk of osteoporosis fracture. There is a known G to T SNP in the promoter region of this gene (Figure 3),[47] which is known to result in a 3-fold difference in activity level.[48] This SNP lies in one of the G-tracts of a putative quadruplex region, and so it is possible that the SNP could have this effect by preventing the formation of a G-quadruplex and hence increasing the level of expression of the gene.

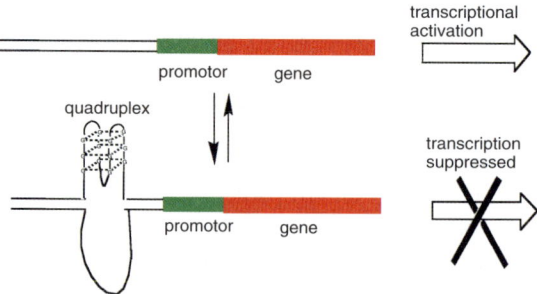

Figure 2 *Possible model for quadruplex formation inhibiting transcription by acting as a steric block, in or near the promoter region*

'S' allele   GGGAATG**G**GGGCGGGATGAGGG

's' allele   T

Figure 3   *Promoter region of the COL1A1 gene, showing the SNP implicated in osteoporosis*

It is also worth highlighting that the COL1A1 sequence is also believed to be a binding site for the transcription factor (TF) Sp1. This TF, as well as a family of other TFs, binds G-rich sequences (or their C-rich converses). Another example is the TF hnRNP K,[49,50] which binds within the *c-myc* promoter sequence. The binding of these factors could clearly be affected strongly by formation of alternative secondary structures such as quadruplexes.

Quadruplexes could also act purely as a structural feature, enabling easy recognition by proteins or other species. This model has been proposed for telomeric repeats, perhaps as a 'beads-on-a-string' model of sequential folded quadruplexes. Other regions could also have such structural behaviours.

Other possible physiological roles for quadruplexes, which have not yet been demonstrated experimentally, include roles in translation initiation or splicing. Many others can be proposed.

An example of the arrangements of G-quadruplexes is shown in Figure 4 in an extract of the 5′ end of the oncogene Ha-ras, which is a G-protein of the ras class involved in signal transduction. It is perhaps noteworthy that all the putative quadruplexes found in this sequences (located using the program *quadparser*[12]), are near the beginning or end of particular regions.

8.6   Genome-Wide Studies on Quadruplexes

To date, there have been three published analyses of genomic information, each based on a different algorithm, as highlighted earlier in this chapter. They all investigate different aspects of the PQS located in the genome. Huppert and Balsubramanian[12] investigated the frequency of PQS in the genome, and then compared that to the number that might be expected to have been found, if there were no selective pressures for or against them. The authors then continued by investigating the prevalence of PQS in RNA-forming regions as compared to the genome as a whole. Lastly, they investigated the distribution of loops in the PQS identified, and how that differed from that expected by chance.

Todd and Neidle[11] examined the sequence of the loops in great details, and demonstrated that some sequences are significantly more prevalent than anticipated. Lastly, d'Antonio and Bagga[13] investigated the relationship between PQS and RNA processing sites. Each of these investigations will be dealt with in more detail in this section.

8.7   Frequency of PQS

One of the most obvious genome-wide questions is to count how many PQS there are in the genome. Using any of the algorithms described above, a count

```
gacctccgcggtgggcggcgccgcgctgccggcgcagggagggcctctggtgcaccggcaccgctgagtc
gggttctctcgccggcctgttcccgggagagcccggggccctgctcggagatgccgccccgggcccccag
acaccggctccctggccttcctcgagcaaccccgagctcggctccggtctccagccaagcccaaccccga
gaggccgcggccctactggctccgcctcccgcgttgctcccggaagccccgcccgaccgcggctcctgac
agacgggccgctcagccaaccggggtggggcggggcccgatggcgcgcagccaatggtaggccgcgcctg
gcagacggacgggcgcggggcggggcgtgcgcaggcccgcccgagtctccgccgcccgtgccctgcgccc
GCAACCCGAGCCGCACCCGCCGCGGACGGAGCCCATGCGCGGGGCGAACCGCGCGCCCCCGCCCCGCCC
CGCCCCGGCCTCGGCCCCGGCCCTGGCCCCGGGGGCAGTCGCGCCTGTGAACG
gtgagtgcgggcagggatcggccgggccgcgcgccctcctcgcccccaggcggcagcaatacgcgcggcg
cgggccgggggcgcggggccggcgggcgtaagcggcggcggcggcggcggcggcggcggtgggtggggccgg
gcggggcccgcgggcacaggtgagcgggcgtcggggctgcggcgggcggggcccttcctccctgggg
cctgcgggaatccgggccccacccgtggcctcgcgctgggcacggtccccacgccgcgtacccgggagc
ctcgggcccggcgccctcacacccggggggcgtctgggaggaggcggccgcggccacggcacgcccgggca
ccccgattcagcatcacaggtcgcggaccaggccgggggcctcagccccagtgccttttccctctccgg
gtctcccgcgcgcgcttctcggcccccttcctgtcgctcagtccctgcttcccaggagctcctctgtcttct
ccagctttctgtggctgaaagatgcccccggttccccgccggggggtgcggggcgctgcccgggtctgccc
tcccctcggcggcggcgcctagtacgcagtaggcgctcagcaaatacttgtcggaggcaccagcgccgcgggg
cctgcaggctggcactagcctgcccgggcacgccgtggcgcgctccgccgtggccagacctgttctggag
gacggtaacctcagccctcgggcgcctccctttagccttctgccgacccagcagcttctaatttgggtg
cgtggttgagagcgctcagctgtcagcctgcctttgagggctgggtccctttttcccatcactggtcat
taagagcaagtgggggcgaggcgacagccctcccgcacgctgggttgcagctgcacaggtaggcacgctg
cagtccttgctgcctggcgttggggcccagggaccgctgtgggtttgcccttcagatggccctgccagca
gctgcctgtgggcctggggctgggcctgggcctggctgagcagggccctccttggcag
GTGGGGCAGGAGACCCTGTAGGAGGACCCCGGGCCGCAGGCCCCTGAGGAGCGATGACGGAATATAAGCT
GGTGGTGGTGGGCGGCCGGCGGTGTGGGCAAGAGTGCGCTGACCATCCAGCTGATCCAGAACCATTTTGTG
GACGAATACGACCCCACTATAGAG
```

Figure 4   *5′ end of the Ha-ras gene. In green is shown the 5′ upstream region. The purple region is the 5′ UTR, which is transcribed but not translated. The black letters are the coding region, beginning with an ATG start codon (AUG as DNA). In red is an intron. Putative quadruplexes are shown inside yellow boxes*

can be made. However, there are some subtleties to the counting. These are perhaps best exemplified by considering the telomeric extract d(GGGTTAG GGTTAGGGTTAGGGTTAGGGTTAGGGTTAGGGTTAGGG).

This sequence consists of eight runs of G_3, and according to the three algorithms could form a variety of different quadruplexes. But should it be treated as 5 quadruplexes (using the 1st–4th G_3 run, the 2nd–5th and so forth). Or should it be considered as two non-overlapping sequences (1st–4th and 5th–8th), or even just as one extra-long PQS?

Both papers which calculated the total numbers of PQS in the genome[11,12] used the 'non-overlapping' definition, which is equivalent to the maximum number of quadruplexes that could form at any given time. They gave very similar results, reporting that there are around 376,000 PQS in the genome (375,157[11] and 376,446[12] exactly).

How many PQS might one expect if these sequences had no physical reality? In order to address this question, it is necessary to develop models for the DNA in these locations. One control that can be used in these models is to consider patterns analogous to the G-rich PQS. Labelling such base-rich sequences as X-patterns, where X is the frequent base, we have G-patterns, which correspond to sequences that form PQS in the strand being considered, and C-patterns, which correspond to sequences that form PQS in the complementary strand.

G-pattern	GGG TTA GGG TTA GGG TTA GGG
C-pattern	CCC AAT CCC AAT CCC AAT CCC
A-pattern	AAA GCG AAA GCG AAA GCG AAA
T-pattern	TTT CGC TTT CGC TTT CGC TTT

Figure 5 *Sample X-patterns*

These can be collectively labelled as GC-patterns. A- and T-patterns (collectively, AT-patterns) do not form a secondary structure like quadruplexes, and so can be used as controls when modelling DNA (Figure 5).

The simplest such model is to treat DNA as a sequence of independent bases (a Bernoulli stream), each occurring with a probability equivalent to their frequency in the human genome. There are approximate methods[51,52] for calculating the expected number of PQS given a certain base frequency, and the problem can be solved explicitly as well. This gives the expected density of PQS, $\rho(PQS)$, as a function of p, the probability of any individual base being guanine:

$$p(PQS) = 343p^{12} - 882p^{13} + 756p^{14} - 1098p^{15} + 2835p^{16} - 3357p^{17} + 2484p^{18}\ldots$$

However, applying this solution to the entire human genome gives predicted frequencies for GC-patterns of 8300 and for AT-patterns of 304,000. These results are clearly significantly lower than those actually found (376,000 and 3,259,000, respectively), by more than an order of magnitude. Since the discrepancy between real genomic data and prediction arises for both 'real' GC-patterns and 'control' AT-patterns[12] this is suggestive of shortcomings in the model used for DNA.

There are two reasons why this simple Bernoulli model may be oversimplistic. Firstly, DNA exhibits considerable structure on the diad base level, shown in Table 1,[12] and as described by previous researchers.[53,54] This means that some bases are more likely to occur after others. As an example, there is only a 5% chance of finding a G after a C, although there is a 21% chance in general of finding a G in any particular position. In particular, homodiads are relatively frequent, which means that quadruplex-forming sequences will be considerably more frequent. Secondly, DNA is not homogenous with regard to base composition, and has regions that are relatively rich in each base. Since the number of quadruplexes found is a very strong function of base density, this factor will have a large impact on the number of PQS found.

In order to address these two issues, a windowed Markov model can be used.[12] This model uses windows of the genome in order to address the variability of DNA composition in different areas of the genome. Comparing the results found using this Markov windowed method with those found in the real human genome, it was found that for window sizes from 150 and above,

Table 1   *Diad analysis of every human chromosome*[12]

Base	Previous base								Overall
	G		C		A		T		
G	0.26	+0.05	0.05	−0.16	0.24	+0.03	0.25	+0.04	0.21
C	0.21	-	0.26	+0.05	0.17	−0.04	0.20	-	0.21
A	0.29	-	0.35	+0.06	0.33	+0.04	0.22	−0.08	0.29
T	0.25	−0.05	0.34	+0.05	0.26	−0.04	0.33	+0.04	0.29

Notes: Vertical lines show the percentage probabilities of each base following a given base, and then the deviation from the percentage probabilities expected if each base was independent.

Table 2   *Total number of GC- and AT-patterns found in the real human genome and simulates using various methods*[12]

Method	GC-patterns	AT-patterns
Markov, size 50	687 k	4.01 M
Markov, size 75	**514 k**	**3.26 M**
Markov, size 100	420 k	2.81 M
Markov, size 150	320 k	2.29 M
Markov, size 200	269 k	2.02 M
Markov, size 400	185 k	1.56 M
Markov, size 1000	123 k	1.20 M
Markov, size 2000	93 k	1.02 M
Markov, size 4000	75 k	0.89 M
Bernoulli	8 k	0.30 M
Real human genome	376 k	3.26 M

Notes: In the window methods, simulates were generated conserving diad base frequencies in windows of the size shown. Five independent analyses were performed, and the standard deviation was in all cases less than 1%. The 'Bernoulli' method treats DNA as a stream of independent bases, with base frequencies homogenous across each chromosome. The Markov model that correctly predicted the number of AT-patterns (window size 75 bp) is shown in bold.

there are fewer predicted sequences than actually found for both GC- and AT-patterns. For window sizes between 75 and 150, there are more GC-patterns than actually found, but fewer AT-patterns. For a window size of 75, the algorithm correctly predicts the number of AT-patterns, that is 3,260,000, but predicts 37% more GC-patterns than actually found – suggesting that GC patterns are under-represented in the genome. For smaller windows, more of both type of pattern is predicted than found (Table 2).

A 'pseudo-Chargaff' rule was also observed, with equivalent numbers of G- and C-patterns, and A- and T-, in each case. This is an expected corollary of the fact that the two strands are equivalent.

8.8   Location of Putative Quadruplexes

Using the ENSEMBL database, it is possible to classify regions of the genome as related to genes. This information was used[12] to investigate the number of

Table 3   *Relative frequency of X-patterns in exonic regions*

Base X	G	C	A	T
Relative frequency	0.48	0.83	0.93	1

putative quadruplexes in genes, and specifically within the exonic regions. These results show marked differences in terms of base composition and frequency of quadruplex-forming patterns.

The base compositions of these exonic regions behave largely as though the bases were equivalent, both as regards individual base frequencies and diad frequencies, when compared to the genome as a whole, which has significant differences between GC on the one hand and AT on the other. As a result of this, it would be expected (and is confirmed by modeling[12]), that X-patterns would occur with roughly equal frequency. In fact, this is not the case, and G-patterns are very much less frequent than C-patterns, which are less frequent than A or T patterns (the relative frequencies are shown in Table 3).

These results break the 'pseudo-Chargaff' rule observed in the rest of the genome – G and C do not behave similarly, and nor do A and T. This can occur because the two DNA strands are distinct in exonic regions, with only one of them being transcribed to form RNA. However, that does not explain why such a marked decrease in G-patterns is observed.

One possible explanation is the simple observation that the G-rich codons do not code for the same amino acids as C-rich codons, but modelling based on codon frequencies shows that the predicted discrimination from this alone is not sufficient to account for the observation. An alternative explanation is based on the observation that the G-patterns would lead to potential quadruplexes in the mRNA strand in addition to the DNA duplex, whereas C-patterns could only lead to quadruplexes in the (complementary strand of) DNA. This evidence is consistent with an evolutionary pressure to reduce the number of quadruplexes allowed to form in mRNA. This may be particularly strong as it seems that RNA quadruplexes are more stable, both thermodynamically and kinetically, than their DNA counterparts.[55] To date, there has been relatively little work focused on RNA quadruplexes, although it has evoked some interest.[13,36,55,56] However these results suggest that RNA quadruplexes could play a significant physiological role, and should be investigated further.

8.9   Distribution of Loop Lengths

Huppert and Balasubramanian[12] then considered the length of the loops separating the G-tracts. It had been shown previously that the length of the loops linking the runs of oligo-G sequences play a significant role in determining the stability of the resultant quadruplexes[27,57].

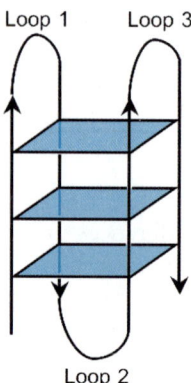

Loop 1   Loop 3

Loop 2

Figure 6 *Schematic structure of a quadruplex, showing the three loops*

By counting the number of PQS observed with loops of different lengths, they demonstrated that the central loop (loop 2, Figure 6) behaved differently from the terminal loops (loops 1 and 3). In particular, they showed that there was a greater tendency for the central loop to be three or four bases long than the others. They also found that there was a significant predominance of very short loop lengths – especially single-base loops – for all three loops. This is shown graphically in Figure 7.

More detailed analysis of the behaviour of the loops required analysis of the co-variance between the loops. All the PQS identified above were examined, and represented as a three-dimensional coordinate set $<i,j,k>$, where i is the length of the first loop, j the second and k the third.

The most common set of loop lengths was $<1,1,1>$, which accounted for 8% of all PQS. Table 4 lists the 20 most common loop length combinations, which between them account for 32% of all PQS. Table 4 also lists the 20 least common loop length combinations (1.7% of all PQS), which generally include longer loops. It is also noteworthy than in none of the least common loop length combinations is the length of loops 1 and 3 the same, thus such sequences are over-represented in the cases of the most common loop length combinations.

These results may be most clearly represented graphically, using a matrix plot, where the area of each segment is proportional to the number of PQS observed with that particular set of loop lengths (Figure 8). It may be clearly seen that there is an interaction between the loops, and in particular a 'spine' of high frequency running along the diagonal.

The best method of modelling this behaviour is using a model called diagonal quasi-independence,[58] and corresponds to a probability mixture model in which with probability a, the loops lengths are constrained to be the same, and with probability $(1-a)$, they are independent. This method gives the relationship shown below, where N_{ik} is the predicted count with first loop length i and third loop length k, β_i and β_k are the two independent distributions,

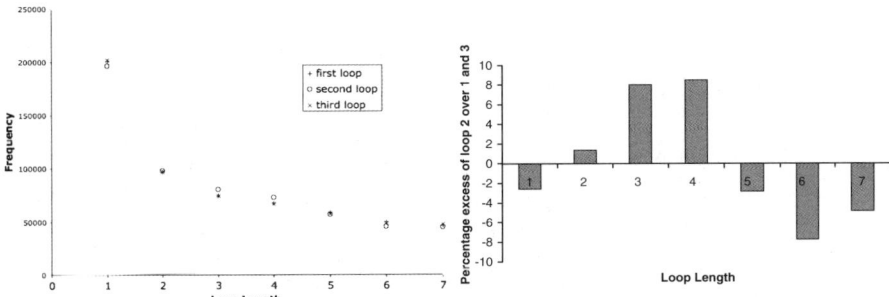

Figure 7 *Left: frequency distributions of loops of lengths 1–7 bases for the entire human genome. Right: percentage excesses of loop 2 counts over the averages of loops 1 and 3 for the entire human genome*

Table 4 *The twenty most common and twenty least common sets of observed PQS loop lengths*

Most common loop lengths				Least common loop lengths			
Loop 1	*Loop 2*	*Loop 3*	*Number*	*Loop 1*	*Loop 2*	*Loop 3*	*Number*
1	1	1	47475	6	5	7	441
1	4	1	11328	7	6	5	441
1	2	1	10656	7	6	3	447
1	1	2	10415	5	6	7	447
2	1	1	10040	6	6	7	449
2	2	2	9411	6	7	7	450
1	3	1	9127	7	5	6	452
1	5	1	7799	5	7	6	484
5	1	1	7379	5	5	7	501
1	1	5	7337	5	6	3	505
3	3	3	6827	7	7	6	505
3	1	1	6458	6	6	5	506
1	1	3	6403	5	6	6	511
1	1	4	6196	3	6	7	521
4	1	1	6189	7	6	6	523
2	2	1	5123	6	7	3	525
1	2	2	5046	7	7	4	528
2	1	2	4780	3	7	6	533
1	6	1	4556	5	7	7	536
6	1	1	4462	6	7	5	538

Notes: Loops are numbered from 5′ to 3′ of the G-rich strand.

a is a constant describing the population of the 'spine' and α_i describes the distribution of sequences along the spine.

$$N_{\{ik\}} = a \cdot \alpha_i + (1 - a)\beta_i\beta_k \quad \text{for } i = k$$
$$N_{\{ik\}} = \qquad (1 - a)\beta_i\beta_k \quad \text{for } i \neq k$$

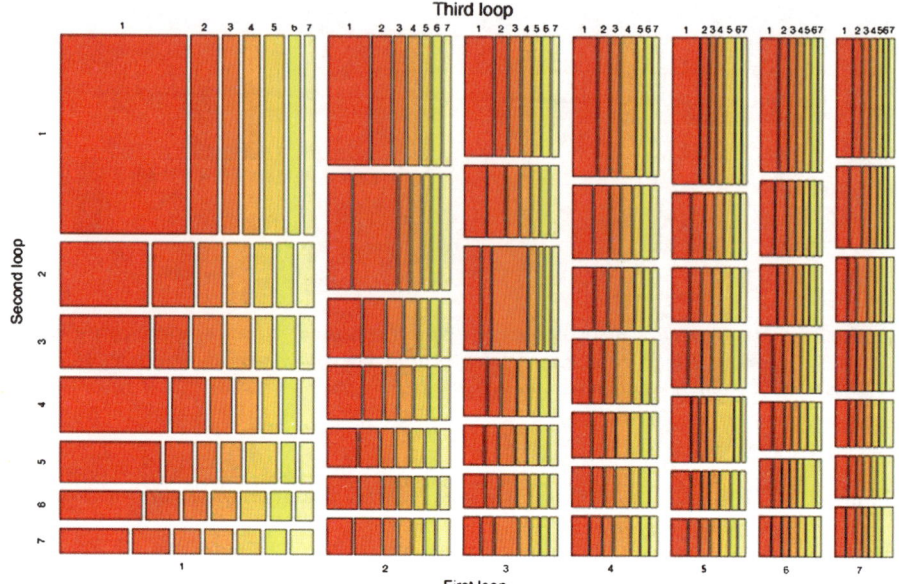

Figure 8 *Mosaic plot representing the loop lengths of all putative quadruplexes found in the human genome. The seven principle columns represent the lengths of the first loop, the seven rows the lengths of the second loop and the seven segments in each box the lengths of the third loop. The area of each box is proportional to the number of sequences found with that combination of loop lengths. The plot was produced using the program R (http://www.r-project.org) using the command mosaicplot*

Detailed analysis of the loops lengths was performed using this method, giving a value for *a* of 0.1, using a variety of different statistical methods.[59] This indicates that approximately 10% of the potential quadruplex patterns show a high correlation between the lengths of loops 1 and 3.

8.10 Loop Sequences

In contrast to the work described above, which examined the *length* of G-quadruplex loops, Todd and Neidle[11] focussed primarily on the *sequence* of the loops. It is known that the sequences of the loops of G-quadruplex can affect their structure and stability.[17,60–62] They firstly showed that there are many potential loop sequences found in PQS (10,551 unique loops sequences), although this does not cover the entire sequence space available; there are 12,289 possible sequences, of which only 85% are actually found.

They examined a variety of properties, focusing on two principle ones – the frequency of various sequences, and the degree to which individual sequences are found in one loop rather than the others.

The most frequent loops are shown in Table 5 below. As described earlier,[12] these results show that single-base loops are very common, and that in general,

Table 5 *Most popular loop sequences found in PQS, separated by which loop the sequence was found in*

Sequence	Population	1st loop	2nd loop	3rd loop
A	193,756	51,361	63,872	78,523
T	121,406	53,234	37,657	30,515
C	44,020	14,983	14,907	14,130
AA	400,026	12,778	13,717	13,531
CT	32,472	11,637	10,554	40,281
CA	32,070	10,781	10,846	10,443
G	29,623	7183	8375	14,065
AT	19,957	6789	7242	5926
AGA	19,144	5377	6919	6848
TT	17,089	7437	5530	4122
TA	12,641	4744	4329	3568
CC	10955	3646	3726	3583
AGT	9896	2767	4447	2682
AGGA	9463	1932	3559	3972
AGGT	9434	1516	6448	1470
TGA	9237	3006	2849	3382
AAA	7839	2393	2970	2476
CCT	7151	2540	2298	2313
TGT	6619	2530	2307	1782
CCA	6269	2105	2048	2116

shorter loops are more frequent than longer loops. However, it also shows significant differences in the frequencies of some particular sequences.

Studying the sequences which showed the greatest distinction between the different loops gave a rather different set of results. These loop sequences are interesting because they may be considered to be the most likely to have some physiological relevance, as there is no obvious other reason why they would be located differentially in the various loops. The results of this analysis are given in Table 6. Interestingly, these results show a significant preponderance of sequences of the form CCTGT* or CCTAT*, especially in the first loop.

Investigating the location of the sequences containing either CCTGTT or CCTGTCA as a loop, Todd and Neidle found that they are significantly repressed within genes, and particularly so on the 'plus' strand. This fact provides further evidence that there is some physiological significance to PQS, as they appear to be disfavoured in regions that form RNA.

8.11 RNA Processing

d'Antonio and Bagga[13] developed tools to search for PQS throughout the genome, but only report on the correlation between the positions of the PQS they find and the positions of RNA processing sites. As a model, they studied the distribution patterns of PQS in 95 alternatively processed mouse transcript sequences. Most of the analysed genes exhibited the presence of PQS near splice sites or poly(A) regions suggesting a role for G-quadruplex elements in RNA processing reactions.

Table 6 *PQS loop sequences showing the greatest discrimination by loop*

Sequence	Ratio	1st loop	2nd loop	3rd loop
CCTGTCA	309	1239	5	4
CCTGTT	140	1266	18	9
CCTGTC	139	836	8	6
CCTGTTA	90	90	1	0
ATCTCCA	74	1	5	74
TGGTCTT	58	3	1	58
CCTATCA	53	53	1	0
TCTGTCA	51	51	3	0
TAGCACA	42	0	5	42
CCTATC	38	38	1	1
CCTATT	37	75	4	2
CCTTTCA	37	37	1	0
CTTGGC	36	13	16	471
TAGCATT	34	1	0	34
CCTGTCC	30	30	0	3
CCTGTTT	29	58	4	2
CTTGTCA	29	58	4	2
CCTGTGA	28	28	8	1
CCAGTC	28	28	1	3
ACCTGTC	27	27	1	2

Notes: Ranked in order of the ratio of the most frequent count in a loop over the least.

They also identified a bias in the occurrence and location of PQS in alternatively processed transcripts. PQS were more likely to be found near RNA processing sites in alternatively processed transcript sequences. Furthermore, G-quadruplex elements were typically associated with processing sites of selective gene products in alternatively spliced genes. Their findings lend supporting evidence that PQS play a regulatory role in differential RNA-processing.

8.12   Conclusions and Future Directions

To date, there has only been preliminary work investigating the general question of quadruplexes within the genome. This work has already revealed a number of interesting results, including strong evidence that the PQS identified are not just random artefacts – they show significant evidence of selective pressures, and are non-random in terms of their frequency, location, loop sequence and loop length. This result is highly encouraging, and supports the earlier suggestions that quadruplexes do indeed play a general physiological role.

Further studies, both theoretical and experimental, will be needed to look at the interactions of other physiological parameters, both modelled and experimentally derived, with these PQS. Some examples include looking at the interaction between PQS and nuclease hypersensitive elements (NHEs), or with single nucleotide polymorphism (SNP) positions, or with any of a huge range of other parameters, such as those being generated by the ENCODE project.[63] Another angle will be to look at how PQS are conserved between species, as such conserved elements will clearly be of great interest.

There is also a need for further base-line studies to support this physiological work. Although there are now some algorithms to identify PQS, none of these are very well developed, and further work is required to refine them. It would be valuable to have a model that could predict both the structure (especially anti-parallel *vs.* parallel folds) and the thermodynamic stability of sequences, in the manner currently possible for duplex DNA. This goal will require more detailed analysis of the role of the sequences, and hence a significant extension of the analysis to date.[27] Another issue is that, to date, almost every biophysical study has looked at G-quadruplex formation from a single strand of DNA. There is a need to better understand the equilibrium between quadruplex and duplex structures in a more native context.

I believe this field will continue to develop, and push ever closer to the holy grail of absolute proof of quadruplex involvement in physiological processes – and to contribute to therapeutic treatments.

References

1. T. Simonsson, G-quadruplex DNA structures – variations on a theme, *Biol. Chem.*, 2001, **382**, 621–628.
2. D. Sen and W. Gilbert, Guanine quartet structures, *Meth. Enzymol.*, 1992, **211**, 191–199.
3. E.H. Blackburn, Structure and function of telomeres, *Nature*, 1991, **350**, 569–573.
4. P. Balagurumoorthy and S.K. Bramachari, Structure and stability of human telomeric sequence, *J. Mol. Biol.*, 1994, **269**(34), 21858–21869.
5. A. Siddiqui-Jain, C.L. Grand, D.J. Bearss and L.H. Hurley, Direct evidence for a G-quadruplex in a promoter region and its targeting with a small molecule to repress *c-MYC* transcription, *Proc. Natl. Acad. Sci. USA*, 2002, **99**(18), 11593–11598.
6. A. Rangan, O.Y. Fedoroff and L.H. Hurley, Induction of duplex to G-quadruplex transition in the c-myc promoter region by a small molecule, *J. Biol. Chem.*, 2001, **276**(7), 4640–4646.
7. J. Seenisamy, E.M. Rezler, T.J. Powell, D. Tye, V. Gokhale, C.S. Joshi, A. Siddiqui-Jain and L.H. Hurley, The dynamic character of the G-quadruplex element in the c-MYC promoter and modification by TMPyP4, *J. Am. Chem. Soc.*, 2004, **126**(28), 8702–8709.
8. D. Sen and W. Gilbert, Formation of parallel four-stranded complexes by guanine-rich motifs in DNA and its implications for meiosis, *Nature*, 1988, **334**, 364–366.
9. International Human Genome Sequencing Consortium, Initial sequencing and analysis of the human genome, *Nature*, 2001, **409**, 860–921.
10. B. Rost and C. Sander, Prediction of protein secondary structure at better than 70% accuracy, *J. Mol. Biol.*, 1993, **232**, 584–599.

11. A.K. Todd, M. Johnstone and S. Neidle, Highly prevalent putative quadruplex sequence motifs in human DNA, *Nucleic Acids Res.*, 2005, **33**(9), 2901–2907.

12. J.L. Huppert and S. Balasubramanian, Prevalence of quadruplexes in the human genome, *Nucleic Acids Res.*, 2005, **33**(9), 2908–2916.

13. L. D'Antonio and P. Bagga, Computational methods for predicting intramolecular G-quadruplexes in nucleotide sequences in 2004 IEEE Computational Systems Bioinformatics Conference, 2004.

14. C. Schaffitzel, I. Berer, J. Postberg, J. Hanes, H.J. Lipps and A. Plückthun, *In vitro* generated antibodies specific for telomeric guanine-quadruplex DNA react with *Stylonychia lemnae* macronuclei, *Proc. Natl. Acad. Sci. USA*, 2001, **98**(15), 8572–8577.

15. M. Gellert, M.N. Lipsett and D.R. Davies, Helix formation by guanylic acid, *Proc. Natl. Acad. Sci. USA*, 1962, **48**, 2013–2018.

16. L.C. Bock, L.C. Griffin, J.A. Latham, E.H. Vermaas and J.J. Toole, Selection of single-stranded DNA molecules that bind and inhibit human thrombin, *Nature*, 1992, **355**, 564–566.

17. I. Smirnov and R.H. Shafer, Effect of loop sequence and size on DNA aptamer stability, *Biochemistry*, 2000, **39**, 1462–1468.

18. M. Fry and L.A. Leob, The Fragile X syndrome d(CGG)$_n$ nucleotide repeats form a stable tetrahelical structure, *Proc. Natl. Acad. Sci. USA*, 1994, **91**, 4950–4954.

19. R.A.J. Darby, M. Sollogoub, C. McKeen, L. Brown, A. Risitano, N. Brown, C. Barton, T. Brown and K.R. Fox, High throughput measurement of duplex, triplex and quadruplex melting curves using molecular beacons and a LightCycler, *Nucleic Acids Res.*, 2002, **30**(9), 39.

20. M.W. da Silva, Association of DNA quadruplexes through G:C:G:C tetrads. Solution structure of d(GCGGTGGAT), *Biochemistry*, 2003, **42**, 14356–14365.

21. N. Escaja, J.L. Gelpí, M. Orozco, M. Rico, E. Pedroso and C. González, Four-stranded DNA structure stabilized by a novel G:C:A:T: tetrad, *J. Am. Chem. Soc.*, 2003, **125**(19), 5654–5662.

22. C. Cáceres, G. Wright, C. Gouyette, G.H. Parkinson and J.A. Subirana, A thymine tetrad in d(TGGGGT) quadruplexes stabilized with Tl^+/Na^+ Ions, *Nucleic Acids Res.*, 2004, **32**(3), 1097–1102.

23. A. Matsugami, K. Ouhashi, M. Kanagawa, H. Liu, S. Kanagawa, S. Uesugi and M. Katahira, An intramolecular quadruplex of (GGA)$_4$ triplet repeat DNA with a G:G:G:G tetrad and a G(:A):G(:A):G(:A):G heptad, and its diametric interaction, *J. Mol. Biol.*, 2001, **313**(2), 255–269.

24. P.K. Patel and R.V. Hosur, NMR observation of T-tetrads in a parallel stranded DNA quadruplex formed by *Saccharomyces cerevisiae* telomere repeats, *Nucleic Acids Res.*, 1999, **27**(12), 2457–2464.

25. Y. Krishnan-Ghosh, D. Liu and S. Balasubramanian, Formation of an interlocked quadruplex dimer by d(GGGT), *J. Am. Chem. Soc.*, 2004, **125**, 11009–11016.

26. M. Crnugelj, P. Sket and J. Plavec, Small change in G-rich sequence, a dramatic change in topology new dimeric: G-quadruplex folding motif with unique loop orientations, *J. Am. Chem. Soc.*, 2003, **125**(26), 7866–7871.

27. P. Hazel, J.L. Huppert, S. Balasubramanian and S. Neidle, Loop-length dependent folding of G-quadruplexes, *J. Am. Chem. Soc.*, 2004, **126**, 16405–16415.

28. M. Crnugelj, P. Sket and J. Plavec, Small change in G-rich sequence, a dramatic change in topology: new dimeric G-quadruplex folding motif with unique loop orientations, *J. Am. Chem. Soc.*, 2003, **125**(26), 7866–7871.

29. M. Fry and L.A. Leob, Human Werner syndrome DNA helicase unwinds tetrahelical structures of the fragile X syndrome repeat sequence d(CGG)$_n$, *J. Biol. Chem.*, 1999, **274**, 12797–12802.

30. P. Fojtik, I. Kejnovska and M. Vorlickova, The guanine-rich fragile X chromosome repeats are reluctant to form tetraplexes, *Nucleic Acids Res.*, 2004, **32**(1), 298–306.

31. T. Saha and K. Usdin, Tetraplex formation by the progressive myoclonus epilepsy type-1 repeat: implications of instability in the repeat expansion diseases, *FEBS Lett.*, 2001, **491**, 184–187.

32. P. Castati, X. Chen, R.K. Moyzis, E.M. Bradbury and G. Gupta, Structure-function correlations of the insulin-linked polymorphic region, *J. Mol. Biol.*, 1996, **264**(3), 534–545.

33. M.N. Weitzmann, K.J. Woodford and K. Usdin, The mouse Ms6-hm hypervariable microsatellite forms a hairpin and two unusual tetraplexes, *J. Biol. Chem.*, 2002, **273**(46), 30742–30749.

34. D. Sun, A. Pourpak, K. Beetz and L.H. Hurley, Direct evidence for the formation of G-quadruplex in the proximal promoter region of the RET protooncogene and its targeting with a small molecule to repress RET protooncogene transcription, *Clin. Cancer Res. (Suppl.)*, 2003, **9**(16), A218.

35. S. Cogoi, F. Quadrifoglio and L.E. Xodo, G-rich oligonucleotide inhibits the binding of a nuclear protein to the Ki-*ras* promoter and strongly reduces cell growth in human carcinoma pancreatic cells, *Biochemistry*, 2004, **43**, 2512–2523.

36. J. Christansen, M. Kofod and F.C. Nielsen, A guanosine quadruplex and two stable hairpins flank a major cleavage site in insulin-like growth factor II mRNA, *Nucleic. Acids Res.*, 1994, **22**(25), 5709–5716.

37. M. Isalan, S.D. Patel, S. Balasubramanian and Y. Choo, Selection of zinc fingers that bind single-stranded telomeric DNA in the G-quadruplex confirmation, *Biochemistry*, 2001, **40**, 830–836.

38. S. Patel, Studies on a designed G-quadruplex binding protein that inhibits human telomerase, PhD Thesis, University of Cambridge, Cambridge, UK, 2000.

39. H. Sun, J.K. Karow, K.I.D. Hickson and N. Maizels, The Bloom's syndrome helicase unwinds G4 DNA, *J. Biol. Chem.*, 1998, **273**, 27587–27592.

40. J.-L. Li, R.J. Harrison, A.P. Reszka, R.M. Brosh Jr., V.A. Bohr, S. Neidle and I.D. Hickson, Inhibition of the Bloom's and Werner's syndrome helicases by G-quadruplex interacting ligands, *Biochemistry*, 2001, **40**(50), 15194–15202.

41. R. Giraldo and D. Rhodes, The yeast telomere-binding protein RP1 binds to and promotes the formation of DNA quadruplexes in telomeric DNA, *The EMBO J.*, 1994, **13**(10), 2411–2420.

42. R. Giraldo, M. Suzuki, L. Chapman and D. Rhodes, Promotion of parallel DNA quadruplexes by a yeast telomere binding protein: a circular dichroism study, *Proc. Natl. Acad. Sci. USA*, 1994, **91**, 7658–7662.

43. L. Laporte and G.J. Thomas Jr., Structural basis of DNA recognition and mechanism of quadruplex formation by the b subunit of the *Oxytricha* telomere-binding protein, *Biochemistry*, 1998, **37**, 1327–1335.

44. G. Sarig, P. Weisman-Shomer, R. Erlitzki and M. Fry, Purification and characterization of qTBP42, a new single-stranded and quadruplex telomeric DNA-binding protein from rat hepatocytes, *J. Biol. Chem.*, 1997, **272**(7), 4474–4482.

45. J. Ying, R.K. Bradley, L.B. Jones, M.S. Reddy, D.T. Colbert, R.E. Smalley and S.H. Hardin, Guanine-rich telomeric sequences stimulate DNA polymerase activity *in vitro*, *Biochemistry*, 1999, **38**, 16461–16468.

46. A.M. Zahler, J.R. Williamson, T.R. Cech and D.M. Prescott, Inhibition of telomerase by G-quartet DNA structures, *Nature*, 1991, **350**, 718–720.

47. S.F.A. Grant, D.M. Reid, G. Blake, R. Herd, I. Fogelman and S.H. Ralston, Reduced bone density and osteoporosis associated with a polymorphic Sp1 binding site in the collagen type I alpha 1 gene, *Nat Genet.*, 1996, **14**(2), 203–205.

48. V. Mann, E.E. Hobson, B. Li, T.L. Stewart, S.F.A. Grant, S.P. Robins, R.M. Aspden and S.H. Ralston, A *COL1A1* Sp1 binding site polymorphism predisposes to osteoporitic fracture by affecting bone density and quality, *J. Clin. Invest.*, 2001, **107**(7), 899–907.

49. T. Tomonaga and D. Levens, Activating transcription from single stranded DNA, *Proc. Natl. Acad. Sci. USA*, 1996, **93**(12), 5830–5835.

50. T.L. Davis, A.B. Firulli and A.J. Kinniburgh, Ribonucleoprotein and protein factors bind to an H-DNA-forming *c-myc* DNA element: possible regulators of the *c-myc* gene, *Proc. Natl. Acad. Sci. USA*, 1989, **86**(24), 9682–9686.

51. R.F. Sewell and R. Durbin, Method of calculation of probability of matching a bounded regular expression in a random data string, *J. Comp. Biol.*, 1995, **2**(1), 25–31.

52. R. Staden, Methods of calculating the probabilities of finding patterns in sequences, *Comput. Appl. Biosci.*, 1989, **5**, 89–96.

53. C. Burge, A.M. Campbell and S. Karlin, Over- and under-representation of short oligonucleotides in DNA sequences, *Proc. Natl. Acad. Sci. USA*, 1992, **89**, 1358–1362.

54. R. Nussinov, Nearest neighbour nucleotide patterns: structural and biological implications, *J. Biol. Chem.*, 1981, **256**(16), 8458–8462.

55. J.-L. Mergny, A. De Cian, A. Ghelab, B. Sacca and L. Lacroix, Kinetics of tetramolecular quadruplexes, *Nucleic Acids Res.*, 2005, **33**(1), 81–94.

56. B. Pan, Y. Xiong, K. Shi and M. Sundaralingam, Crystal structure of a bulged RNA tetraplex at 1.1 Å resolution: Implications for a novel binding site in RNA tetraplex, *Structure*, 2003, **11**, 1423–1430.

57. A. Risitano and K.R. Fox, Influence of loop size on the stability of intramolecular G-quadruplexes, *Nucleic Acids Res.*, 2004, **32**(8), 2598–2606.

58. A. Agresti, *Categorical Data Analysis*, Wiley, Hoboken, NY,2002.

59. M.-R. Wilkinson, *Analysing the Frequencies of Loop Lengths of Genomic G-Quadruplex Structures*, University of Cambridge, Cambridge, UK, 2005.

60. V.M. Marathias and P.H. Bolton, Determinants of DNA quadruplex structural type: sequence and potassium binding, *Biochemistry*, 1999, **38**(14), 4355–4364.

61. V. Dapic, V. Abdomerovic, R. Marrington, J. Peberdy, A. Rodger, J.O. Trent and P.J. Bates, Biophysical and biological properties of quadruplex oligonucleotides, *Nucleic Acids Res.*, 2003, **31**(8), 2097–3107.

62. S. Rankin, A.P. Reszka, J.L. Huppert, M. Zloh, G.H. Parkinson, A.K. Todd, S. Ladame, S. Balasubramanian and S. Neidle, Putative DNA quadruplex formation within the human *c-kit* oncogene, *J. Am. Chem. Soc.*, 2005, 10.1021/ja050823u.

63. The ENCODE Project Consortium, The ENCODE (ENCylopedia Of DNA Elements) Project, *Science*, 2004, **306**, 636–640.

CHAPTER 9

Quadruplexes and the Biology of G-Rich Genomic Regions

NANCY MAIZELS

Departments of Immunology and Biochemistry, University of Washington Medical School, Seattle, WA, USA

When we picture DNA, the image that comes immediately to mind is the iconic double helix. However, the double helix is only the storage form for genetic information, and DNA takes on very different structures when actively used during transcription, replication, and recombination. G-rich genomic regions have unusual structural potential, as they readily form G-quadruplex structures. This chapter discusses how regulated formation of G-quadruplexes contributes to key cellular processes.

9.1 Structural Potential of G-Rich DNA

DNA containing runs of guanines readily forms structures stabilized by hydrogen bonding between guanines.[1] The basic unit of these structures is the G-quartet, a planar array of four guanines, in which each guanine pairs with two neighbors by hydrogen bonding[2] (Figure 1). G-quartets can promote inter- or intra-molecular interactions, between parallel or anti-parallel DNA strands.[1,3,4] DNA structures stabilized by G-quartets are variously referred to as G-quadruplex, G-tetraplex, and G4 DNA. For simplicity, in this chapter we use the terms G4 DNA and G-quadruplexes to include all structures stabilized by G-quartets. The details of G4 structure are discussed in Chapter 1.

9.1.1 G-Rich Nucleic Acids Spontaneously Form G4 DNA or G4 RNA *in Vitro*

Formation of G-quadruplex structures occurs spontaneously in synthetic DNAs which contain at least four runs of guanines, each at least 3 nt in length (GGG). A minimalist example is (TTAGGG)$_4$, a 4-mer repeat of the vertebrate telomeric sequence, which readily forms G4 DNA.[5] Some sequences containing shorter G-runs will form G-quadruplexes, although this may depend upon the number and identity of bases in the loops between the G-planes, and the

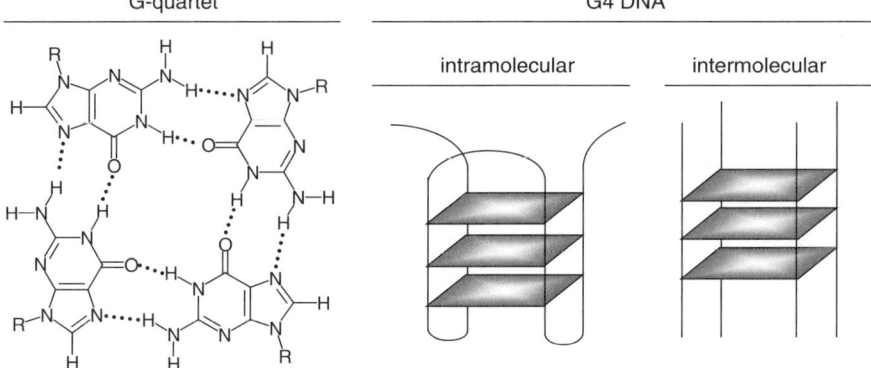

Figure 1   *G-quartet and G4 DNA. Left: G-quartet, a planar array of four guanines, in which each guanine pairs with two neighbors by hydrogen bonding. Right: G4 DNA. G-quartets can promote intra- or intermolecular interactions between DNA strands, as shown in the figure. Parallelograms represent G-quartets*

structures may be only marginally stable, if stable at all. Sequences like the telomeric repeat can be described as "G-rich," to indicate strand bias as well as sequence composition. This contrasts with "G/C-rich," which describes overall base composition without regard to strand bias.

9.1.2   The Uses and Limitations of Chemical Probing in Assaying G4 DNA Formation

The presence of G4 DNA can be assayed by footprinting DNA with dimethyl sulfate (DMS).[6] DMS attacks the N7 of guanine, which is accessible in single-stranded and duplex DNA, but hydrogen-bonded with the exocyclic amino group of a neighboring guanine in a G-quartet (Figure 1). Methylation of guanine renders the glycosidic bond unstable to heating. DMS modification is assayed by heating the DNA, which creates abasic sites, followed by treatment with mild alkali to break the phosphodiester backbone at the abasic sites. Fragments are then resolved by gel electrophoresis, either directly or following PCR amplification.

DMS footprinting is very useful for assaying the presence of G4 DNA in a short defined region of sequence, especially a synthetic oligonucleotide. However, DMS footprinting can only measure the average protection of any specific guanine. If a G-rich region forms G4 DNA but the structure is heterogeneous rather than homogenous (see Figure 2A), DMS footprinting may produce an apparently negative result. This limits the utility of DMS footprinting for identifying G-quadruplexes formed within long G-rich sequences, since a different structure may form within each individual molecule.

9.1.3   G4 DNA is Stable

G4 DNA is very stable once formed. Stability derives from hydrogen bonding between guanines; from stacking of the hydrophobic G-quartets; and from the

A   **Transcription**

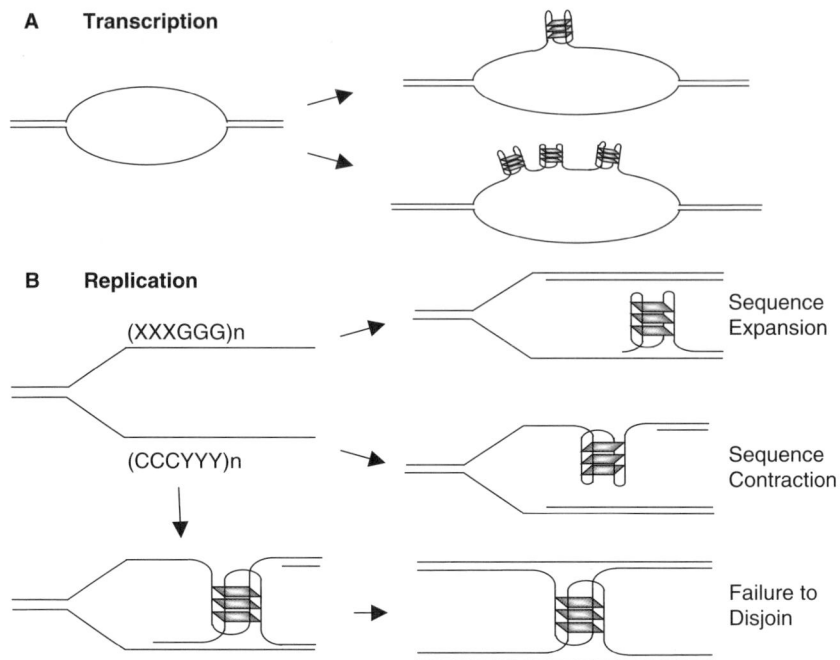

B   **Replication**

(XXXGGG)n   Sequence Expansion

(CCCYYY)n   Sequence Contraction

                                                               Failure to Disjoin

Figure 2 *Formation of G4 DNA during transient DNA denaturation that accompanies transcription or replication. (A) Transcription results in formation of intramolecular G4 DNA. Within a long G-rich region, structure formation may occur at different positions in different molecules, as diagrammed in the two examples shown. (B) Replication results in formation of G4 DNA. If this occurs on a single strand, it may lead either to sequence expansion or contraction. If intermolecular G4 DNA forms, it may result in failure of newly replicated daughter molecules to disjoin (below)*

presence of a monovalent cation within the channel formed by the stacked G-quartets.[7] Potassium in this channel confers particular stability. The intracellular potassium ion concentration in a mammalian cell is 120 mM, creating an intracellular ionic environment which will maintain G-quadruplex structures once formed. G4 DNA resists thermal denaturation, but it can be denatured by treatment with alkali or by removal of potassium.

9.1.4   G4 DNA has a Distinctive Structure and Enzymology

G-quadruplex and duplex DNA are structurally distinct. The G-quartets produce a core of a very large diameter, encircled by four phosphodiester backbones separated by grooves narrower than the minor groove of *B*-form DNA.[8] This distinctive structure enables G4 DNA to resist attack by even the most potent of the enzymes that target duplex DNA. Specific enzymes are therefore required to maintain G-rich genomic regions. Enzymes have been identified that bind, unwind, and cleave G4 DNA with great affinity and specificity. In addition, some factors can promote formation of G4 DNA within a single-stranded G-rich region, while other factors disrupt G4 DNA. Dynamic formation and disruption of G-quadruplexes confers specialized properties upon G-rich sequences in living cells.

9.1.5 G4 DNA Formation *In vivo* Could Threaten Genomic Stability

The vast majority of genomic DNA is normally maintained as a Watson–Crick duplex, in which G–C pairing prevents formation of G4 DNA. However, guanines within a G-rich region have the potential to form G4 DNA during transient denaturation accompanying transcription, replication, and recombination. Figure 2 illustrates how G4 DNA can form upon transcription or replication of a G-rich region. If intramolecular G4 DNA is not eliminated, it could lead to expansion or contraction of the G-rich region, analogous to the instability documented for triplet repeats.[9] Intermolecular G4 DNA formation could interfere with disjunction of daughter DNA molecules following replication. As discussed elsewhere in this chapter, cells have mechanisms to prevent this instability.

9.2 G-Rich Genomic Regions

Eukaryotic cells contain many G-rich regions capable of forming G4 DNA. Among these are repetitive and functionally essential chromosomal domains, including the telomeres, the rDNA, and the immunoglobulin heavy chain switch regions of higher vertebrates. Repeats comprise over half the human genome, and many minisatellite and microsatellite repeats are G-rich and have the potential to form G4 DNA. G-rich regions are also found within specific single-copy genes.

9.2.1 Telomeres

In almost all eukaryotes, the telomeric repeat contains runs of G's. Some examples include the yeast *S. cerevisiae*, TG_{1-3}; the ciliates *Tetrahymena*, TTTTGGGG, and *Oxytricha nova*, TTGGGG; the plant, Arabidopsis, TTTAGGG; and the vertebrate repeat, TTAGGG. Telomeres contain a duplex region, in which a G-rich strand is base-paired to the complementary C-rich strand, and they terminate with a 3′ single-stranded overhang on the G-rich strand. G-rich telomeric overhangs readily form G-quadruplex structures *in vitro*, which are characterized not only by the presence of G-quartets but also by distinctive loops determined by the sequence of bases between the G-runs.[5,10] The notion that regulated formation of G4 DNA may be central to telomere biology received considerable attention at the time G4 DNA was first described,[1] and it has received considerable recent support, as described in a later section.

9.2.2 rDNA

Eukaryotic rDNA is G-rich, with G-runs concentrated on the non-template strand and abundant in both coding and spacer regions. The rDNA is heavily transcribed, which could enhance G4 DNA formation by prolonging the DNA denaturation that normally accompanies transcription.

9.2.3 Immunoglobulin Heavy Chain Switch Regions

In mammals, the immunoglobulin heavy chain switch (S) regions comprise a third G-rich chromosomal domain. S regions are repeats from 2 to 10 kb in length, which consist of reiterations of highly degenerate consensus sequences. The S regions are critical for class switch recombination, a region-specific recombination process which joins an expressed variable region to a new constant region, deleting many kb of DNA.[11]

9.2.4 Hypervariable Repeats

The human genome is replete with repetitive regions, including both short minisatellite repeats (repeat unit less than 14 bp) and larger microsatellite repeats. Many of them exhibit pronounced instability, and among the most unstable are G-rich sequences. These include the MS1 repeat (D1S7), AGGGTGGAG; D4S43, GGGGAGGGGGAAGA; the insulin-linked hyper-variable repeat, ACAGGGGTGTGGGG; MS32 (29 bp), CEB1 (37–43 bp), D1Z2 (40 bp); and MS205 (45–54 bp), to name a few.[12–15] In contrast to repeats which are unstable as a result of replication slippage,[16,17] these G-rich sequences cannot form stable hairpin structures; however, they almost certainly can form G4 DNA. G4 DNA formation has been directly confirmed for two G-rich repeats, D4S43, and the insulin-linked hypervariable repeat.[15]

9.2.5 G-Rich Single-Copy Genes

G-richness is not restricted to repeated sequences. Many mammalian single-copy genes contain G-rich regions, and the potential importance of regulated G-quadruplex formation to genomic biology has stimulated development of algorithms for determining whether a sequence can form G DNA, and analyses of genomic sequences to identify regions with structural potential. Tellingly, searches of the human genome sequence using algorithms designed to detect potential for intramolecular G4 DNA formation has shown that such sequences are typically depleted in coding exons.[18] Consistent with the view that G-richness confers specific properties, specific sequence motifs are prevalent in the "loops" connecting the G-barrels.[19]

9.3 G4 DNA Forms in Living Cells

Despite the readiness with which G4 DNA forms *in vitro*, skepticism about the biological importance of G4 DNA persisted for many years because it had not been directly identified in living cells. Two lines of experiments have recently shown that G4 DNA formation occurs intracellularly.

9.3.1 G-Loops and G4 DNA Formation in Transcribed G-Rich Regions

G4 DNA has been directly identified within transcribed G-rich regions in experiments that used electron microscopy to visualize structures formed in individual molecules.[20] Transcription of G-rich templates was shown to cause formation of characteristic loops (Figure 3), which contained G4 DNA interespersed with single-stranded regions on the G-rich strand. A stable RNA/DNA hybrid formed on the C-rich strand, as predicted by experiments which used either gel electrophoresis[21] or atomic force microscopy[22] to analyze products of transcription on G-rich templates. The presence of G4 DNA was demonstrated by probing the transcribed molecules with two reagents specific for G4 DNA. A recombinant derivative of nucleolin with very high affinity for G4 DNA[23] was shown to decorate one strand of the loops; and GQN1, a mammalian nuclease which specifically cleaves G4 DNA[24] was shown to cleave G-rich strand of the loops.

The loops formed upon transcription of G-rich regions are called "G-loops" to emphasize that their structure depends upon G-richness of the transcribed region, and they contain G4 DNA.[20] G-loop formation is transcription-dependent, and occurs efficiently upon transcription either *in vitro* or intracellularly. The RNA/DNA hybrid in a G-loop forms cotranscriptionally, not by invasion of one DNA molecule by RNA synthesized on another template. Stability of the RNA/DNA hybrid in a G-loop is determined in part by base pairing. Inosine (I) forms a stable base pair with cytosine, but in contrast to the G/C pair, which contains three hydrogen bonds, an I/C pair contains only two hydrogen bonds, and G-loops do not form in transcription reactions in which GTP has been replaced by ITP. G-loop formation also depends at least in part on the unusual stability of rG/dC base pairs,[25] because G-loop formation occurs only in one transcriptional orientation, with a C-rich strand as template for RNA synthesis. This is the physiological orientation of transcription in both the G-rich rDNA and immunoglobulin heavy-chain S regions.

The potential for formation of alternative structures is a common property of G-rich regions: G-rich sequences derived from the mammalian telomeric

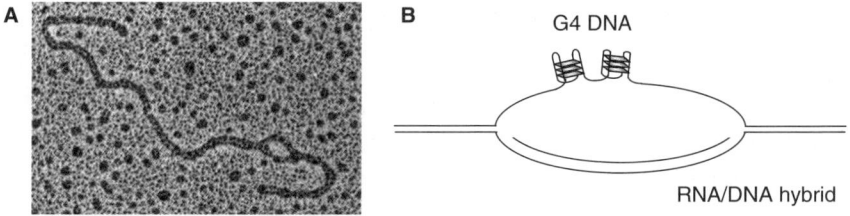

Figure 3 *G-loops form in transcribed G-rich regions. (A) Electron micrograph of transcribed G-rich plasmid template (courtesy of Michelle Duquette). (B) Diagram of G-loop, showing the RNA/DNA hybrid on the template strand and G4 DNA interspersed with single-stranded regions on the G-rich strand*

repeat, the immunoglobulin class switch regions, synthetic G-rich repeats, and the c-*myc* gene have all been shown to form G-loops upon transcription.[20,26] The readiness with which G-loops and G4 DNA form provides support to the notion that biology takes advantage of regulated G4 DNA formation at G-rich genomic regions.

9.3.2 G4 DNA in Ciliate Macronuclei

Ciliates offer an extraordinary model for studies of telomere biology. Following sexual reproduction, genomic DNA becomes fragmented into millions of gene-size chromosomes, each of which carries its own short telomeres.[27] The minichromosomes are contained within the macronucleus, which can be readily visualized by microscopy. A high-affinity single-chain antibody which recognizes G4 DNA has been shown to decorate the macronucleus but not the micronucleus or the replication band in *Stylonychia*, and staining is specific for antiparallel G4 DNA.[28] This suggested that regulated formation of antiparallel G4 DNA occurs at telomeres of ciliate minichromosomes as a result of interactions between telomeric overhangs on two minichromosomes (see Figure 4A). As discussed in greater detail below, recent experiments show that telomere end-binding proteins regulate formation of G4 DNA not only at ciliate telomeres, but also at human telomeres.

9.4 Critical Roles of G-Rich Telomeres in Health and Disease

Telomeres are specialized structures at the ends of linear chromosomes, which are essential for maintenance of chromosomal and genomic integrity. They are composed of telomeric DNA bound to specific and dynamic protein complexes. Telomeric DNA in essentially all organisms consists of G-rich repeated duplex sequence, terminating with a single-stranded 3′ G-rich overhang. In living cells, telomeres can form lariat-shaped structures, called "t-loops," in which the G-rich overhang at the very end of the chromosome interacts with a distant region of telomeric DNA.[29]

Telomere length is maintained by telomerase, an enzyme composed of one polypeptide subunit and one RNA subunit.[30,31] The catalytic subunit of telomerase bears homology to retroviral reverse transcriptases, consistent with its role in catalyzing synthesis of DNA on an RNA template. The 3′ end of a telomeric G-rich overhang is the primer for addition of new telomeric sequence, which is templated by the complementary sequence in the telomerase RNA subunit. *In vitro*, G-rich telomeric sequences readily form G-quadruplexes with distinctive structural features determined by telomere sequence and strand orientation.[10] G4 DNA formation could in principle promote telomere–telomere interaction [Figure 4(A)]; protect the 3′ end from extension or nucleolytic attack [Figure 4(B)]; or stabilize t-loops [Figure 4(C)].

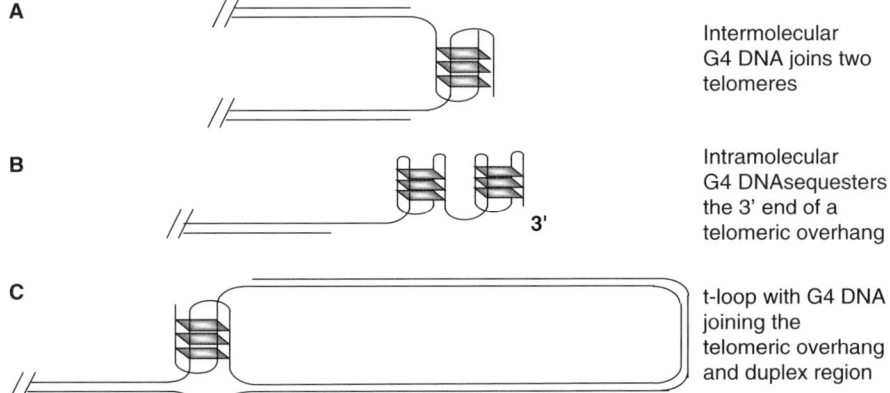

A Intermolecular G4 DNA joins two telomeres

B Intramolecular G4 DNAsequesters the 3' end of a telomeric overhang

3'

C t-loop with G4 DNA joining the telomeric overhang and duplex region

Figure 4 *Formation of G4 DNA at telomeres. (A) Intermolecular G4 DNA promotes telomere–telomere interactions. (B) Intramolecular G4 DNA sequesters the 3' end of a telomere to prevent extension by telomerase. (C) The G-rich overhang invades telomeric sequence to form G4 DNA, stabilizing a t-loop*

9.4.1 Telomeres in Aging and Cancer

Telomere length is a critical determinant of cellular lifespan. In primary cultured human cells, and in most human tissues, telomere length diminishes with each successive cell division. In humans, telomerase activity is high in germ cells, developing embryos, and very young children, but it is downregulated in most tissues starting at approximately age four.[32,33] Diminished telomerase activity is due to downregulation of expression of the catalytic subunit of telomerase. In cells that must proliferate, such as B and T lymphocytes, reactivation of telomerase accompanies cell activation. The possibility that telomeres do not simply reflect but actually control cellular lifespan was articulated in the "Hayflick hypothesis," which proposed that telomeres are a clock that counts the number of mitotic divisions allowed to each cell.[34] This hypothesis was first borne out by experiments which showed that ectopic expression of telomerase permits cells to avoid senescence and proliferate indefinitely in culture.[35]

Tumor cells must maintain telomere length in order to divide indefinitely. Most tumors (85–90%) contain telomerase activity, which supports cell proliferation. The importance of telomerase to tumor proliferation has been established in a simple experiment, which demonstrated that transfection with a construct expressing a dominant-negative telomerase catalytic subunit would halt cell proliferation.[36] Telomerase is therefore an exciting and promising target for therapeutics.[37] In a minority of tumors, telomerase activity is absent, and recombination-dependent mechanisms are responsible for telomere maintenance.[38] This pathway for telomering lengthening, called the alternative lengthening of telomeres, or "ALT" pathway, substitutes for telomerase in about 10–15% of human cancers. This raises the possibility that therapeutics directed against telomerase in cancer cells might inadvertently select for a population of cells, which use the ALT pathway, and are therefore therapy-resistant.

9.4.2 Telomeric G4 DNA as a Therapeutic Target

Hypothesizing that drugs that interact with G4 DNA might interfere with telomere replication and thus inhibit cell proliferation, a number of laboratories have worked to identify or design small molecule ligands, which specifically bind to G4 DNA. Candidate molecules have been identified which have high affinity for G4 DNA and inhibit cell proliferation, and some of these compounds have been shown to inhibit telomerase or to affect telomere length.[39–44] Evidence that regulated G-quadruplex formation occurs at telomeres, described below, will undoubtedly provide further impetus to development of drugs, which interact with or promote formation of G4 DNA. It will of course be important that compounds be designed and tested for target specificity. The mammalian telomeric repeat is known to form a distinctive G-quadruplex structure.[5] This should allow design of ligands that can specifically target telomeric G4 DNA.[45]

9.4.3 Genetic Diseases Due to Impaired Telomerase Activity

Human genetic diseases can result from impaired telomerase activity.[46] These diseases are clinically heterogeneous but typically characterized by failure in blood cell production, leading to aplastic anemia. This reflects the importance of cell proliferation to development of blood cells, which are a constantly renewing tissues. Other tissues affected are hair, nails, and skin. Telomerase deficiency can result from mutations in *TERC*, which encodes the RNA component of telomerase; to mutations in *TERT*, the gene encoding the catalytic protein subunit of telomerase; or from an absence of the nucleolar enzyme dyskerin, which converts uracil to pseudouracil and is essential for posttranscriptional base modification of telomerase RNA.[47,48] These mutations show "anticipation:" effects become increasingly severe with successive generations, as telomeres from a normal ancestor gradually erode. Moreover, levels of telomerase are limiting, so disease can develop in haploids despite the presence of some functional telomerase.

9.4.4 Animal Models for Human Telomerase Deficiencies

Mouse models for telomerase deficiency have been extensively studied and shown to recapitulate human disease.[49–51] Inbred mice have longer telomeres than *Homo sapiens*, so the effects of diminished telomerase activity are not apparent in the first few generations following deletion of the m*Terc* gene, which encodes murine telomerase RNA, but become apparent in the third or fourth generation as diminished fertility, chromosomal anomalies, alopecia (baldness), and bone-marrow failure. Chromosomal aberrancies documented in cells lacking telomerase are analogous to those evident in cancer cells, suggesting that telomere dysfunction contributes to genomic destabilization during malignant progression.[52] The availability of robust animal models permits therapies to be tested in a physiological setting. This is especially important with a factor like telomerase, which is regulated in tissue-specific fashion and which displays profound tissue-specificity when expression is impaired by mutation.

9.5 Formation of G4 DNA Regulates Telomere Extension by Telomerase

The G-richness of telomeric repeats in a great variety of organisms suggested that G4 DNA formation could contribute to telomere function. Biochemical analysis has shown that formation of G4 DNA can prevent extension of a 3' end by telomerase.[53–55] Recent experiments have now shown that conserved telomere end-binding proteins regulate G4 DNA formation at telomeres.[55,56] Results of these experiments have convincingly established that formation of G-quadruplexes at telomere ends could be key to regulation of telomere length and, in turn, cellular lifespan.

9.5.1 Telomere End-Binding Proteins

Proteins that bind specifically to telomeric overhangs have been identified in essentially every species. One of these proteins, POT1, is conserved in fission yeast through humans.[57] In humans, POT1 is found at telomeres in a six-member complex which includes the duplex telomere binding proteins, TRF1 and TRF2.[58] POT1 binds cooperatively and with considerable specificity to single-stranded DNA containing the telomeric repeat, with preferential binding at the 3' end of DNA.[59] This suggests that POT1 sequesters the 3' end of the telomere, acting as a cap to protect it from degradation.

This picture appeared to be confirmed by structural analysis. The N-terminal region of POT1 contains conserved oligonucleotide/oligosaccharide-binding (OB) folds homologous to those of the telomere end binding proteins (TEBPs) from ciliates and *S. cerevisiae*.[60] A high resolution (1.73 Å) crystal structure of the N-terminal region of human POT1 complexed to a 10 nt sequence of the telomeric repeat, d[TTAGGGTTAG], revealed that one of two OB folds binds to 6 nt of telomeric DNA, and the other protects the very 3' end.[61] Similar results had been obtained from analyses of the N-terminal half of *S. pombe* POT1 complexed to telomeric DNA;[62] and from a 2.8 Å resolution crystallographic analysis of the more distantly-related telomere end-binding complex from the ciliate, *Oxytricha nova*, complexed to d[$G_4T_4G_4$].[63] These structures led to the view that POT1 and its homologs binds to unstructured DNA ends to repress telomere extension and prevent telomere erosion.

The cocrystal structures cited above provided no evidence of G-quartets or G4 DNA formed within the G-rich telomeric DNA complexed to the protein. While a higher resolution structure (1.86 Å) of the *Oxytricha* telomere end-binding complex identified G-quartets stabilizing antiparallel interactions between two telomeric oligonucleotides, protein/G4 DNA interactions were not extensive, and it was not clear whether G4 DNA formation was biologically relevant or reflected constraints imposed by crystallization.[64] The absence of G-quartets or G-quadruplexes in the cocrystal structures was taken by many as evidence that, despite the propensity of the G-rich telomeric overhangs to form G4 DNA *in vitro*, telomere binding proteins do not recognize this DNA structure. This view was recently overturned by two papers demonstrating that regulated G-quadruplex formation can determine the availability of the 3′ single-stranded end of the telomere to prime extension by telomerase.

9.5.2   POT1 Disrupts G-Quadruplexes to Promote Telomere Extension by Telomerase

Evidence for POT1 function in regulating formation of G4 DNA at human telomeric repeats comes from recent biochemical analysis.[55] Telomerase had earned a reputation as a relatively promiscuous enzyme, able to extend a great variety of short telomeric and non-telomeric primers. Extension typically produced a characteristic 6 nt ladder, reflecting cycles of copying the 6 nt template sequence, followed by translocation. However, longer telomeric oligonucleotides prove to be poor primers for recombinant human telomerase.[55] Inability to prime telomerase extension correlate with the ability of an oligonucleotide to form G4 DNA; an example of a poor primer is d[GGG(TTAGGG)]₃, which represents a sequence found in telomeric tails. Addition of hPOT1 was shown to restore processive elongation, in stoichiometric fashion, to long telomeres capable of forming G-quadruplex structures. This same effect could be achieved by adding a complementary oligonucleotide, which base-paired with the 5′ end of the primer to prevent formation of intramolecular G4 DNA. Taken together with evidence that hPOT1 can bind G4 DNA, these results lead to a model in which formation of G4 DNA at telomeric G-rich overhangs prevents extension by telomerase, and thereby limits telomere length; and POT1 binding disrupts G4 DNA, to allow the 3′ end to function as a primer for telomerase [Figure 5(A)].

This model for POT1 function [Figure 5(A)] provides mechanistic understanding of an earlier observation that ectopic expression of POT1 stimulates telomere elongation in cultured human cells, but only if telomerase activity is present in those cells.[65] This model is consistent with the hypothesis that stabilization of telomeric G4 DNA could prevent telomere extension and limit cell proliferation, which is central to current development of therapeutics, as described elsewhere. Nonetheless, it is important to remember that regulation of telomere length in living cells is almost certainly more complex. The proteins which interact with telomere ends, including POT1, are likely to themselves be targets of regulation in human cells, as they are known to be in ciliates (see below), and could respond dynamically to small molecule-induced stabilization of telomeric G-quadruplexes.

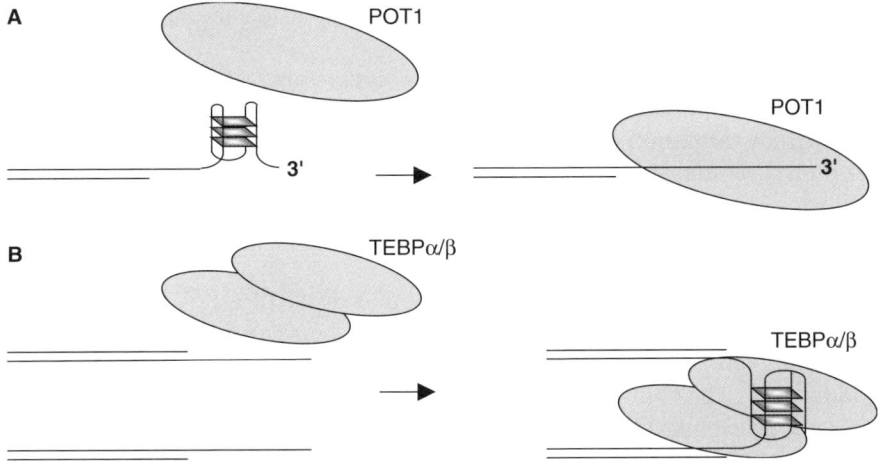

Figure 5 *Interactions of end-binding proteins with telomeric G-rich overhangs. (A) Human POT1 binding disrupts G-quadruplexes to allow extension by telomerase at the 3' end of a telomeric overhang. (B) The ciliate end-binding complex TEBPα/β promotes G-quadruplex formation to prevent telomere extension*

9.5.3 TEBPα/β Regulate G4 DNA Formation at Ciliate Telomeres

Telomere end-binding proteins regulate G4 DNA formation not only in human cells, but also in ciliates. The macronuclear DNA of these unicellular eukaryotes consists of literally millions of short chromosomes, with short telomeres containing both duplex regions and a short single-stranded overhang. A heterodimeric factor, TEBPα/β, recognizes and binds to the single-stranded telomeric overhang, and TEBP binding prevents extension by telomerase.[66] This could in principle reflect formation of G4 DNA, which is a poor substrate for telomerase;[53,54] or inaccessibility of the 3' end of the tail due to protection by bound protein.

A compelling case for regulated formation of G-quadruplexes at telomeres has been made in recent experiments which used RNAi to silence TEBPα or TEBPβ in ciliate macronuclei.[56] Silencing of either subunit of the TEBP complex caused loss of telomeric DNA, and cell death, accompanied by the loss of detectable macronuclear G4 DNA as assayed by antibody-staining. TEBPβ had long been known to promote formation of G4 DNA by the single-stranded telomeric repeat, in a reaction which is sequence-specific and first-order with respect to DNA concentration.[67] These observations lead to a model in which TEBPα/β binds telomeric overhangs to stabilize G4 DNA [Figure 5(B)]. Despite the structural homology of ciliate TEBPs and POT1, there appears to be an important difference in the roles these proteins play in regulated formation of G4 DNA. Binding by POT1 disrupts G-quadruplexes to allow extension by telomerase; while binding by TEBPα/β promotes

G-quadruplex formation to prevent telomere extension (compare Figure 5A and B).

Why was G4 DNA not apparent in the cocrystals of telomeric DNA with ciliate TEBP complexes?[63,64] An explanation for this apparent paradox was provided by mutational analysis of TEBPβ, which showed that the basic C-terminal region of TEBPβ is necessary to promote G4 DNA formation.[56] Following the common "divide and conquer" strategy for structural analysis, this region had been truncated in the TEBPα/β heterodimer analyzed in the cocrystals.[63,64]

Telomerase is not active on telomeres bound by TEBPα/β.[66] The complex of TEBPα/β with telomeric DNA must therefore be released to allow telomere extension. This appears to be regulated by cell cycle-dependent protein modification. The C-terminal of TEBPβ has been found to carry two conserved sites for phosphorylation by the Cdk2 kinase, which promotes cell-cycle progression.[56] The phosphorylated protein cannot interact with TEBPα to form a telomere binding complex. Cell-cycle dependent phosphorylation could thereby promote dissociation of the TEBPα/β complex from the telomere, allowing G-quadruplexes to unfold and thereby permitting extension by telomerase.

9.6   G4 DNA in Immunoglobulin Class Switch Recombination

Immunoglobulin heavy chain class switch recombination provides a dramatic example of regulated function of G-rich regions in genomic rearrangement. Class switch recombination is a regulated, irreversible, essential process of DNA deletion which alters genomic structure at the immunoglobulin heavy chain locus.[11] The C region determines how antigen is removed from the body, and recombination modifies how an Ig molecule removes antigen without affecting specificity of antigen-recognition by literally *switching* C regions, thus causing a change in the isotype (or "class") of the expressed antibody. Switch recombination joins a rearranged and expressed variable (V) region to a new downstream constant (C) region, deleting the DNA between as an excised switch circle (Figure 6).

9.6.1   G-Rich Switch (S) Regions are Targets for Recombination

Switch recombination is targeted to repetitive and G-rich regions, called switch (S) regions, which are the sites of chromosomal and extrachromosomal junctions formed upon recombination. S regions are 2–10 kb in length and located in the intron upstream of those C regions that participate in switch recombination: Cμ, Cγ, Cε, and Cα (Figure 6). There is no S region upstream of Cδ, and RNA processing rather than DNA recombination regulates Cδ expression. Switch recombination joins two S regions to produce junctions which display limited microhomology (<4 nt), and which are heterogeneous in sequence and in the sites of breakpoints within both the donor and the

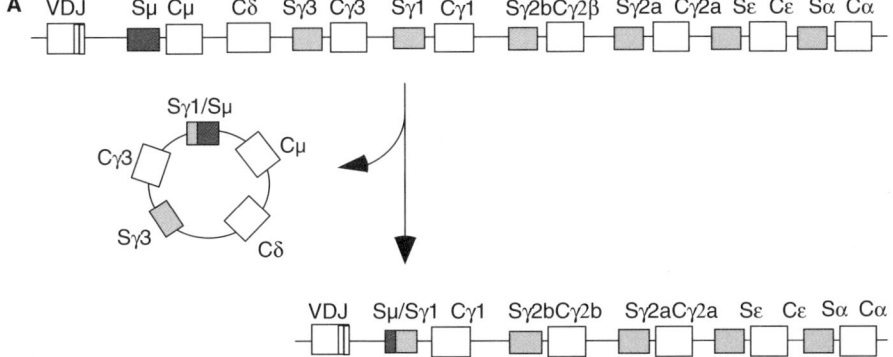

B <u>Murine IImmunoglobulin Switch   Region Consensus Repeats</u>

Sμ   GCTGAGCT<u>GGGG</u>TGAGCTGA

Sγ3   <u>GGGG</u>ACCA<u>GG</u>CT<u>GGG</u>CAGCTCT<u>GGGGG</u>AGCT<u>GGGG</u>TA<u>GG</u>TT<u>GGG</u>AGTGT

Sγ1   ACCCA<u>GG</u>CAGAGCAGCTCCA<u>GGGG</u>AGCCA<u>GG</u>ACA<u>GG</u>T<u>GG</u>AAGTGT<u>GG</u>TG

Sα   <u>GGG</u>ATGAGCTGAGCTA<u>GG</u>CT<u>GG</u>AATA<u>GG</u>CT<u>GGG</u>CT<u>GGG</u>CT<u>GG</u>TGTGAGCT<u>GGG</u>TT

Figure 6   *Immunoglobulin heavy chain class switch recombination. (A) The murine heavy chain locus is shown before and after recombination from μ to γ1, which results in expression of IgG1 antibodies. Deleted sequences can be recovered in circular DNA molecules from B cells which have recently completed switch recombination. V, variable region; C, constant region; and S, switch region. (B) Examples of consensus G-rich repeats from some murine S regions*

acceptor S region.[68] Switch junctions are located within introns, so the imprecision of the DNA recombination event leaves no mark on the heavy chain polypeptide.

9.6.2   Structure-Specific Recombination at Transcribed S Regions

Switch recombination is activated by transcription of the targeted S regions, driven by dedicated promoters/enhancers upstream of each S region. S region transcription *per se* is essential, not the transcript or translation products. Targeted deletion of an S region promoter affects switching only *in cis*, abolishing recombination only on the mutant allele. Transcription alters the structure of S region DNA, creating an active molecular partner for recombination. Transcribed S regions form G-loops, which contain a stable, cotranscriptional RNA/DNA hybrid on the C-rich template strand,[20–22,69–72] and intramolecular G4 DNA interspersed within single-stranded regions on the G-rich strand.[20,26] As shown above [Figure 3(A)], G-loops can be readily imaged by electron microscopy transcribed of S region sequences. G4 DNA in G-loops is the target for at least two factors genetically linked to switch recombination, BLM helicase and MutSα.

9.6.2.1 *BLM Helicase in Immunoglobulin Class Switch Recombination*

Bloom's syndrome, the genetic disease resulting from deficiency in BLM helicase, is characterized by a predisposition to malignancy and also immunodeficiency. In individuals affected with Bloom's syndrome, production of immunoglobulin-producing plasma cells and class switch recombination is impaired.[73] BLM helicase actively unwinds G4 DNA.[74,75] Deficient unwinding of G4 DNA that forms at transcribed switch regions may contribute to or cause the immunodeficiency associated with Bloom's syndrome.

9.6.2.2 *MutSα Targets G4 DNA in G-loops to Promote S Region Synapsis*

In classical mismatch repair, MutSα (the MSH2/MSH6 heterodimer) recognizes DNA mismatches and short loops, and stimulates excision of the mismatch or structured region, followed by DNA replication.[76] MutSα has also been proven to recognize G-quadruplex structures.[77] This possibility was raised by genetic analysis showing that, in *Msh2*- and *Msh6*-deficient mice, levels of switch recombination decrease, and switch junctions lose their characteristic sequence heterogeneity.[78] Binding assays with recombinant human MutSα, expressed in and purified from insect cells, identified G4 DNA as a high affinity substrate (apparent $k_D = 1$ nM) of MutSα. MutSα is enriched at transcribed S regions in switching B cells, consistent with participation in the recombination mechanism. Moreover, EM imaging showed that purified recombinant MutSα not only binds to G4 DNA within G-loops, but also promotes interactions between G-loops, effectively causing synapsis. These results lead to a model for MutSα function in the synapsis step of class switch recombination: MutSα binds G4 DNA, and MutSα molecules interact to synapse the transcribed S-regions.

9.7 Special Properties of G-Rich Single-Copy Genes: c-*myc*

The human genome contains many non-repetitive regions that are G-rich and have the potential for G4 DNA formation.[18,19] A key current challenge is to determine whether and how this affects gene function. At present, one fascinating gene, c-*myc*, provides tantalizing insights into the potential impact of G4 DNA formation on genetic stability and gene function.

The c-*myc* gene encodes a transcription factor that is a key regulator of cell proliferation. Deregulated c-*myc* expression contributes to tumorigenesis in many different tissues. A variety of mechanisms contribute to deregulation of c-*myc*, including translocation, mutation, and amplification. The c-*myc* gene contains three exons and three introns; and c-*myc* expression can be driven by several upstream promoters (Figure 7). Firing of these promoters depends on

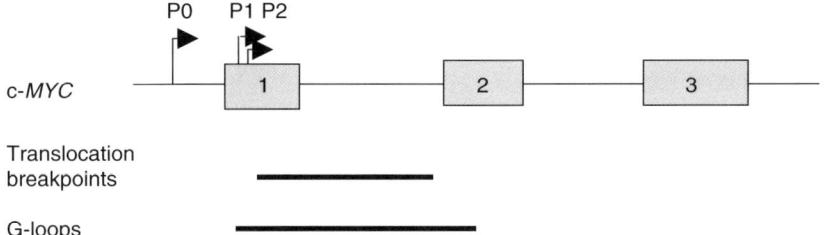

Figure 7   *G4 DNA forms in regions of c-myc where translocation breakpoints cluster. The human c-myc locus is diagrammed, with exons 1, 2, and 3 indicated as boxes, and start sites of transcription from promoters P0, P1, and P2 marked. Heavy lines indicate translocation breakpoints identified in B cell lymphomas, and region, which forms G-loops upon transcription*

tissue type and whether c-*myc* expression occurs in a normal or transformed cell. c-*myc* is unusual in that the first AUG start codon is within intron 2. The promoter, exon 1 and intron 1 of c-*myc* all contain strikingly G-rich regions, raising the possibility that G-quadruplex formation is a key determinant of gene regulation or genetic stability.

9.7.1 Quadruplex Formation by the c-*myc* Promoter

Oligonucleotides bearing a G-rich region from the c-*myc* promoter readily form G-quadruplex structures *in vitro*.[79,80] Structure formation appears to impair transcription, suggesting that small molecules that stabilize G4 DNA could prove useful as antitumor drugs. Porphyrins can interact with and stabilize G4 DNA, and derivatives can be generated with considerable selectivity for the structure formed at the c-*myc* promoter.[81] Sequences in the promoter of another oncogene, c-*kit*, will similarly form G-quadruplex structures, suggesting that this may be a general strategy for therapeutic intervention.[82] Relatively little is yet known about the mechanisms by which G-rich sequences in promoters contribute to regulation of gene expression. Greater understanding of these mechanisms, and identification of the entire subset of genes at which G-quadruplex formation contributes to regulation, should aid in establishing biological specificity of such compounds.

9.7.2 c-*myc* Translocations in B Cell Lymphomas

Translocations of c-*myc* to the immunoglobulin loci are common in B cell lymphomas, a tumor of activated B cells.[83] The most common translocation, t(8;14), moves c-*myc* to the immunoglobulin heavy chain locus and deregulates c-*myc* expression by juxtaposing this oncogene with the intensely active transcriptional regulatory elements that drive immunoglobulin gene expression. Two competing hypotheses could explain frequent translocation of c-*myc*. c-*myc* is a potent oncogene, and if translocation increased c-*myc* expression or activity, this would contribute to cell proliferation and tumorigenesis.

Alternatively, c-*myc* may itself be a target for factors that promote class switch recombination in activated B cells. Like the immunoglobulin heavy chain switch regions, c-*myc* is rapidly transcribed in activated B cells, and the first exon and intron of c-*myc*, where translocation junctions map, contains numerous G-runs with potential for formation of G-quadruplexes.

The hypothesis that c-*myc* is a translocation target has been supported by experiments showing that transcription of c-*myc* results in formation of G-loops which are indistinguishable from those formed upon S region transcription.[26] Moreover, the G-loops map to the zone that is targeted for translocation in B cell lymphomas (Figure 7). These results provide strong support for the hypothesis that c-*myc* is specifically targeted by factors that promote recombination of the G-rich S regions, and that this contributes to the genetic instability of c-*myc* in B cell malignancies. It will be of interest to learn whether potential for G4 DNA formation characterizes other proto-oncogenes, which undergo frequent translocation.

9.8   G4 DNA and Genomic Stability and Instability

G-quadruplex formation provides a powerful mechanism for altering DNA structure in response to biological regulation, but there is a strong potential downside. As illustrated in Figure 2, formation of G4 DNA during replication could lead to expansion or contraction of the G-rich regions, analogous to the instability documented for triplet repeats. Intermolecular G4 DNA could interfere with separation of DNA molecules following replication. To prevent this, mechanisms exist for elimination of G4 DNA. We are only beginning to understand how these mechanisms function, but what is currently known suggests some directions for future research.

9.8.1   Specialized Helicases Maintain G-Rich Genomic Regions

The RecQ helicases comprise a small, conserved enzyme family which is essential for maintenance of genomic stability in organisms from *E. coli* to human.[84,85] There is only one RecQ family member in *E. coli* (RecQ), one in *S. cerevisiae* (Sgs1), one in *S. pombe* (Rqh1), and five in *Homo sapiens*. Three human members of the RecQ family are associated with genomic instability and predisposition to malignancies: WRN with Werner's syndrome, BLM with Bloom's syndrome, and RTS (or RECQ4) with Rothmund–Thompson syndrome. RecQ helicases are unusual in their ability to unwind G4 DNA,[74–75,86–88] and these enzymes have specialized functions in G-rich regions.

9.8.1.1   Nucleolar Functions of RecQ Family Members

Eukaryotic ribosomal RNA (rRNA) is transcribed from a family of repeated DNA sequences, ribosomal DNA, rDNA. rDNA repeats are G-rich on the non-template strand, not only within the regions that encode ribosomal RNA but also in the spacer regions. rDNA transcription and rRNA biogenesis occur in a specialized compartment within the nucleus, called the nucleolus. In actively proliferating cells, the rDNA is very rapidly transcribed, with polymerases separated by as little as 100 bp.[89] DNA must undergo denaturation to allow transcription to occur. Passage of each polymerase could be accompanied by denaturation and reannealing of the duplex, but this would be likely to diminish the overall efficiency of transcription; and it is possible that regulated formation of G4 DNA by the G-rich non-template strand could enable an extended region to become denatured, leaving the template strand exposed for rapid transcription. This possibility was first suggested by the discovery that the abundant nucleolar protein, nucleolin, binds tightly to G4 DNA.[23] Consistent with G4 DNA formation occurring during rDNA transcription, the eukaryotic RecQ helicases Sgs1, BLM, and WRN all localize to the nucleolus. Moreover, in *S. cerevisiae*, Sgs1-deficiency results in reduced proliferative capacity, popouts of rDNA circles, nucleolar fragmentation, and impaired rDNA transcription and replication.[90–93]

9.8.1.2 WRN Helicase Unwinds G4 DNA to Promote Mitotic Replication of Telomeres

The possibility that the human RecQ helicase, WRN, may function to maintain telomeres has drawn considerable attention because premature aging is the hallmark of Werner syndrome. Affected individuals develop normally until their teens, then rapidly develop characteristics of aging, including graying hair, alopecia, osteoporosis, type II diabetes and cataracts, and premature death, at an average age of 40.[94] This is accompanied by chromosomal instability and cancer.

Telomere maintenance has recently been shown to depend upon WRN helicase activity.[95,96] WRN is active not only as a helicase, but also as an exonuclease, and it is a very large protein (180 kD), which could in principle function as a scaffold for other factors. However, telomere maintenance has been shown to depend specifically on the DNA unwinding activity of WRN, as a point mutation that abolished helicase activity causes telomeric sequence loss. Most tellingly, loss of telomeric sequence in the absence of WRN helicase activity is specific to only one of the telomeric strands: replication of the G-rich template is impaired.[96] Coupled with evidence that WRN associates with telomeres specifically in S phase, when replication occurs, these results lead to a model in which WRN helicase activity unwinds G4 DNA formed within the G-rich template strand, to allow complete replication of telomeric sequences (Figure 8).

9.8.1.3 RTEL and DOG-1: Putative Helicases Critical to G-Rich Regions

Other helicases may also be critical to maintenance of G-rich regions. Genetic screens have identified a gene, *dog-1*, that is necessary to maintain polyguanine

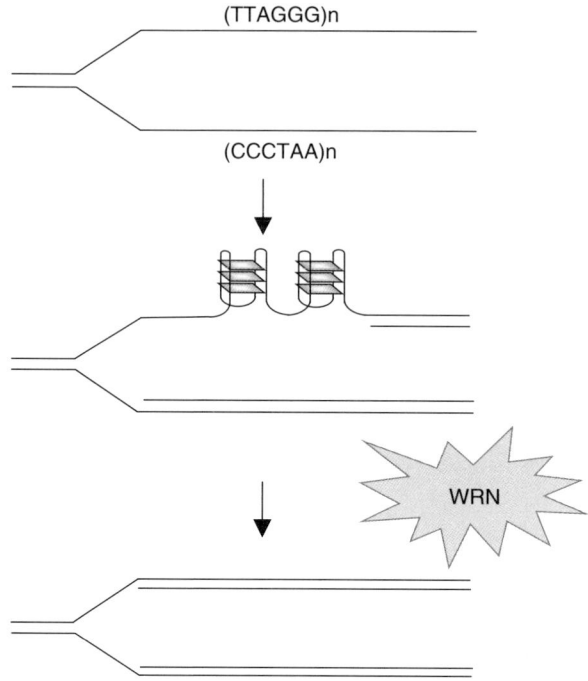

Figure 8 *WRN helicase unwinds G4 DNA to maintain telomere sequence. Top, replication fork in duplex telomeric DNA. Center, the G-rich strand forms G4 DNA, which impairs replication. Bottom, WRN unwinds G4 DNA, allowing complete replication of telomeric sequence*

tracts in the nematode.[97] Ablation of *dog-1* resulted in genome-wide destabilization of polyguanine tracts more than 18 nt in length, but did not have an obvious effect on shorter homonucleotide tracts or on telomere function. In contrast, the murine homolog of *dog-1*, named *Rtel*, has proven to have a key role in maintenance of telomere stability.[98] Mutation of the *Rtel* gene resulted in embryonic lethality in a murine model, but the functions of this gene could be analyzed in mutant cell lines. *Rtel-/-* cells displayed profound chromosomal instability, evident as end-to-end chromosomal fusions, broken chromosomes, chromosome fragments, and telomere-less chromosomes, similar to the abnormalities typical of cells lacking telomerase. In contrast to *dog-1*, *Rtel* does not appear to be required for stabilization of non-telomeric G-rich repeats.

Both *Rtel* and *dog-1* encode proteins which contain a conserved helicase motif as well as a domain for interaction with proliferating cell nuclear antigen (PCNA), part of the replication apparatus. This suggests that these proteins may function by unwinding G4 DNA that could otherwise impair replication.[97,98] Biochemical analysis of the activities of RTEL, DOG-1, and related factors should prove very interesting.

9.8.2 Potential Functions of MutSα in Recognition and Repair of G4 DNA

MutSα functions in canonical mismatch repair to identify irregularities in duplex DNA, such as base mismatches, modified bases, and short loops, and then recruit factors to excise and repair these discontinuities.[76] Recruitment depends upon ATP hydrolysis by MutSα, which is thought to result in a conformation change necessary for recruitment of downstream factors. Binding of MutSα to these substrates is sensitive to the presence of ATP, which is thought to reflect this same conformational change.

MutSα binds to G4 DNA with affinity even higher than for the best of its documented substrates in canonical mismatch repair (apparent $k_D = 1$ nM).[77] This raises the possibility that MutSα directs a pathway for G4 DNA removal. However, downstream factors in this pathway may very well be distinct from those that typically participate in mismatch repair, because binding to G4 DNA is not sensitive to addition of ATP. Interestingly, the RecQ family helicase BLM interacts with and is stimulated by MutSα.[99,100] This raises the possibility that MutSα might stimulate G4 DNA unwinding, rather than excision of quadruplex structures.

9.8.3 Nucleases that Cleave G4 DNA

Nucleases specific for G4 DNA have been identified in yeast and human cells. The KEM1 nuclease of *S. cerevisiae*, was the first nuclease shown to be active on G4 DNA [Liu, 1994, p. 148]. KEM1 cleaves a single nucleotide from the 5' end of G4 DNA formed from synthetic oligonucleotides. KEM1 is also active on G4 RNA, and its primary function appears to be in RNA processing.[101]

Another G4 DNA cleavage activity, GQN1, has been identified in human cells. GQN1 cleaves within the single-stranded region 8–12 nt 5' of the barrel formed by stacked G-quartets, to generate single-stranded 5' fragments. GQN1 cleaves either intramolecular or intermolecular G4 DNA, but does not cleave G4 RNA, free single-stranded DNA, duplex DNA, or Holliday junctions. Thus, GQN1 and KEM1 are distinct in both specificity and cleavage pattern.

9.8.4 Cotranscriptional RNA/DNA Hybrids and Genomic Instability

Transcribed G-rich templates form two distinct structures: G4 DNA, and a stable, cotranscriptional RNA/DNA hybrid.[20] Cotranscriptional hybrid formation occurs when the nascent transcript reanneals with the DNA template. RNA polymerases appear designed to prevent cotranscriptional hybrid formation by forcing the nascent RNA to exit through a topological tunnel which separates the RNA transcript from the DNA template strand.[102,103] However, separation is not completely efficient, as shown by experiments which identify stable hybrids in transcribed genes in living cells.[20,104,105] Factors associated with RNA processing have recently been shown to prevent cotranscriptional hybrid formation.

Two different factors have so far been shown to prevent formation of cotranscriptional RNA/DNA hybrids, and the absence of either results in genomic instability. The THO/TREX complex, identified in yeast, is associated with the transcription apparatus and functions to prevent formation of stable cotranscriptional hybrids.[104,106] Deficiencies in subunits of the THO/TREX complex result in increased genomic instability, suggesting that the RNA/DNA hybrid itself or structures formed within the DNA strand may be targets for recombination and repair pathways. Expression of G/C-rich transcripts is especially dependent upon THO/TREX. The factor ASF/SF2, initially identified with pre-mRNA splicing, is also critical to preventing cotranscriptional hybrid formation. ASF/SF2 interactions with the CTD domain of RNA polymerase to prevent stable hybrid formation.[105] In cells lacking ASF/SF2, DNA fragmentation leads to a hypermutation phenotype. In the absence of ASF/SF2 activity, DNA breaks can be specifically detected within a region of the β-actin gene that is G-rich and has potential for G4 DNA formation.[105]

The mechanisms that promote genomic instability upon cotranscriptional hybrid formation have not been defined. Nonetheless, the propensity of transcribed G-rich regions to form cotranscriptional hybrids suggests that there may be overlap between factors and pathways that remove or resolve G4 DNA and promote instability at sites of cotranscriptional hybrid formation.

9.9 Perspective and Future Directions

Until very recently there has been a gulf between chemical and biological analysis of G-quadruplexes. We knew much about structures and ligands, but relatively little about functions in living cells. Evidence for regulated formation of G4 DNA, and its importance in key cellular processes, particularly telomere maintenance, has overcome lingering skepticism about whether DNA in a living cell ever abandons the security of the double helix to form G-quadruplex structures, and the future should witness a flood of new experiments. Some of the questions we can hope will be answered include: What mechanisms regulate dynamic G4 DNA formation? What cellular processes depend upon formation of this alternative structure? What is the structural basis for high affinity interactions of proteins and G-quadruplexes? How does G4 DNA formation affect the function and regulation of single genes? And how does G4 DNA formation impact genomic stability?

9.10 Acknowledgments

I thank members of my laboratory and my colleagues for valuable discussions, particularly Michelle Duquette, Michael Huber, and Erik Larson. We are grateful to the US National Institutes of Health for supporting our research on G4 DNA (GM65988).

References

1. D. Sen and W. Gilbert, *Nature*, 1988, **334**, 364.
2. M. Gellert, M.N. Lipsett and D.R. Davies, *Proc. Natl. Acad. Sci. USA*, 1962, **48**, 2014.
3. D. Sen and W. Gilbert, *Nature*, 1990, **344**, 410.
4. D. Sen and W. Gilbert, *Biochemistry*, 1992, **31**, 65.
5. G.N. Parkinson, M.P. Lee and S. Neidle, *Nature*, 2002, **417**, 876.
6. D. Sen and W. Gilbert, *Meth. Enzymol.*, 1992, **211**, 191.
7. D.E. Gilbert and J. Feigon, *Curr. Opin. Struct. Biol.*, 1999, **9**, 305.
8. G. Laughlan, A.I. Murchie, D.G. Norman, M.H. Moore, P.C. Moody, D.M. Lilley and B. Luisi, *Science*, 1994, **265**, 520.
9. C.E. Pearson, K.N. Edamura and J.D. Cleary, *Nat. Rev. Genet.*, 2005, **6**, 729.
10. S. Neidle and G.N. Parkinson, *Curr. Opin. Struct. Biol.*, 2003, **13**, 275.
11. N. Maizels, *Annu. Rev. Genet.*, 2005, **39**, 23.
12. N. Buroker, R. Bestwick, G. Haight, R.E. Magenis and M. Litt, *Hum. Genet.*, 1987, **77**, 175.
13. Z. Wong, V. Wilson, I. Patel, S. Povey and A.J. Jeffreys, *Ann. Hum. Genet.*, 1987, **51**(Pt 4), 269.
14. A.J. Jeffreys, N.J. Royle, V. Wilson and Z. Wong, *Nature*, 1988, **332**, 278.
15. M.N. Weitzmann, K.J. Woodford and K. Usdin, *J. Biol. Chem.*, 1997, **272**, 9517.
16. S.M. Heale and T.D. Petes, *Cell*, 1995, **83**, 539.
17. L.P. Ranum and J.W. Day, *Curr. Opin. Genet. Dev.*, 2002, **12**, 266.
18. J.L. Huppert and S. Balasubramanian, *Nucleic Acids Res.*, 2005, **33**, 2908.
19. A.K. Todd, M. Johnston and S. Neidle, *Nucleic Acids Res.*, 2005, **33**, 2901.
20. M.L. Duquette, P. Handa, J.A. Vincent, A.F. Taylor and N. Maizels, *Genes Dev.*, 2004, **18**, 1618.
21. M.E. Reaban and J.A. Griffin, *Nature*, 1990, **348**, 342.
22. R. Mizuta, K. Iwai, M. Shigeno, M. Mizuta, T. Ushiki and D. Kitamura, *J. Biol. Chem.*, 2003, **278**, 4431.
23. L.A. Hanakahi, H. Sun and N. Maizels, *J. Biol. Chem.*, 1999, **274**, 15908.
24. H. Sun, A. Yabuki and N. Maizels, *Proc. Natl. Acad. Sci. USA*, 2001, **89**, 12444.
25. N. Sugimoto, S. Nakano, M. Katoh, A. Matsumura, H. Nakamuta, T. Ohmichi, M. Yoneyama and M. Sasaki, *Biochemistry*, 1995, **34**, 11211.
26. M.L. Duquette, P. Pham, M.F. Goodman and N. Maizels, *Oncogene*, 2005, **24**, 5791.
27. C.L. Jahn and L.A. Klobutcher, *Annu. Rev. Microbiol.*, 2002, **56**, 489.
28. C. Schaffitzel, I. Berger, J. Postberg, J. Hanes, H.J. Lipps and A. Pluckthun, *Proc. Natl. Acad. Sci. USA*, 2001, **98**, 8572.
29. J.D. Griffith, L. Comeau, S. Rosenfield, R.M. Stansel, A. Bianchi, H. Moss and T. de Lange, *Cell*, 1999, **97**, 503.
30. T.R. Cech, *Cell*, 2004, **116**, 273.

31. A. Smogorzewska and T. de Lange, *Annu. Rev. Biochem.*, 2004, **73**, 177.

32. D.L. Aisner and W.E. Wright, and J.W. Shay, *Curr. Opin. Genet. Dev.*, 2002, **12**, 80.

33. Y.S. Cong, W.E. Wright and J.W. Shay, *Microbiol. Mol. Biol. Rev.*, 2002, **66**, 407.

34. J.W. Shay and W.E. Wright, *Nat. Rev. Mol. Cell. Biol.*, 2000, **1**, 72.

35. A.G. Bodnar, M. Ouellette, M. Frolkis, S.E. Holt, C.P. Chiu, G.B. Morin, C.B. Harley, J.W. Shay, S. Lichtsteiner and W.E. Wright, *Science*, 1998, **279**, 349.

36. W.C. Hahn and R.A. Weinberg, *N. Engl. J. Med.*, 2002, **347**, 1593.

37. C.B. Harley, *Curr. Mol. Med.*, 2005, **5**, 205.

38. A.A. Neumann and R.R. Reddel, *Nat. Rev. Cancer*, 2002, **2**, 879.

39. S. Neidle and G. Parkinson, *Nat. Rev. Drug Discov.*, 2002, **1**, 383.

40. E.M. Rezler, D.J. Bearss and L.H. Hurley, *Annu. Rev. Pharmacol. Toxicol.*, 2003, **43**, 359.

41. J. Cuesta, M.A. Read and S. Neidle, *Mini. Rev. Med. Chem.*, 2003, **3**, 11.

42. D. Gomez, T. Lemarteleur, L. Lacroix, P. Mailliet, J.L. Mergny and J.F. Riou, *Nucleic Acids Res.*, 2004, **32**, 371.

43. G. Pennarun, C. Granotier, L.R. Gauthier, D. Gomez, F. Hoffschir, E. Mandine, J.F. Riou, J.L. Mergny, P. Mailliet and F.D. Boussin, *Oncogene*, 2005, **24**, 2917.

44. C. Granotier, G. Pennarun, L. Riou, F. Hoffschir, L.R. Gauthier, A. De Cian, D. Gomez, E. Mandine, J.F. Riou, J.L. Mergny, P. Mailliet, B. Dutrillaux and F.D. Boussin, *Nucleic Acids Res.*, 2005, **33**, 4182.

45. S.M. Haider, G.N. Parkinson and S. Neidle, *J. Mol. Biol.*, 2003, **326**, 117.

46. P.M. Lansdorp, *Ann. N.Y. Acad. Sci.*, 2005, **1044**, 220.

47. M. Armanios, J.L. Chen, Y.P. Chang, R.A. Brodsky, A. Hawkins, C.A. Griffin, J.R. Eshleman, A.R. Cohen, A. Chakravarti, A. Hamosh and C.W. Greider, *Proc. Natl. Acad. Sci. USA*, 2005.

48. H. Yamaguchi, R.T. Calado, H. Ly, S. Kajigaya, G.M. Baerlocher, S.J. Chanock, P.M. Lansdorp and N.S. Young, *N. Engl. J. Med.*, 2005, **352**, 1413.

49. M.A. Blasco, H.W. Lee, M.P. Hande, E. Samper, P.M. Lansdorp, R.A. DePinho and C.W. Greider, *Cell*, 1997, **91**, 25.

50. H.W. Lee, M.A. Blasco, G.J. Gottlieb, J.W. Horner 2nd, C.W. Greider and R.A. DePinho, *Nature*, 1998, **392**, 569.

51. K.L. Rudolph, S. Chang, H.W. Lee, M. Blasco, G.J. Gottlieb, C. Greider and R.A. DePinho, *Cell*, 1999, **96**, 701.

52. R.A. DePinho and K. Polyak, *Nat. Genet.*, 2004, **36**, 932.

53. A.M. Zahler, J.R. Williamson, T.R. Cech and D.M. Prescott, *Nature*, 1991, **350**, 718.

54. T.M. Fletcher, D. Sun, M. Salazar and L.H. Hurley, *Biochemistry*, 1998, **37**, 5536.

55. A.J. Zaug, E.R. Podell and T.R. Cech, *Proc. Natl. Acad. Sci. USA*, 2005, **102**, 10864.

56. K. Paeschke, T. Simonsson, J. Postberg, D. Rhodes and H.J. Lipps, *Nat. Struct. Mol. Biol.*, 2005.
57. P. Baumann and T.R. Cech, *Science*, 2001, **292**, 1171.
58. T. de Lange, *Genes Dev.*, 2005, **19**, 2100.
59. M. Lei, P. Baumann and T.R. Cech, *Biochemistry*, 2002, **41**, 14560.
60. D.L. Theobald and D.S. Wuttke, *Structure (Camb)*, 2004, **12**, 1877.
61. M. Lei, E.R. Podell and T.R. Cech, *Nat. Struct. Mol. Biol.*, 2004, **11**, 1223.
62. M. Lei, E.R. Podell, P. Baumann and T.R. Cech, *Nature*, 2003, **426**, 198.
63. M.P. Horvath, V.L. Schweiker, J.M. Bevilacqua, J.A. Ruggles and S.C. Schultz, *Cell*, 1998, **95**, 963.
64. M.P. Horvath and S.C. Schultz, *J. Mol. Biol.*, 2001, **310**, 367.
65. L.M. Colgin, K. Baran, P. Baumann, T.R. Cech and R.R. Reddel, *Curr. Biol.*, 2003, **13**, 942.
66. S.J. Froelich-Ammon, B.A. Dickinson, J.M. Bevilacqua, S.C. Schultz and T.R. Cech, *Genes Dev.*, 1998, **12**, 1504.
67. G. Fang and T.R. Cech, *Cell*, 1993, **74**, 875.
68. W. Dunnick, G.Z. Hertz, L. Scappino and C. Gritzmacher, *Nucleic Acids Res.*, 1993, **21**, 365.
69. M.E. Reaban, J. Lebowitz and J.A. Griffin, *J. Biol. Chem.*, 1994, **269**, 21850.
70. G.A. Daniels and M.R. Lieber, *Nucleic Acids Res.*, 1995, **23**, 5006.
71. M. Tian and F.W. Alt, *J. Biol. Chem.*, 2000, **275**.
72. K. Yu, F. Chedin, C.L. Hsieh, T.E. Wilson and M.R. Lieber, *Nat. Immunol.*, 2003, **4**, 442.
73. N. Kondo, F. Motoyoshi, S. Mori, N. Kuwabara, T. Orii and J. German, *Acta Pediatr.*, 1992, **81**, 86.
74. H. Sun, J.K. Karow, I.D. Hickson and N. Maizels, *J. Biol. Chem.*, 1998, **273**, 27587.
75. M.D. Huber, D.C. Lee and N. Maizels, *Nucleic Acids Res.*, 2002, **30**, 3954.
76. T.A. Kunkel and D.A. Erie, *Annu. Rev. Biochem.*, 2005, **74**, 681.
77. E.D. Larson, M.L. Duquette, W.J. Cummings, R.J. Streiff and N. Maizels, *Curr. Biol.*, 2005, **15**, 470.
78. M.R. Ehrenstein and M.S. Neuberger, *EMBO J.*, 1999, **18**, 3484.
79. A. Rangan and O.Y. Fedoroff, and L.H. Hurley, *J. Biol. Chem.*, 2001, **276**, 4640.
80. A. Siddiqui-Jain, C.L. Grand, D.J. Bears and L.H. Hurley, *Proc. Natl. Acad. Sci. USA*, 2002, **99**, 11593.
81. J. Seenisamy, S. Bashyam, V. Gokhale, H. Vankayalapati, D. Sun, A. Siddiqui-Jain, N. Streiner, K. Shin-Ya, E. White, W.D. Wilson and L.H. Hurley, *J. Am. Chem. Soc.*, 2005, **127**, 2944.
82. S. Rankin, A.P. Reszka, J. Huppert, M. Zloh, G.N. Parkinson, A.K. Todd, S. Ladame, S. Balasubramanian and S. Neidle, *J. Am. Chem. Soc.*, 2005, **127**, 10584.
83. R. Kuppers and R. Dalla-Favera, *Oncogene*, 2001, **20**, 5580.
84. I.D. Hickson, *Nat. Rev. Cancer*, 2003, **3**, 169.
85. V.A. Bohr, *Mutat. Res.*, 2005, **577**, 252.

86. H. Sun, R.J. Bennett and N. Maizels, *Nucleic Acids Res.*, 1999, **27**, 1978.
87. X. Wu and N. Maizels, *Nucleic Acids Res.*, 2001, **29**, 1765.
88. P. Mohaghegh, J.K. Karow, R.M. Brosh Jr., V.A. Bohr and I.D. Hickson, *Nucleic Acids Res.*, 2001, **29**, 2843.
89. Y. Osheim, E.B. Mougey, J. Windle, M. Anderson, M. O'Reilly, O.L. Miller, A. Beyer and B. Sollner-Webb, *J. Cell Biol.*, 1996, **133**, 943.
90. D.A. Sinclair and L. Guarente, *Cell*, 1997, **91**, 1033.
91. D.A. Sinclair, K. Mills and L. Guarente, *Science*, 1997, **277**, 1313.
92. S.K. Lee, R.E. Johnson, S.L. Yu, L. Prakash and S. Prakash, *Science*, 1999, **286**, 2339.
93. G. Versini, I. Comet, M. Wu, L. Hoopes, E. Schwob and P. Pasero, *EMBO J.*, 2003, **22**, 1939.
94. P.L. Opresko, W.H. Cheng, C. von Kobbe, J.A. Harrigan and V.A. Bohr, *Carcinogenesis*, 2003, **24**, 791.
95. Y. Bai and J.P. Murnane, *Hum. Genet.*, 2003, **113**, 337.
96. L. Crabbe, R.E. Verdun, C.I. Haggblom and J. Karlseder, *Science*, 2004, **306**, 1951.
97. I. Cheung, M. Schertzer, A. Rose and P.M. Lansdorp, *Nat. Genet.*, 2002, **31**, 405.
98. H. Ding, M. Schertzer, X. Wu, M. Gertsenstein, S. Selig, M. Kammori, R. Pourvali, S. Poon, I. Vulto, E. Chavez, P.P. Tam, A. Nagy and P.M. Lansdorp, *Cell*, 2004, **117**, 873.
99. G. Pedrazzi, C.Z. Bachrati, N. Selak, I. Studer, M. Petkovic, I.D. Hickson, J. Jiricny and I. Stagljar, *Biol. Chem.*, 2003, **384**, 1155.
100. Q. Yang, R. Zhang, X.W. Wang, S.P. Linke, S. Sengupta, I.D. Hickson, G. Pedrazzi, C. Perrera, I. Stagljar, S.J. Littman, P. Modrich and C.C. Harris, *Oncogene*, 2004, **23**, 3749.
101. A.W. Johnson, *Mol. Cell. Biol.*, 1997, **17**, 6122.
102. Y.W. Yin and T.A. Steitz, *Science*, 2002, **298**, 1387.
103. K.D. Westover, D.A. Bushnell and R.D. Kornberg, *Science*, 2004, **303**, 1014.
104. P. Huertas and A. Aguilera, *Mol. Cell*, 2003, **12**, 711.
105. X. Li and J.L. Manley, *Cell*, 2005, **122**, 365.
106. K. Strasser, S. Masuda, P. Mason, J. Pfannstiel, M. Oppizzi, S. Rodriguez-Navarro, A.G. Rondon, A. Aguilera, K. Struhl, R. Reed and E. Hurt, *Nature*, 2002, **417**, 304.

CHAPTER 10

The G-Quartet in Supramolecular Chemistry and Nanoscience[†]

MARK S. KAUCHER, WILLIAM A. HARRELL JR. AND
JEFFERY T. DAVIS

*Department of Chemistry and Biochemistry, University of Maryland, College
Park, Maryland, USA*

10.1 Guanosine Self-Assembly in Supramolecular Chemistry

Self-assembly is central to many processes in biology and chemistry. The G-quartet, a hydrogen-bzonded macrocycle formed upon the cation-templated self-association of guanosine analogs, was first identified in the early 1960s as the basic building block for hydrogels formed by 5'-GMP.[1–3] Since those early days, many different nucleosides, oligonucleotides and synthetic derivatives have been shown to form a rich array of G-quadruplex structures. This chapter summarizes some of the many recent studies of G-quartets in the general areas of supramolecular chemistry and nanoscience.

10.1.1 Guanosine Self-Assembly in Water

With its complementary hydrogen-bonding edges and its polarized aromatic surfaces, it is not at all surprising that guanosine analogs have a high propensity to self-associate. The guanine nucleobase has two hydrogen bond acceptors (N7 and O6 atoms) on its Hoogsteen edge and two hydrogen bond donors (N1 amide and N2 amino hydrogens) on its Watson-Crick edge [Figure 1(a)]. Depending on the environmental conditions, guanosine derivatives can self-associate into hydrogen-bonded dimers, extended ribbons, or stacks of discrete macrocycles.[1–3]

The cyclic G-quartet structure was first identified in 1962 as the basic building block for formation of hydrogels by 5'-GMP **1**.[4] Using fiber

---

[†] Dedicated to Prof. Giovanni Gottarelli on the occasion of his retirement.

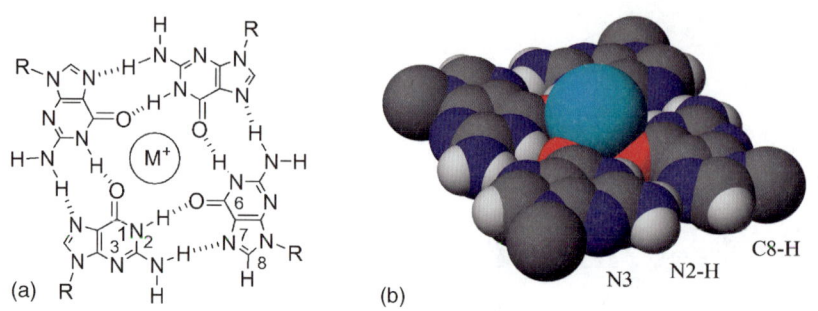

Figure 1 *(a) The G-quartet. (b) A space filling model from the crystal structure of [G 5]₁₆·3K⁺/Cs⁺·4Pic⁻ showing a G-quartet, without the sugars, with a K⁺ cation bound above the plane of the G-quartet*
(Adapted from ref 3 with permission).

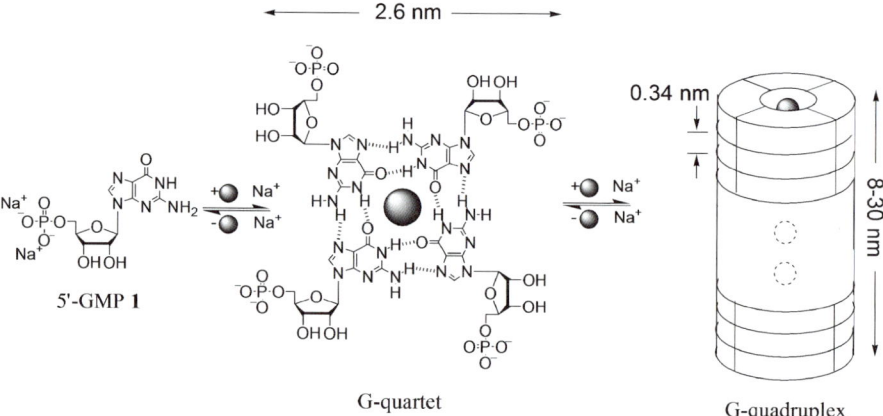

Figure 2 *Depiction of the G-quadruplex cylinder formed by the self-assembly of 5′-GMP*
(See ref 6)

diffraction data, Gellert and colleagues proposed that a square planar G-quartet was formed by eight intermolecular hydrogen bonds between the Hoogsteen and Watson-Crick edges of neighboring nucleobases [Figure 1(a)]. Pinnavaia and colleagues later showed that these G-quartets are stabilized by Na⁺ and K⁺ ions that coordinate to the four inward directed carbonyl oxygens in stacked G-quartets [Figure 1(b)]. They found that 5′-GMP **1** forms diastereomeric G₈-K⁺ octamers by sandwiching two G-quartets around the central cation.[5] More recently, Wu and colleagues used a combination of data from diffusion NMR and dynamic light scattering measurements to determine the size of nanostructures formed by sodium 5′-GMP **1** at pH 8 (Figure 2).[6] Wu's group identified two major species in solution: stacked 5′-GMP monomers and stacked G-quartets. For 5′-GMP concentrations in the 18–34 wt% range, these structures had an average length between 8 and 30 nm, corresponding to a cylinder composed of 24–87 stacked G-quartets. The impressive length of

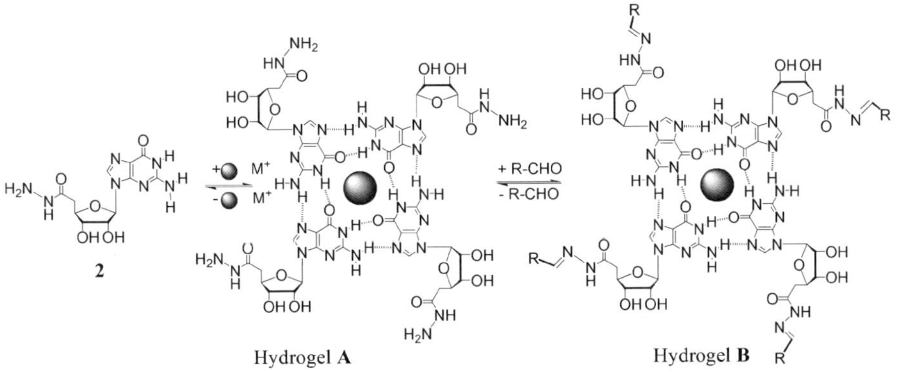

Figure 3 *Dynamic hydrogels synthesized using a G-quartet scaffold*
(Adapted from ref 7 with permission).

G-quadruplexes formed from 5′-GMP in water underscores the highly cooperative participation of hydrogen bond, ion-dipole, and π–π stacking interactions inherent to these G-quartet based assemblies.

Such highly organized assemblies may also be used as templates for the synthesis of new nanostructures and biomaterials. For example, Sreenivasachary and Lehn[7] recently prepared dynamic hydrogels by covalent modification of the 5′-sidechains that extend from stacked G-quartets. Reaction of a hydrogel A made from 5′-hydrazido G **2** with a mixture of aldehydes produced a family of acylhydrazone G-quartets (Figure 3). This dynamic library of acylhydrazones demonstrated preferential synthesis of the most stable hydrogel B. As described in more detail later in this chapter, Lehn's study is a unique example of using dynamic covalent chemistry (DCC), with G-quartets as a scaffold, to control phase-organization.

10.1.2 Self-Assembly of Lipophilic Guanosine Analogs

Until about 10 years ago it was believed that G-quartet assemblies only formed in water, as G derivatives were believed to give mostly poorly organized structures in organic solvents. Since G nucleosides and nucleotides are not soluble in most organic solvents, it really was not until the ribose hydroxyl groups were modified with protecting groups that chemists recognized that lipophilic guanosine nucleosides could indeed self-associate into discrete assemblies in organic solvents. In 1995, Gottarelli and his colleagues[8] reported that 3′,5′-didecanoyl-2′-dG **3** extracts K^+ picrate from water into chlorinated organic solvents to give a discrete and highly stable $[dG\ 3]_8 \cdot K^+$ octamer. As is the case in water, the K^+ cation was absolutely essential for templation and stability of the G-quartets (Figure 4).

In the absence of stabilizing cations, dG **3** instead organized into two types of hydrogen-bonded ribbons.[9] Changing either the sugar substituents or the solvent can modulate the specific hydrogen-bonding pattern (either ribbon A or B in Figure 5). These ordered ribbons have some potential applications in

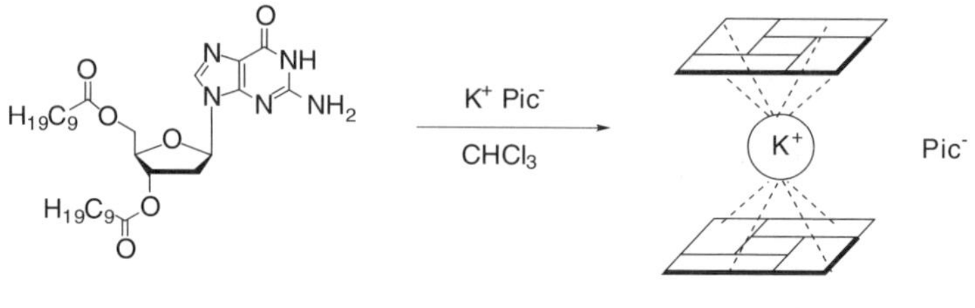

Figure 4   *Lipophilic [dG 3]₈ · K⁺ octamer formed by extraction of K⁺ picrate from water into CHCl₃*

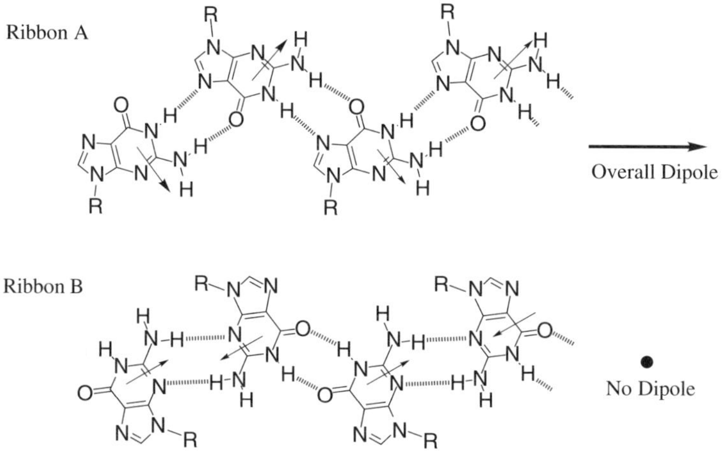

Figure 5   *Two different H-bonded ribbons formed by lipophilic dG 3 self-assembly in the absence of cations. Ribbon A has a net dipole, whereas ribbon B contains no dipole*

(Adapted from ref 3 with permission).

the molecular electronics field. Thus, Gottarelli and colleagues made an organic semiconductor using dG **3**. Asymmetric *I–V* curves, characteristic of molecular rectifiers, were attributed to the dipole that is inherent to the supramolecular structure of ribbon A.[10]

More recently, Gottarelli's group described another unique self-assembled structure built from the self-assembly of a lipophilic 8-oxoG derivative **4**.[11] This 8-oxoG **4** self-assembled to give extended helices in solution *via* solvophobic interactions and hydrogen bonds (Figure 6). This self-assembly pattern for 8-oxoG **4** was clearly different from the pattern provided by dG **3** that gave hydrogen-bonded ribbons.

As mentioned above, hydrogen-bonded G ribbons can be converted to stacked cyclic G-quartets upon addition of a templating cation. To better understand the organization of these lipophilic G-quartets within higher ordered G-quadruplex structures, our group collaborated with Gottarelli's group to solve the NMR structure of [dG **3**]₈ · KI in CDCl₃.[12] Remarkably, this discrete octamer

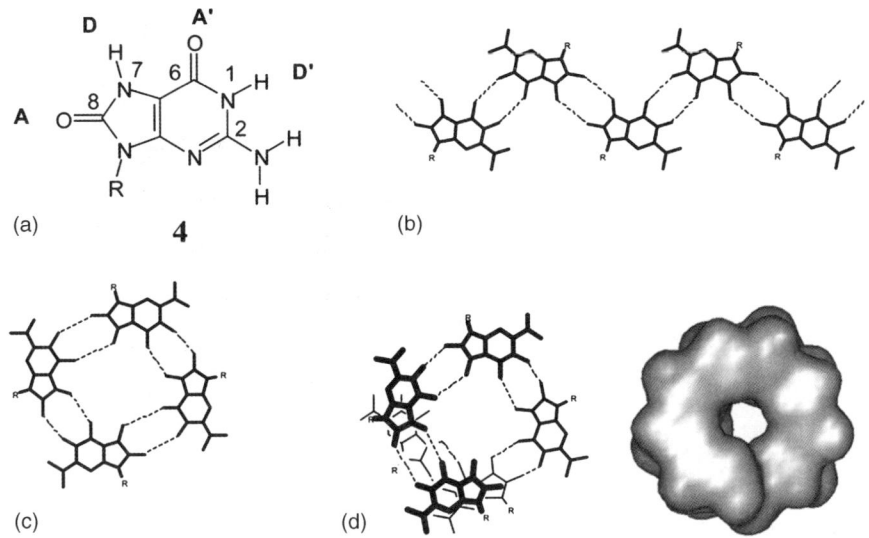

Figure 6 *(a) 8-oxoguanine, (b) ribbon structure, (c) 8-oxoG-quartet, and (d) 8-oxoG-helical structure*
(Adapted from ref 11 with permission).

[dG **3**]$_8$ · KI existed as a single supramolecular diastereomer with K$^+$ sandwiched between an all-*anti* G-quartet and an all-*syn* G-quartet. In 2000, an X-ray crystal structure definitively proved that larger lipophilic G-quadruplexes are formed in high diastereoselectivity, even from nonpolar organic solvents.[13] Thus, the lipophilic G-quadruplex [G **5**]$_{16}$ · 3K$^+$ · Cs$^+$ · 4pic$^-$ is a complex composed of four stacked G-quartets and is generated when 5′-silyl-2′,3′-isopropylidene G **5** is used to extract K$^+$ picrate from water into CH$_2$Cl$_2$ (Figure 7). Crystals of the G-quadruplex were grown from acetonitrile. This G-quadruplex can be best described as a pair of head-to-tail [G **5**]$_8$ octamers; each octamer uses eight carbonyl oxygens to coordinate sandwiched K$^+$ ions. A third K$^+$ ion holds the two [G **5**]$_8$ octamers together and a solvated Cs$^+$ ion caps the structure. The four hydrogen-bonded G-quartets within [G **5**]$_{16}$ · 3K$^+$ · Cs$^+$ · 4pic$^-$ all show pi-stacking separations of 3.3–3.4 Å. Additionally, four picrate anions hydrogen bond to the exposed N2 amino groups that extend from the two central G-quartets, providing the G-quadruplex structure with the appearance of a cation channel that has an anionic belt wrapped around its periphery. Recently, we used diffusion NMR techniques to show that the hexadecamer [G **5**]$_{16}$ · 4K$^+$ · 4pic$^-$ observed in the solid state is also the predominant species in CD$_3$CN solution.[14] These stable lipophilic G-quadruplex structures, held together by numerous noncovalent interactions, can be used either as biomimetic models for DNA G-quadruplex structures or for the development of functional supramolecular systems, including synthetic ion channels.[3]

The bound cations in these lipophilic G-quadruplexes have been directly observed using solid-state NMR spectroscopy. Wu and colleagues used ^{23}Na

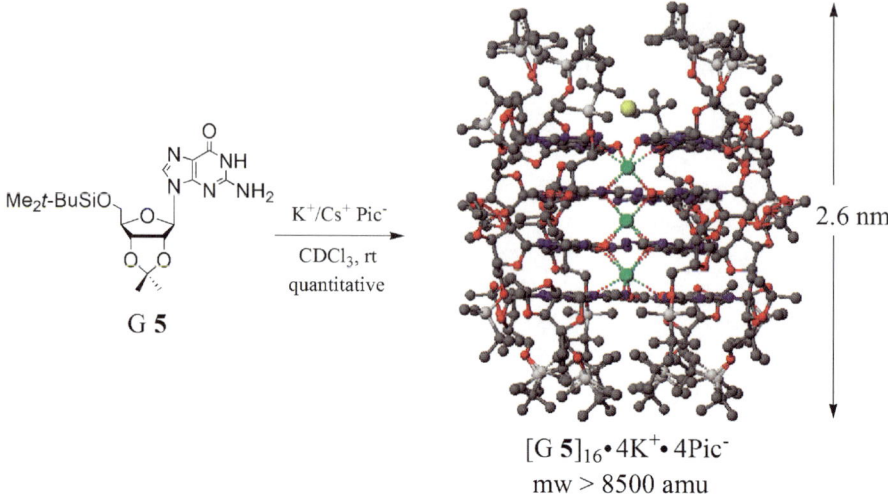

$$[G\ 5]_{16} \bullet 4K^+ \bullet 4Pic^-$$
$$mw > 8500\ amu$$

Figure 7 *Crystal structure shows that the cation-templated self-assembly of 16 equiva-lents of G **5** gives a lipophilic G-quadruplex [G **5**]$_{16}$·3K$^+$/Cs$^+$·4Pic$^-$. This G-quadruplex is prepared quantitatively by extracting salts from water with a CHCl$_3$ solution of G **5***
(Adapted from ref 13 with permission).

and ^{39}K NMR to identify specific channel cations.[15,16] This solid-state NMR work was an important development, as these lipophilic G-quadruplexes could then be reliably used as models to help clarify some ambiguous issues about how Na$^+$ and K$^+$ cations bind to DNA G-quadruplexes. The identity of the bound cations also helps control the solution properties of these G-quadruplex assemblies. For example, G-quadruplexes containing divalent cations such as Pb^{2+}, Ba^{2+}, or Sr^{2+} are both thermodynamically and kinetically more stable than are the corresponding G-quadruplex assemblies that contain monovalent Na$^+$ or K$^+$.[17] We attributed this enhanced stability in the presence of the divalent cations to the stronger ion-dipole interactions between the bound cations and the coordinating carbonyl oxygen, as well as to a cooperative enhancement of the strength of the G-quartet's hydrogen bonds.

Like the bound cations down the central ion channel of [G **5**]$_{16}$·4K$^+$, the phenolate anions bound to the G-quadruplex surface also help control the solution properties of these assemblies (Figure 8). The base strength and structure of the bound phenolate anions influences the exchange rates for both the bound cations and for the G subunits that make up the hexadecamer. For example, the rate of supramolecular isomerization of G$_8$ octamers in solution depends on the identity of the bound anions. Anions that hydrogen bond strongly to the central two G-quartets completely inhibited subunit exchange in CD$_2$Cl$_2$, presumably by increasing the kinetic stability of the complex and making subunit dissociation more difficult.[18] Both the cation and anion clearly influence the stability of these lipophilic G-quadruplexes. Indeed, with the proper combinations of cations and anions to stabilize these structures, we can direct

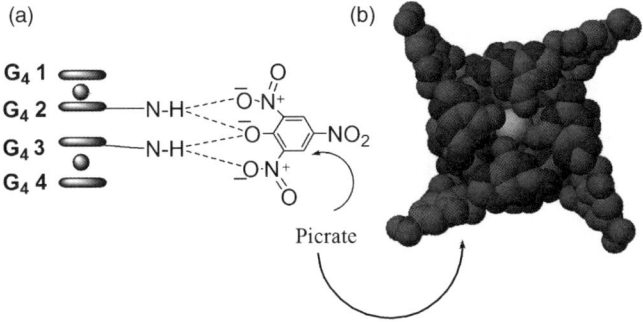

Figure 8 *(a) A schematic showing the nucleobase-picrate hydrogen bonds in the hexade-camer $[G\ 5]_{16}\cdot 2Sr^{2+}\cdot 4Pic^{-}$. Amino protons are from the guanine's exocyclic N2 groups. (b) Top view of the X-ray structure of the G-quadruplex with the sugars removed. Due to the nucleobase-anion hydrogen bonds, four picrate anions form an anionic belt around the G-quadruplex between G-quartet layers 2 and 3* (Adapted from ref 13 with permission).

post-assembly modifications to only the outer quartet in these G-quadruplexes, a result that has important consequences in noncovalent synthesis.[19]

10.1.3 Controlling Supramolecular Structure with Monomer Building Blocks

Purine nucleobases other than guanosine can also self-assemble into discrete supramolecular structures. IsoG **6** is an isomer of G **5**, differing only by transposition of the nitrogen and oxygen atoms at the purine's C2 and C6 positions. This subtle structural difference causes these isomers to self-assemble into unique arrangements. Whereas G derivatives form hydrogen-bonded tetramers, self-assembly of isoG **6** leads to a decamer composed of two hydrogen-bonded isoG$_5$ pentamers that sandwich a templating Cs$^+$ cation.[20,21] This difference in self-assembly and macrocycle size for G and isoG has been rationalized by considering the hydrogen-bonding geometries for the two nucleobases (Figure 9).[3] Donor and acceptor sites in G **5** are separated by 90°, an optimal orientation for a cyclic tetramer with linear hydrogen bonds. For isoG **6**, the angle between the donor and acceptor groups (about 110°) is best for formation of a pentagon. The larger size of the isoG$_5$ pentamer, relative to the G quartet, also helps explain the different ion binding selectivity shown by isoG and G derivatives. IsoG **6** is highly selective for coordinating the largest alkali cation, Cs$^+$ ($r = 1.67$ Å), whereas G-quartets are selective for K$^+$ ion ($r = 1.33$ Å).[20,21]

To clearly illustrate that the cation dictates the self-assembly pattern for G **5** and isoG **6**, we conducted a "self-sorting" study in CDCl$_3$.[22] Thus, an equimolar mixture of the two isomers in CDCl$_3$ formed a variety of hydrogen-bonded species, as monitored by ^{1}H NMR. Addition of Ba (pic)$_2$ to this mixture gave immediate and quantitative formation of two discrete hydrogen-bonded complexes, namely the G-quadruplex $[G\ 5]_{16}\cdot 2Ba^{2+}$ and the decameric [isoG **6**]$_{10}\cdot Ba^{2+}$ (Figure 10). This self-sorting study illustrated that an appropriate

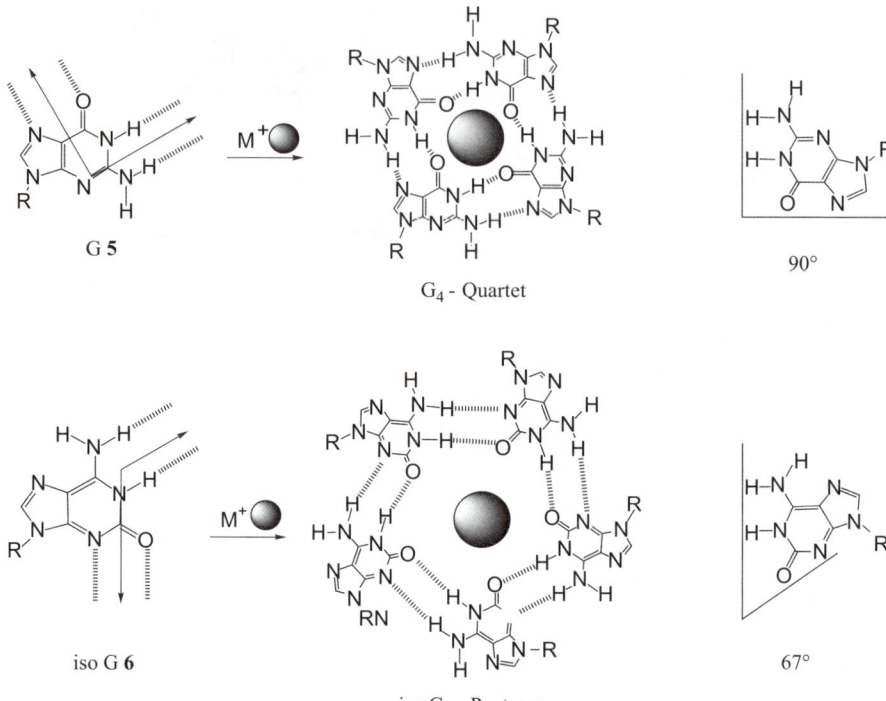

Figure 9 *Lipophilic nucleosides G 5 and isoG 6 self-associate in the presence of cations to give hydrogen-bonded G4-quartets or isoG₅-pentamers. The relative orientation of the nucleoside's hydrogen bond donor and acceptor groups determines assembly size* (Adapted from ref 3 with permission).

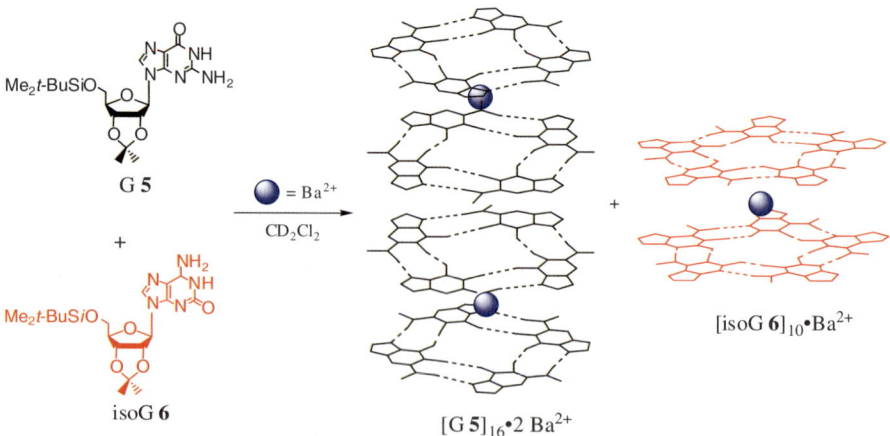

Figure 10 *The isomers G 5 and isoG 6 "self-sort" in the presence of barium picrate to give discrete complexes, namely [G 5]₁₆ · 2Ba²⁺ · 4Pic⁻ and [isoG 6]₁₀ · Ba²⁺ · 2Pic⁻* (Adapted from ref 22 with permission).

cation is a requirement for triggering formation of distinct supramolecular structures in solution from a mixture of nucleoside building blocks, a prime example of equilibrium shifting in dynamic noncovalent chemistry.

10.1.4 Enantiomeric Self-Association of Lipophilic Nucleosides

The cation's central role in controlling nucleoside self-assembly is also well illustrated by the expression of supramolecular stereochemistry upon addition of cations to solutions containing these optically active ligands. For G **5**, the cation's identity (Ba^{2+} *vs.* K^+) has a significant influence on the level of diastereoselectivity in the self-association process.[23] When K^+ was added to a solution of racemic (*D, L*)-G **5** the resulting G-quadruplexes were a mixture of heterochiral diastereomers, as determined by ^1H NMR spectroscopy. The divalent Ba^{2+}, however, directed almost complete enantiomeric self-recognition of a mixture of (*D, L*)-G (**5**), giving homochiral G-quadruplexes (Figure 11). To explain this ion-dependent diastereoselectivity we suggested that the enhanced enthalpy inherent to the divalent cation–oxygen interaction might help overcome the unfavorable entropy associated with enantiomeric self-sorting.

Cation-templated self-association of isoG **6** is also highly diastereoselective. A crystal structure showed that (*D, L*)-isoG **6** undergoes enantiomeric self-recognition in the presence of a Cs^+ template. The x-ray structure revealed that the decamer, $[(D)\text{-isoG }\mathbf{6}]_5 \cdot Cs^+ \cdot [(L)\text{-isoG }\mathbf{6}]_5 \cdot Ph_4B^-$ had one isoG$_5$ pentamer unit composed of only (*D*)-isoG **6** and the other pentamer was made up of (*L*)-isoG **6** (Figure 12).[21] This so-called "meso" diastereomer was also clearly the

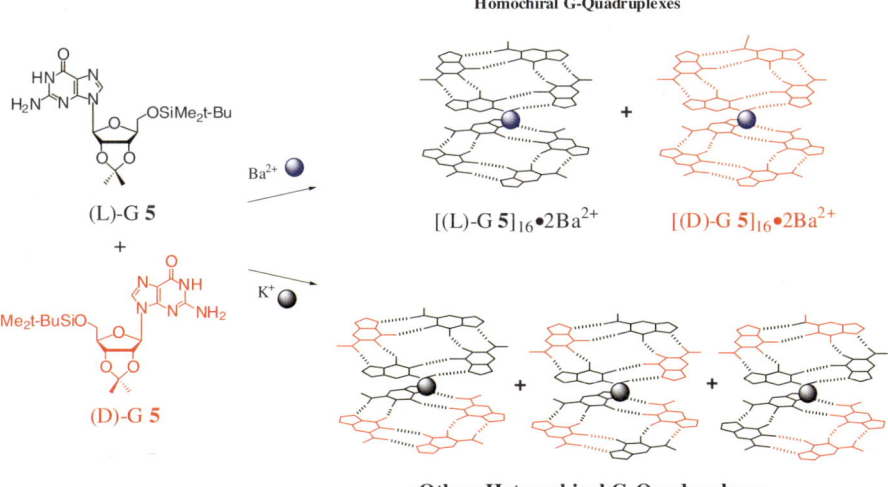

Figure 11 *G* **5** *undergoes cation dependent enantiomeric self-association. Racemic (D, L)-G* **5** *self-assembles in the presence of* Ba^{2+} *to give homochiral G-quadruplexes* $[(D)\text{-G }\mathbf{5}]_{16} \cdot 2Ba^{2+} \cdot 4Pic^-$ *and* $[(L)\text{-G }\mathbf{1}]_{16} \cdot 2Ba^{2+} \cdot 4Pic^-$. *Addition of* K^+ *to G* **5** *gave a diastereomeric mixture of heterochiral assemblies* (Adapted from ref 23 with permission).

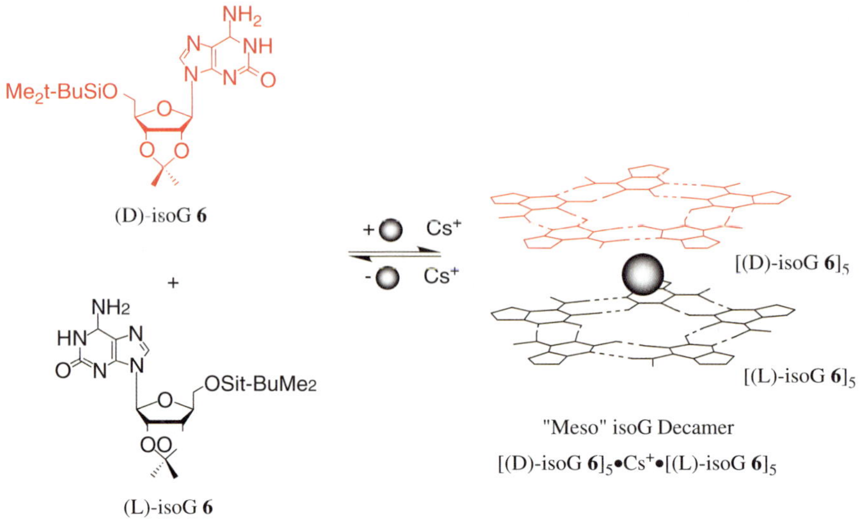

(D)·isoG **6**

+

(L)-isoG **6**

[(D)-isoG **6**]₅

[(L)-isoG **6**]₅

"Meso" isoG Decamer

[(D)-isoG **6**]₅•Cs⁺•[(L)-isoG **6**]₅

Figure 12   *Addition of Cs⁺ to racemic (D, L)-isoG **6** resulted in formation of homochiral hydrogen-bonded pentamers through enantiomeric self-association. The major species in solution and in the solid state is a "meso" decamer, [(D)-isoG **6**]₅·Cs⁺·[(L)-isoG **6**]₅*
(Adapted from ref 21 with permission).

major species in solution, as judged by NMR. The x-ray structure showed that, within a pentamer, each ribose formed sugar-base hydrogen bonds with its neighbor. We proposed that these sugar-base hydrogen bonds help transmit stereochemical information from a chiral sugar to its base-paired neighbor, thus leading to a homochiral hydrogen-bonded pentamer.

Eventually, lipophilic G-quadruplexes might well be useful as chiral resolving agents or as enantioselective catalysts. In this vein, Gottarelli and colleagues discovered that G-quartet structures formed from dG **7** are modestly enantio-selective in their ability to extract chiral anions from water into organic solvents. Thus, dG **7** extracted a K⁺ N-dinitrophenyl-(L)-tryptophan salt from water into CDCl₃ with a 3:1 enantioselectivity over the (D)-Trp enantiomer, indicating that there must be significant interactions between anions and the chiral G-quadruplex.[24]

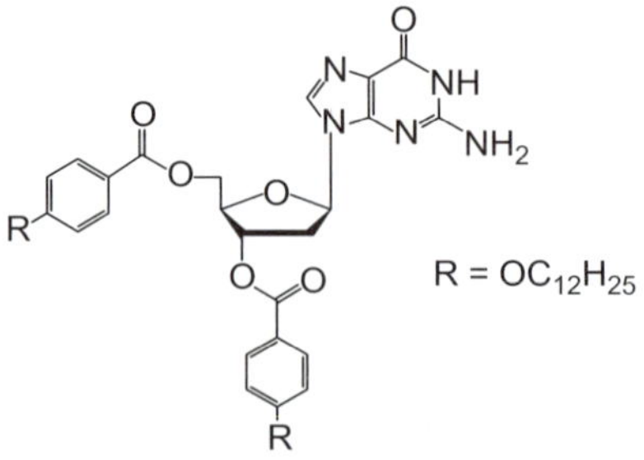

$R = OC_{12}H_{25}$

dG **7**

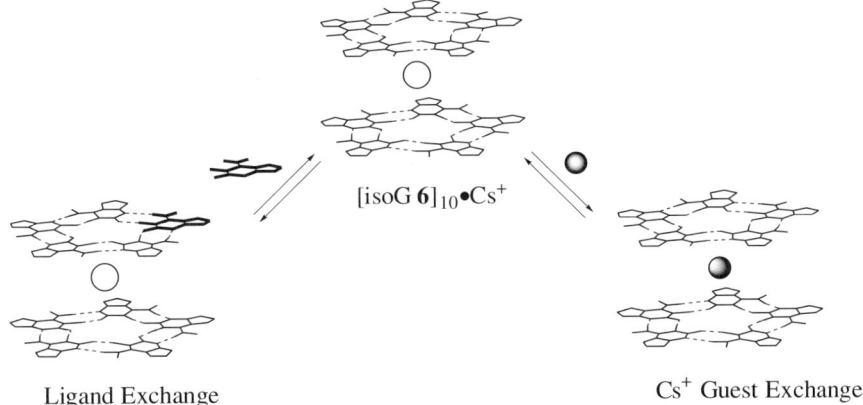

Ligand Exchange Cs^+ Guest Exchange

Figure 13   *Dynamic exchange processes in the decamer [isoG **6**]$_{10} \cdot Cs^+$. Both the isoG ligand **6** and the cationic guest Cs^+ exchange with free species in solution. NMR data shows that Cs^+ exchange is over 40,000 faster than ligand exchange of isoG **6***
     (Adapted from ref 19 with permission).

10.1.5 Self-Assembled Ionophores as Selective Metal Ion Extractants

Nucleoside assemblies also show promise for effecting demanding separations of environmentally important cations. A daunting challenge in nuclear waste remediation involves separating radioactive $^{137}Cs^+$ from excess Na^+ and K^+ ions in solution. Our group has shown that isoG **6** is highly selective for binding Cs^+, *vs.* other alkali metal cations.[20,25] Significantly, even though [isoG **6**]$_{10} \cdot Cs^+$ is thermodynamically stable, it readily exchanges its bound cation, a key issue for developing a practical extractant that can be used in an industrial clean-up application. We showed that the Cs^+ guest in [isoG **6**]$_{10} \cdot Cs^+$ exchanges with "free" Cs^+ in solution about 40,000 times faster than the rate for the exchange of the free isoG ligand **6** (Figure 13).[19] These much different exchange rates suggest that the isoG decamer complex does not dissociate during ion exchange.

Another radioactive ion of significant environmental concern is the cancer causing $^{226}Ra^{2+}$, a naturally occurring species. In collaboration with the Reinhoudt group in the Netherlands, we reported that the self-assembled ionophore [isoG **6**]$_{10}$ extracts $^{226}Ra^{2+}$ with excellent selectivity and affinity from both simulated and real wastewater.[26,27] Thus, even in the presence of higher concentrations of the other divalent alkaline earth metals, isoG **6** showed a remarkable selectivity for binding the radium cation ($>$ 10,000 to 1 for $^{226}Ra^{2+}$ *vs.* Ba^{2+}).

10.1.6 "Empty" G-Quartets

Without a templating cation, most guanosine analogs form hydrogen-bonded dimers or ribbons. But, this is not always the case. Thus, Sessler and colleagues

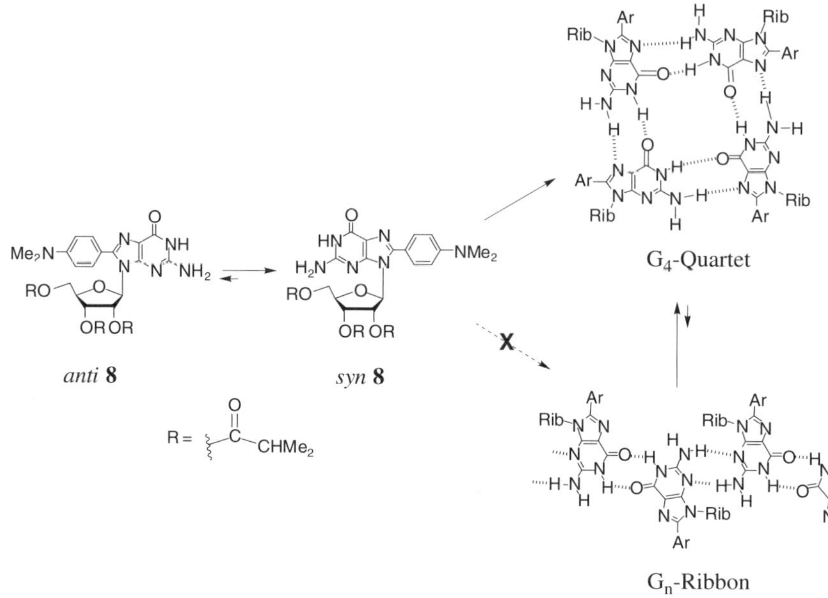

Figure 14 *Conformationally constrained G* **8** *forms a G-quartet without presence of a templating cation*

synthesized a G analog **8** that self-assembles into a G-quartet even without the assistance of a templating cation.[28] Attachment of a dimethylaniline unit to C8 of the guanine ring gave a conformationally constrained nucleoside that prefers to adopt a *syn* glycosidic bond conformer in both the solid state and solution. This *syn* conformation precludes the nucleoside from any hydrogen-bonded ribbon formation and thus favors formation of the macrocyclic G-quartet (Figure 14). Sessler's study showed the power of synthetic chemistry in designing unnatural nucleobases for making well-defined supramolecular structures. The implications of using such designer bases to build discrete assemblies are clearly important in supramolecular chemistry.

Our group showed that a calixarene-guanosine analog forms a hydrogen-bonded dimer (cG **9**)$_2 \cdot$ (H$_2$O)$_n$ in wet CDCl$_3$, with water presumably taking the cation's place within the center G-quartet cavity (Figure 15).[29] This finding was consistent with a prediction, made by Gellert and colleagues in the original 1962 paper on 5′-GMP self-association, that a G-quartet "...*would contain a hole in the middle in which it might be possible to place one water molecule.*"[4]

More recently, Besenbacher and colleagues[30] showed that guanine **10** can form a kinetically stable "empty" G-quartet on a gold surface (Figure 16). Using AFM they found that the empty G-quartet was not the thermodynamic minimum, as annealing the deposited G-quartet network led to rearrangement into a hydrogen-bonded ribbon. In this case, the available N9-H and the neighboring N3 positions of guanine **10** seem crucial for stabilizing the network of connected G-quartets. The Besenbacher paper is, to our knowledge, the first

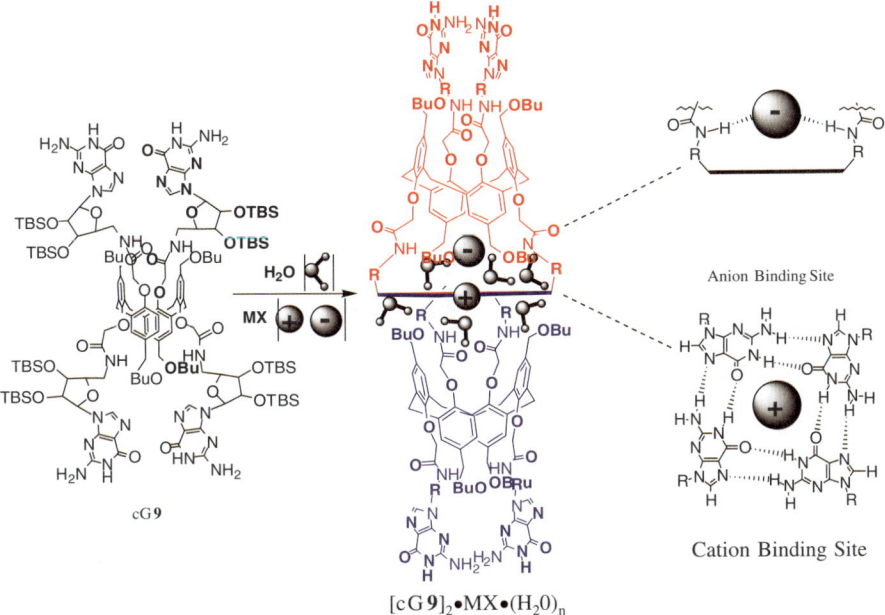

$[cG\,9]_2 \cdot MX \cdot (H_2O)_n$

Figure 15 *A schematic representation of $[cG\,9]_2 \cdot MX \cdot (H_2O)_n$ that shows the anion and cation binding sites*
(Adapted from ref 29 with permission).

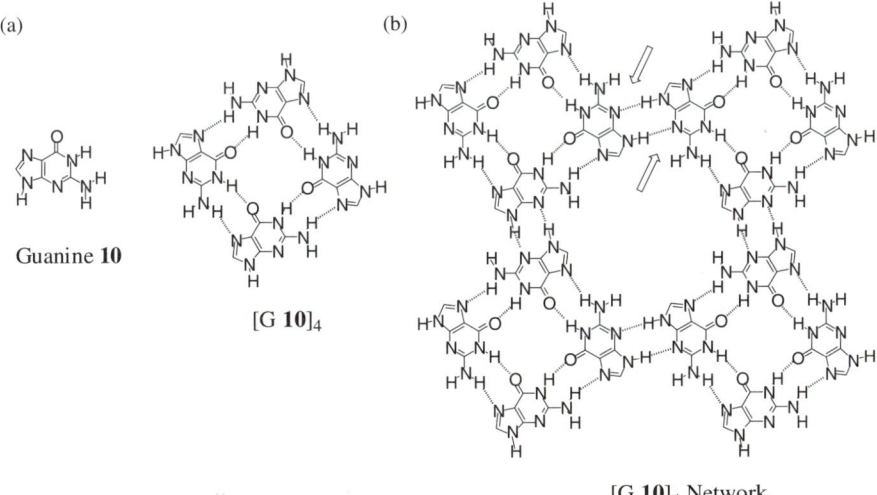

Figure 16 *(a) An empty G-quartet formed by guanine **10**. (b) A hydrogen bound network of empty G-quartets. Each G-quartet can form up to eight additional hydrogen bonds with neighboring G-quartets (arrows)*
(See ref 30).

demonstration that guanine itself forms cyclic quartets, as other G-quartets have always involved N9-substituted G nucleobases.

10.1.7 New Hydrogen-Bonded Assemblies form Other Nucleoside Analogs

Rivera and co-workers reported another way to stabilize G-quartet units by using 8-aryl-dG analogs such as dG **11**.[31] By adding a hydrogen bond acceptor to the C8 position, they hoped to involve the exocyclic N2 amino hydrogen that does not normally participate in the G-quartet's hydrogen bonding (Figure 17).

Variable temperature and dilution NMR experiments on the G-quadruplex [dG **11**]$_{16}$·3K$^+$ showed an increased stability when compared with assemblies from the unsubstituted G derivatives. Rivera proposed that the stability of the 8-aryl-dG analog **11** was due to a combination of factors. First, C8 substitution forces the derivative into the *syn* conformation, again precluding formation of hydrogen-bonded ribbon structures. Second, the four additional aromatic rings attached to C8 provide larger surface for π–π interactions between the stacked G-quartets. Third, the C8 substituent in dG **11** enables four additional hydrogen bonds per G-quartet, as depicted by the arrows in Figure 17.

Sessler and co-workers[32] designed a guanosine-cytidine dinucleoside **12** that self-assembles into a cyclic trimer in organic solvents (Figure 18). They used the potent GC hydrogen-bonding association to direct structure. An ethylene bridge separates the guanosine and cytidine moieties in **12** and orients these complementary groups so that they are preorganized for formation of the macrocycle by making three sets of GC basepairs. This well-defined supramolecular structure suggests that such analogs may enable construction of self-assembled dendrimers and other nanostructures.

Fenniri and co-workers have formed intriguing supramolecular structures from compounds that also combine the complementary hydrogen-bonding

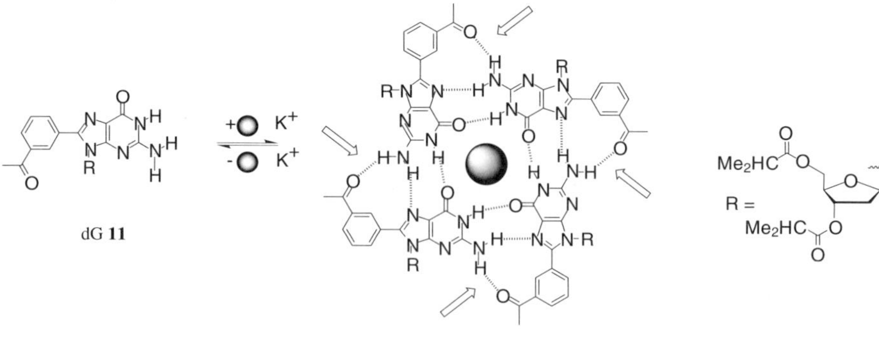

dG 11

[dG 11]$_4$·K$^+$

Figure 17   *A G-quartet formed from dG **11**, a modified nucleobase with an expanded Hoogsteen hydrogen-bonding face. Note the additional hydrogen bonds, depicted by arrows, thought to be a reason for increased stability*
(See ref 31).

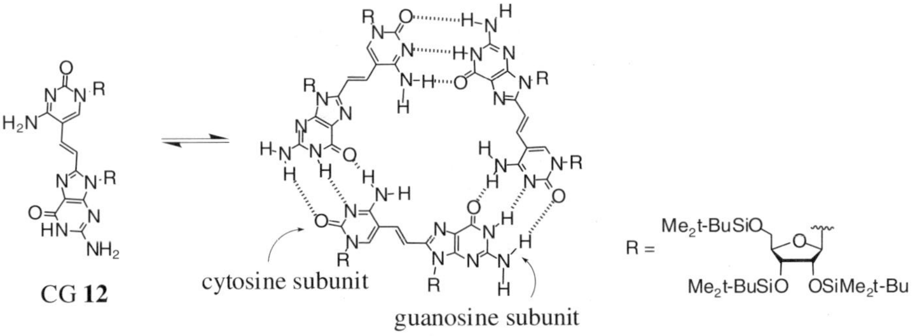

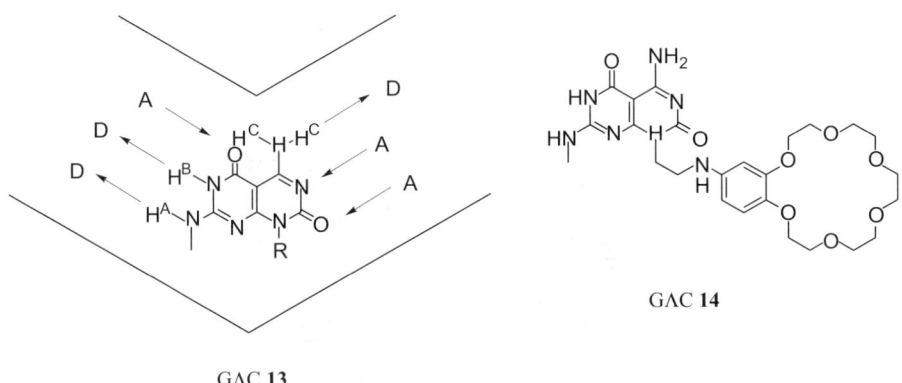

Figure 18 *Self-assembly of the lipophilic dinucleoside CG **12** into cyclotrimer [CG **12**]₃* (See ref 32).

Figure 19 *The G∧C **13** hydrogen-bonding motif uses the donor-acceptor-acceptor hydrogen-bonding face of cytosine and the acceptor-acceptor-donor face of guanosine. Crown-ethers are attached to the G∧C **13** scaffold to give G∧C **14*** (Adapted from ref 34 with permission).

patterns of cytidine and guanosine.[33–36] They synthesized the bicyclic unit, G∧C **13**, which combines the Watson-Crick donor and acceptor faces of G and C, respectively. The G∧C **13** monomers self-assemble in water to give a six-member rosette (Figure 19). These macrocycles stack to form helical rosette nanotubes (HRN). Although a number of synthetic nucleobase analogs self-assemble in organic solvents, Fenniri's G∧C **13** monomer is one of the first synthetic analogs to form such discrete assemblies in water. G∧C **13** has an amino acid moiety attached to the heterocycle *via* a spacer. This amino acid also dictates the supramolecular chirality of the system.

Investigation of rosettes as scaffolds for self-assembly of multichannel nanotubes led to synthesis of a G∧C analog **14** with a crown ether sidechain.[34] When these rosettes stack into an HRN, the crown ethers form channels around the main pore. This compound, G∧C **14**, has tunable chiroptical

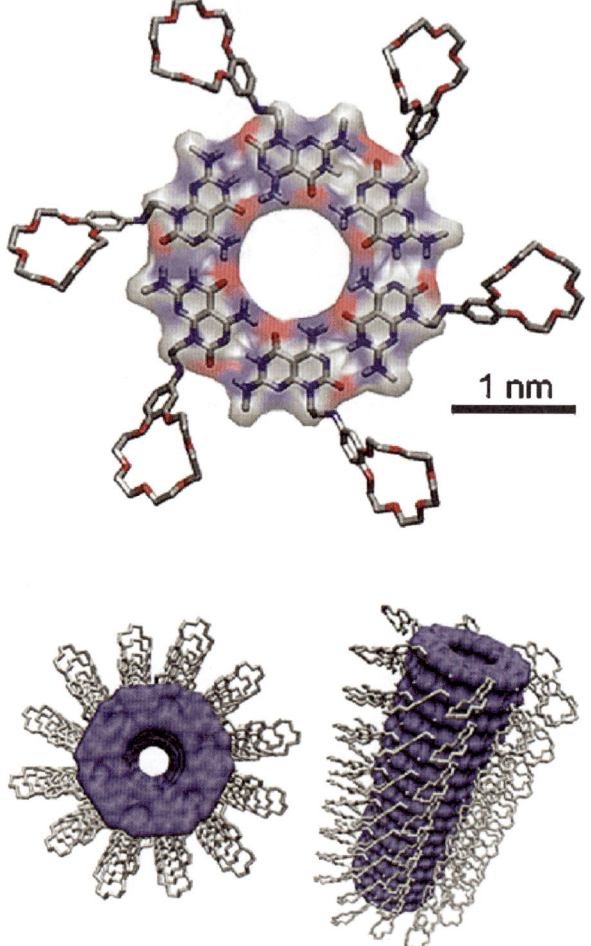

Figure 20 *G* ∧ *C* **14** *self-assembles into a six-member macrocycle called a rosette. These*
rosettes stack forming a helical rosette nanotube (HRN)
(Adapted from ref 35 with permission).

properties.[35] When formed in low concentrations, HRNs exist as racemic left-
and right-handed helices. Addition of chiral amino acid results in complexation
by the peripheral crown ethers of the HRNs, leading to a homochiral helix.
This process occurs in an "all-or-none" manner, as 93–99% of the crown ethers
must be occupied by an amino acid guest for a complete transition to the
homochiral nanotubes (Figure 20).[35]

10.1.8 The G-Quartet and Dynamic Covalent Chemistry

DCC is now a major synthesis strategy in supramolecular chemistry, enabling
amplification of select compounds from a dynamic combinatorial library

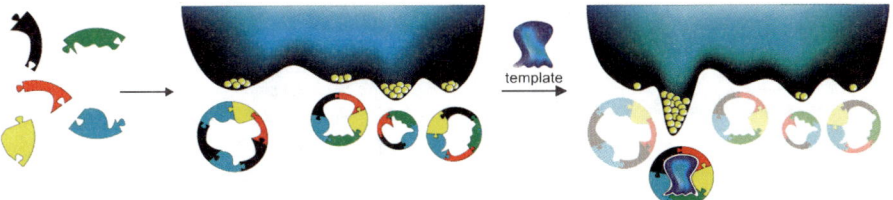

Figure 21 *The concept of dynamic covalent chemistry: the presence of a template shifts the equilibrium and amplifies a specific member of the DCL*
(Adapted from ref 38 with permission).

(DCL) of equilibrating compounds.[37,38] In this approach, building blocks that form reversible covalent bonds are used to build a DCL (Figure 21). Stabilization of a library member upon addition of a template results in a new equilibrium. The end result in accord with Le Chatelier's principle, is amplification of stabilized products in the mixture.

This DCC strategy has been used to produce new ligands that bind to G-quadruplexes. Previous studies have shown that (i) acridone ligands (A) stack on the terminal tetrad of a G-quadruplex and that (ii) certain peptides (P) interact with the grooves formed by the DNA backbone of the tetraplex.[39–42] Balasubramanian and colleagues[43] used a disulfide exchange reaction, carried out in the presence of glutathione disulfide and a G-quadruplex template, to identify new G-quadruplex interactive ligands that combine both the acridone and peptide recognition motifs. Disulfide exchange under aqueous conditions is popular for DCC applications, as the reaction is relatively rapid at pH >7 but not under acidic conditions (pH <5). Thus, disulfide exchange can be carried out under reversible conditions at moderate pH but then the reaction can be acid quenched to determine product composition. Using the deoxyoligonucleotide 5'-biotin(GTTAGG)$_5$, which contains the human telomeric sequence as a template, Balasubramanian's team demonstrated that there was a 400% increase in the formation of the heterodimeric disulfide AssP, a compound containing both the acridone (from A **15**) and peptide (from P **16**) domains, when compared to control experiments carried out in the absence of a G-quadruplex (Figure 22). In addition, the authors made the surprising discovery that the peptide dimer PssP was formed in 5-fold greater amount in the presence of the G-quadruplex template. Quantitative binding studies using surface plasmon resonance showed that the complexes formed by these new ligands and the human telomere G-quadruplex have dissociation constants of $K_d = 30$ and 22.5 µM for AssP and PssP, values that are far lower than the dissociation constant for the AssA–DNA complex ($K_d > 2.5$ mM). This same research group also used disulfide exchange to identify pyrrole-amide dimers formed with modest amplification in the presence of a G-quadruplex template.[44] These two studies established that DCC could provide new G-quadruplex ligands, a potentially important endeavor in the search for effective telomerase inhibitors.

In addition to the discovery of new ligands that interact with tetraplex structures, the DCC concept has also been used by the Balasubramanian, Lehn,

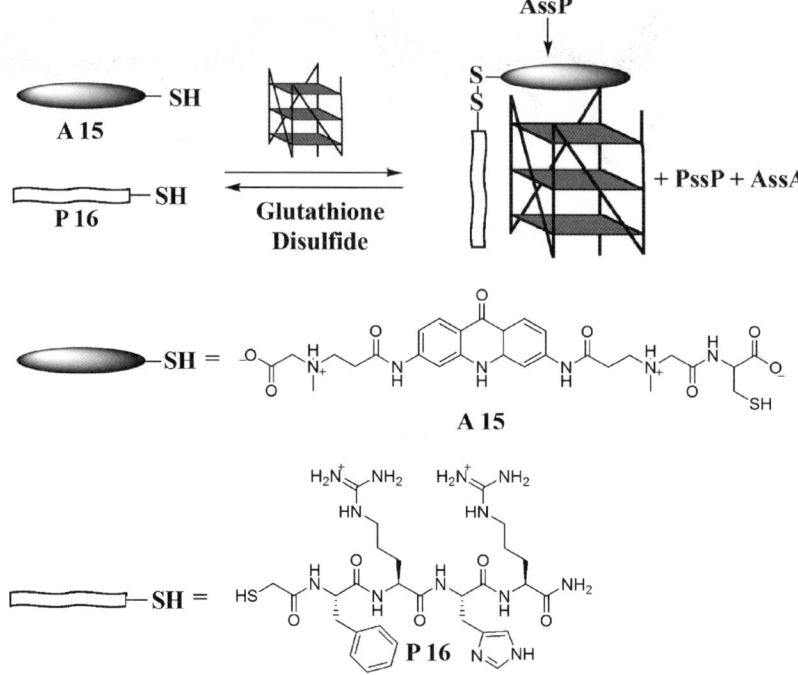

Figure 22 *The AssP disulfide product is selectively amplified in the presence of a G-quadruplex template*
(See ref 43).

and Davis groups to form new G-quadruplex structures, each with its own unique properties.[7,45–47] Thus, Balasubramanian and colleagues reported that a G-rich PNA, modified to allow for covalent bond formation between strands, underwent "self-templation" to form a bimolecular G-quadruplex.[45] In this study, they showed that formation of a noncovalent PNA G-quadruplex preceded covalent bond formation. The authors first demonstrated that an equimolar mixture of PNA-peptide strands, namely Lys-TGGG-GlyGlyCys-SH (G_S) and Lys-TTTT-GlyGlyCys-SH (T_S) gave a 1:2:1 statistical mixture of the three possible disulfides $G_{SS}G$, $G_{SS}T$, and $T_{SS}T$ when oxidized with sodium perborate (Figure 23). In contrast, the slower air oxidation of a mixture of the same two PNA strands gave a 2:1:2 ratio of $G_{SS}G$, $G_{SS}T$, and $T_{SS}T$ indicating that $G_{SS}G$ was somehow stabilized under these oxidation conditions. Mass spectrometry and UV melting experiments indicated that the $G_{SS}G$ dimer formed a bimolecular G-quadruplex $(G_{SS}G)_2$, presumably a bimolecular hair-pin structure wherein the Gly-Cys-Cys-Gly tetrapeptide forms the loop regions. Other measurements indicated that the G_S PNA strands were preorganized into a G-quadruplex prior to formation of the disulfide bond that gave the $G_{SS}G$ product. Significantly, formation of the $G_{SS}G$ disulfide bond depended on the cation template, being most effective with K^+, the cation that best stabilizes a G-quadruplex.

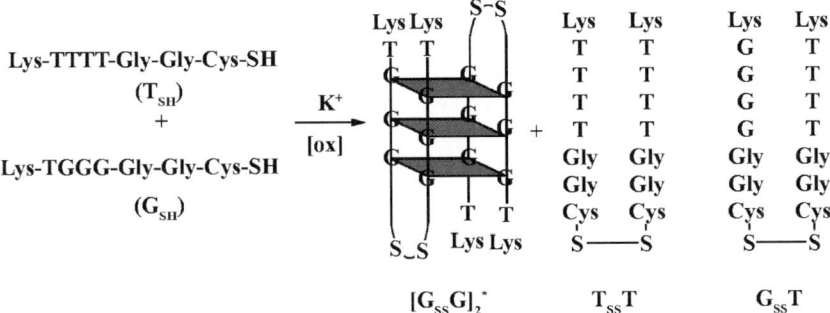

Figure 23 *Oxidation of the PNA strands T_{SH} and G_{SH} provides disulfides. In the presence of K^+, $G_{SS}G$ is amplified. *The structure depicted for $(G_{SS}G)_2$ represents just one possible orientation of a bimolecular G-quadruplex*
(See ref 45).

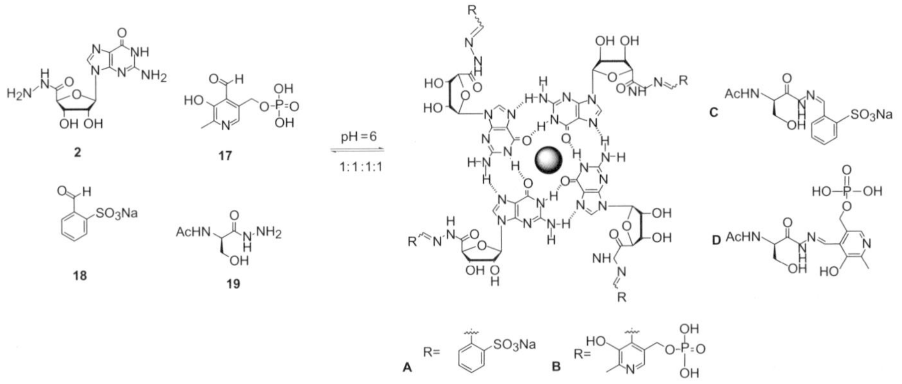

Figure 24 *The stability of G-quartet hydrogel B altered the dynamic equilibrium of acylhydrazones and directed reaction of the G hydrazide 2 with aldehyde 18*
(See ref 7).

As mentioned earlier, Lehn and Sreenivasachary[7] have recently described a G-quartet based system wherein component selection from a DCL is driven by the product's physical properties. They first showed that guanosine hydrazide **2** formed viscous, thermally reversible gels at moderate pH in the presence of Na^+ and K^+. These gels arose from the stacking and crosslinking of cation stabilized G-quartets. The 5′-hydrazide group in G-quartet gels formed by G **2** was then reacted with various aldehydes to form acylhydrazone bonds, allowing the authors to study the effects of sidechain modification on gel properties. While addition of some aldehydes destroyed the hydrogels, other aldehydes (including **17**) formed acylhydrazone gels that were even stronger than the parent gel from hydrazide G **2**. These findings prompted the authors to explore whether the stability of the gel phase might drive component selection in a DCL (Figure 24). A dynamic mixture composed of 4 acylhydrazones, from reaction of aldehydes **6** and **18** with hydrazides G **2** and serine **19**, was generated under conditions where the 5′-acylhydrazones could equilibrate by undergoing reversible bond cleavage and reformation. The resulting product distribution, measured by ^{1}H NMR, was sensitive to temperature. At 80°C, well above the gel transition temperature, the solution distribution of products was statistical, indicating that the four acylhydrazones (A D) were of similar stability. Between 25 and 55°C acylhydrazone B, in its gel state, and C in solution were significantly favored over acylhydrazones A and D. In this case, self-assembly of G hydrazide **2** was driven by selection of the components that gave the most stable hydrogels. The stability of G-quartet hydrogel B altered the dynamic equilibrium of acylhydrazones and directed reaction of the G hydrazide **2** with aldehyde **18**. Lehn explained that "...*(t)he process amounts to gelation-driven self-organization with component selection and amplification ... based on G-quartet formation and reversible covalent connections.*" This DCC approach may well have broad applications in medicinal chemistry and material science.

Ghoussoub and Lehn[46] have recently described another dynamic sol–gel interconversion process, in this case triggered by the reversible binding and release of K^+ by a G-quartet hydrogel. Supramolecular hydrogels formed by the ditopic monomer G–G **20** were readily converted to soluble $(G–G)_n$ polymers upon addition of [2.2.2]-cryptand **21**, an ionophore able to extract the stabilizing K^+ from the G-quartet hydrogel. The gel state could be regenerated upon expelling K^+ from the $[K^+\ 2.2.2]$-cryptate complex by protonating the cryptand's bridgehead nitrogen atoms to give $[2H^+\ 2.2.2]$ **21**. In this way, gel–sol interconversion was affected over multiple cycles by simply controlling the equilibrium of the bound K^+ between the G-quartet and the [2.2.2] cryptand, a ligand whose cation binding properties can be modulated by the solution pH (Figure 25).

In another example of the power of the DCC approach, our group recently reported on the templated synthesis of a unimolecular G-quadruplex that functions as a transmembrane Na^+ transporter.[47] The strategy combined noncovalent synthesis and postassembly modification of the noncovalent G-quadruplex. We used reversible olefin metathesis to cross-link sub-units that had been preorganized within a G-quadruplex (Figure 26).

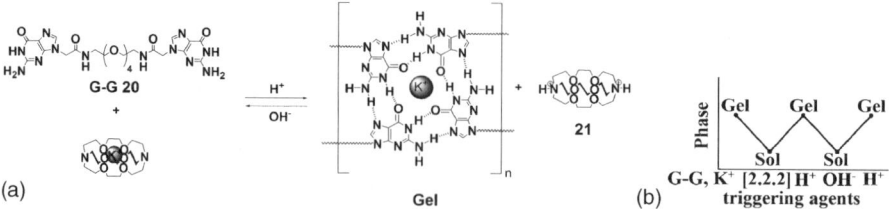

Figure 25   *(a) Structure of G-G **20** and schematic of the reversible formation of polymeric G-quartet based hydrogels. Changing pH in the presence of [2.2.2 cryptand] could modulate the sol–gel equilibrium. (b) Representation of the modulation of the gel–sol status induced by the sequence of triggering agents*
          (See ref 46).

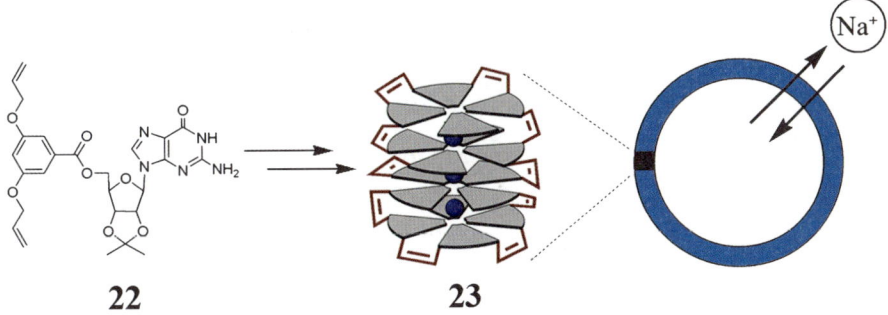

Figure 26 *Olefin metathesis was used to cross-link sub-units in the lipophilic guanosine 22. The resulting unimolecular G-quadruplex 23 was shown to transport Na$^+$ ions across phospholipid bilayer membranes*
(See ref 47).

The precursor, 5′-(3,5-*bis*(allyloxy)benzoyl)-2′,3′-isopropylidene G **22** was outfitted with two meta-substituted allyl ethers to enable olefin metathesis to be carried out within an individual G-quartet and between layers. A variety of techniques, including mass spectrometry, NMR, and CD spectroscopy, confirmed that the analog G **22** formed a hexadecameric G-quadruplex [G **22**]$_{16}$·4K+·4DNP$^-$. Olefin metathesis of this self-assembled G-quadruplex (8 mM), using Grubb's catalyst, provided quantitative formation of the metathesis product **23**. The unimolecular G-quadruplex **23** apparently folds into a conformation that allows it to transport Na$^+$ ions across phospholipid bilayer membranes. Direct evidence for the ability of the lipophilic G-quadruplex **23** to transport Na$^+$ across EYPC liposomes (200 nm) was obtained using ^{23}Na NMR spectroscopy. Shortly after the report on the ion transport properties of this unimolecular G-quadruplex **23**, the Matile and Kato[48] groups definitively showed that hydrogen-bonded rosettes prepared from the lipophilic folate **24**, structures closely related to the G-quartet, could also function as synthetic ion channels (Figure 27).

10.2 Biosensors and Nanostructures Based on DNA G-Quadruplex Structures

10.2.1 The Thrombin Binding Aptamer

The thrombin binding aptamer (TBA) is a 15-residue DNA oligonucleotide with the sequence d(5′-GGTTGGTGTGGTTGG-3′) that binds with high affinity and selectivity to the protease thrombin.[49] Nanomolar concentrations of this DNA aptamer can inhibit formation of the fibrin clots that result from thrombin activation. Shortly after its discovery, the groups of Bolton and Feigon used NMR spectroscopy to determine TBA's solution structure in the presence of K$^+$.[50,51] The single-stranded d(5′-GGTTGGTGTGGTTGG-3′) can form a unimolecular G-quadruplex that is shaped like a chair, with two

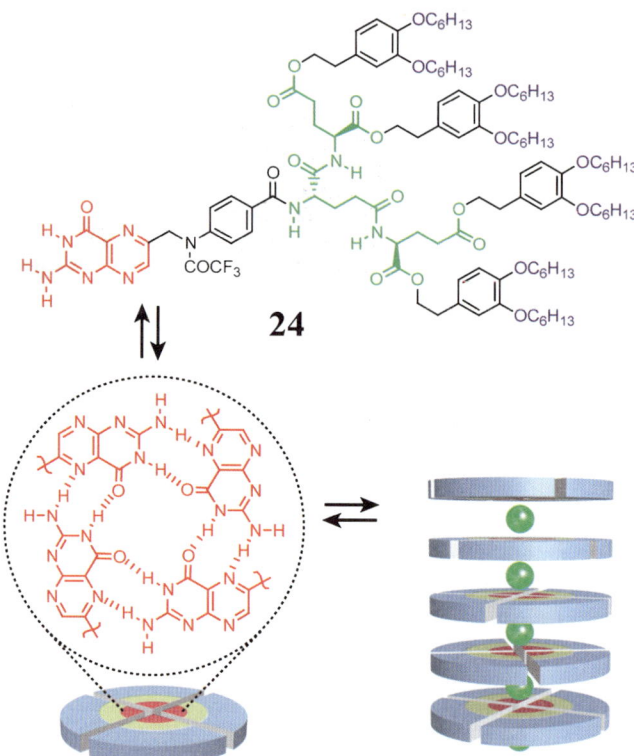

Figure 27   *A synthetic ion channel formed by self-association of folate derivative* **24**. *The folic acid tetramer is quite similar to a G-quartet*
(Adapted from ref 48 with permission).

stacked G-quartets connected by two TT loops and a central 3-base TGT loop (Figure 28). Potassium cation is essential for the templation and stabilization of the chair-type G-quadruplex by TBA, and both solution NMR spectroscopy and mass spectrometry have shown that the TBA G-quadruplex has a pronounced selectivity for coordination of K^+ over Na^+.[52,53] An X-ray crystal structure of a thrombin–TBA complex confirmed TBA's chair-like structure and suggested that this G-quadruplex DNA bound to the fibrinogen exosite, an anion binding location distinct from the protease's active site.[54] Later experiments have shown that thrombin has two distinct binding epitopes that recognize different G-quadruplex ligands.[55] By using thrombin mutants, competitive binding assays, and chemical crosslinking, Tasset and colleagues confirmed that the 15-mer TBA binds to the fibrinogen exosite, whereas another 29-mer oligonucleotide, one that folds into a different G-quadruplex topology, binds tightly to thrombin's heparin-binding exosite. A number of thrombin biosensors have been developed based on the simultaneous use of these two distinct G-quadruplex recognition sites. Although the TBA aptamer originally gained notoriety for its potential as a therapeutic antithrombolytic

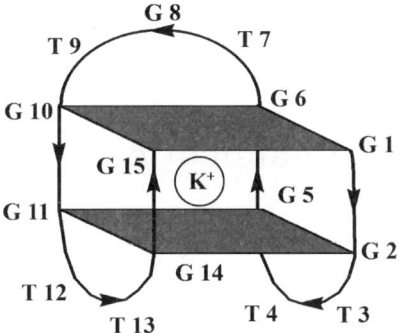

Figure 28 *Schematic of the thrombin binding aptamer (TBA)*

agent, this oligonucleotide has also been important in the supramolecular chemistry of G-quadruplexes, serving as the primary model for development of a range of sensors and nanomachines. As described below, the knowledge that this 15-mer oligonucleotide will fold into a stable G-quadruplex under well-defined conditions has been exploited to develop optical and electronic sensors, for analytes ranging from K^+ ion to proteins to nucleic acids.

10.2.2 Potassium Ion Sensors

The use of the TBA sequence as the basis for a biosensor is nicely demonstrated in a recent study by Takenaka's group.[56] They used a modified TBA as a fluorescent indicator for detecting K^+ in water. Attachment of pyrene groups to the 5′ and 3′-ends of the DNA gave a probe coined "PSO-py" for potassium sensing oligonucleotide-pyrene. This PSO-py is a promising sensor for the real-time detection of K^+ in biological and environmental samples. One challenge in developing an optical K^+ sensor is achieving selectivity in the presence of high Na^+ concentrations. Another challenge is to obtain a fast response that allows for real-time monitoring of the cation. PSO-py used the excimer formation from pi-stacked pyrenes to signal K^+ binding. In the absence of K^+, PSO-py is primarily unfolded and provides little excimer emission. In the presence of K^+, the 5′ and 3′ ends of the folded DNA stack pyrenes in a face-to-face geometry to give a new excimer band (Figure 29). Importantly, the presence of other cations gave little interference as only K^+ binds with high affinity to the TBA G-quadruplex. The fluorescence spectrum of PSO-py in the absence of K^+ showed a weak monomer emission at 390 nm. Addition of K^+ gave a strong excimer band at 480 nm, accompanied by quenching of monomer emission. Changes in excimer fluorescence indicated that the K^+ and Na^+ complexes of PSO-py had dissociation constants (K_d) of 7.33 and 272 mM, respectively. This K^+/Na^+ selectivity coefficient of 37 for PSO-py is higher than for many other previous K^+ sensors. Independent CD measurements of PSO-py, in the presence and absence of K^+, confirmed that the excimer fluorescence corresponded to a structural shift from a random coil to a chair-like G-quadruplex. The dynamics of the fluorescence response for the PSO-py/K^+ system also

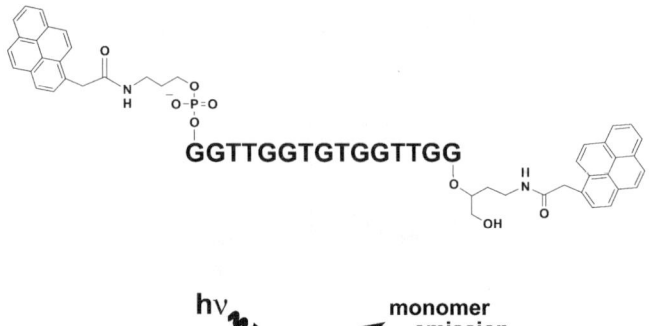

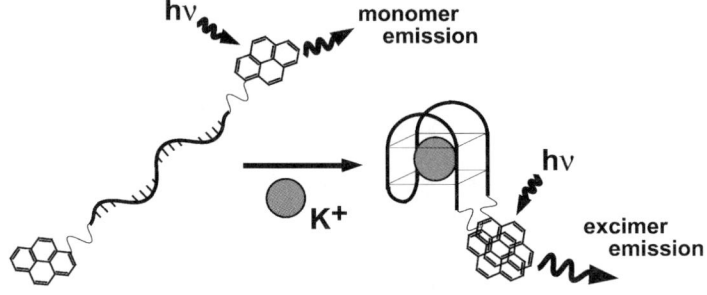

Figure 29   *Chemical structure of the PSO-py and the expected G-quadruplex induced by*
K⁺ binding. Pyrene excimer emission occurs in the presence of K⁺
(Adapted from ref 56 with permission).

showed a short response time (within seconds) upon variation in ion concentration. Moreover, this dynamic excimer fluorescence was both reversible and reproducible. The PSO-py oligonucleotide, well suited for real-time monitoring of K^+ in water, is representative of a range of bioprobes that have been rationally designed by using knowledge of G-quadruplex structure and properties.

The PSO-py oligonucleotide, which uses excimer emission as an optical signal, is actually a second-generation sensor. Takenaka's prototype, described in 2002, was a modified DNA oligonucleotide that underwent efficient fluorescence resonance energy transfer (FRET) upon folding into an intramolecular G-quadruplex.[57] This original PSO with the sequence d(5′-GGGTTAGGGT-TAGGGTTAGGG-3′) had a 6-carboxyfluorescein donor group attached to its 5′-end and a rhodamine acceptor linked to the 3′-terminus. When folded into a G-quadruplex, the two chromophores are located close enough together to undergo efficient energy transfer (Figure 30). Importantly, G-quadruplex formation by this PSO, as measured by FRET, was again highly selective for K^+ over Na^+.

Ho and Leclerc described another interesting method for the optical detection of K^+, based on formation of colored complexes between a cationic polythiophene and negatively charged DNA (Figure 31).[58,59] Because of changes in the conformation of its conjugated backbone, this flexible polymer senses different DNA topologies. Ho and Leclerc showed that this polythiophene distinguishes the single-stranded and G-quadruplex forms of TBA,

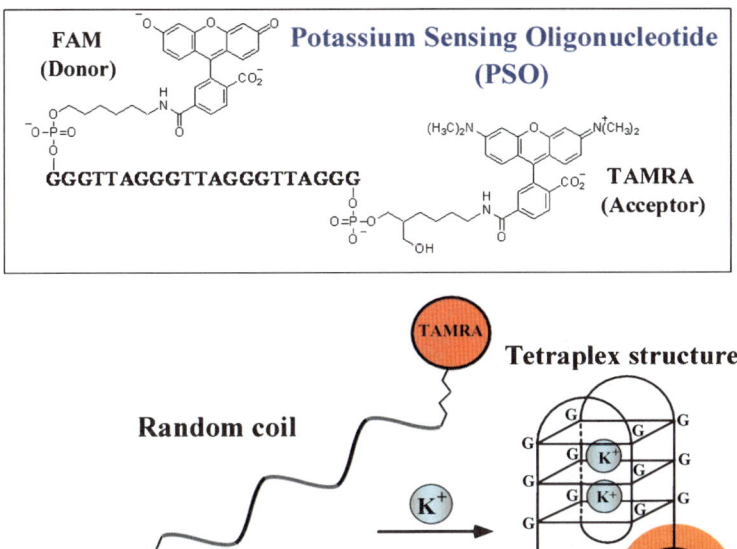

Figure 30 *Chemical structure of the PSO and the expected G-quadruplex induced by K^+ binding. In this case FRET occurs in the presence of K^+* (Adapted from ref 57 with permission).

enabling the polymer to be used as a selective probe for K^+, since that specific ion is required for folding TBA. This simple "staining" method for detection of the TBA G-quadruplex (or for any species that templates or stabilizes G-quadruplex structure) has the obvious advantage that it does not require chemical labeling of the DNA. Ho and Leclerc[59] have also shown that their method is useful for the selective and sensitive (femtomolar range) measurement of the thrombin protein and for the highly enantioselective detection of L-adenosine. Leclerc's biosensor strategy should also be ideal for identification of small molecules that bind to the G-quadruplex, thus providing a new method for screening potential antitelomerase drugs.

Wang and co-workers recently developed a related polymer-based assay for K^+ detection that benefits from the sensitivity that is available from the FRET process. In their case, energy transfer was observed from a cationic conjugated polymer to a TBA oligonucleotide labeled at its 5'-end with a fluorescein acceptor. Notably, they observed a significant increase in emission at 518 nm for the polymer-labeled TBA complex only when in the presence of relatively low concentrations of K^+. The magnitude of the FRET signal, which has a $1/r^6$ dependence on the distance between donor and acceptor, was attributed to the stronger electrostatic interactions that hold the cationic polymer closer to the compact and charge-dense G-quadruplex form of the TBA (Figure 32). In this

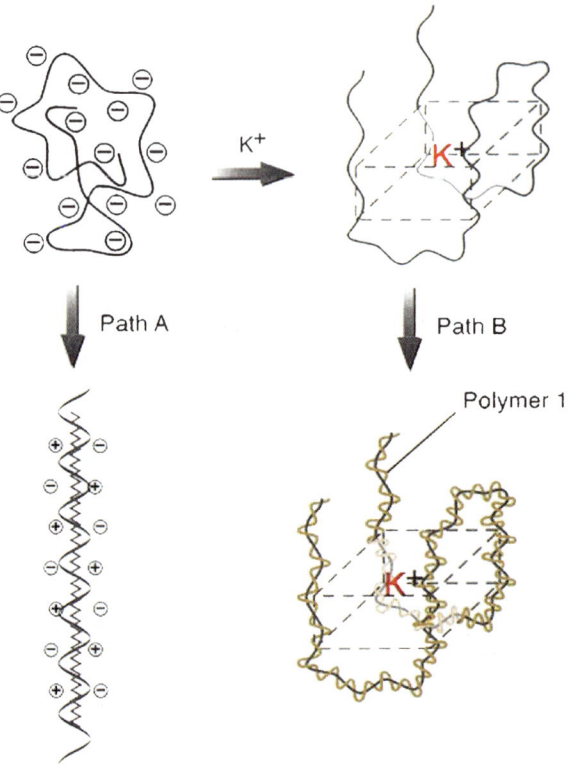

Figure 31   *An optical K^+ sensor based on a complex formed between G-quadruplex DNA and a conjugated cationic polymer*
(Adapted from ref 59 with permission).

way, K^+ ion was readily detected in water at low concentrations, even when other monovalent and divalent cations were present in excess.[60]

10.2.3   G-Quadruplexes as Optical Sensors for Proteins

In 1998, Hieftje and colleagues[61] described the first example of a protein sensor formed by the TBA sequence. They prepared a DNA conjugate that had the TBA labeled at its 5′-end with fluorescein and modified at its 3′-end by an amino siloxane linker, enabling covalent attachment of the oligonucleotide to a glass surface (Figure 33). Once the modified DNA had been tethered to glass, they used evanescent-wave-induced detection of fluorescence anisotropy to detect the specific binding of thrombin in solution to the immobilized TBA ligand. The resulting protein–DNA complex, being much larger than the DNA probe, showed a significant change in its rotational diffusion rate, as detected by the change in fluorescence anisotropy. The change in fluorescence aniso-tropy was specific to both the TBA and the protein analyte. Thus, scrambled DNA sequences that do not form G-quadruplexes did not show any enhanced

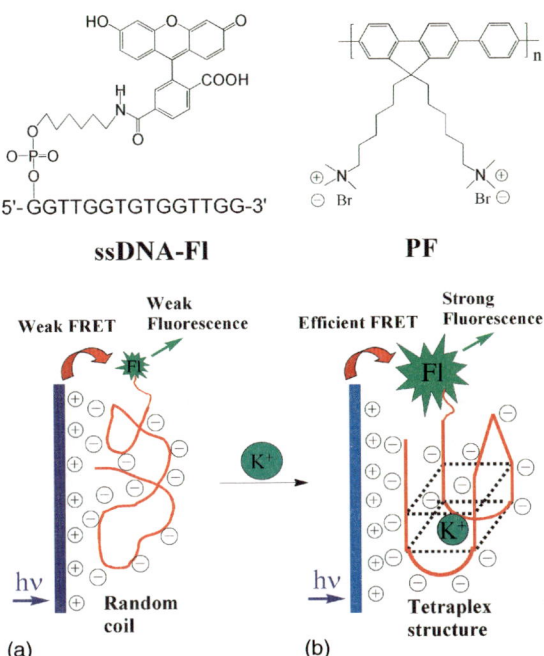

5'-GGTTGGTGTGGTTGG-3'

ssDNA-Fl

PF

Figure 32 *Schematic representation of an optical K^+ sensor based on G-quadruplex–polymer interactions that lead to FRET*
(Adapted from ref 60 with permission).

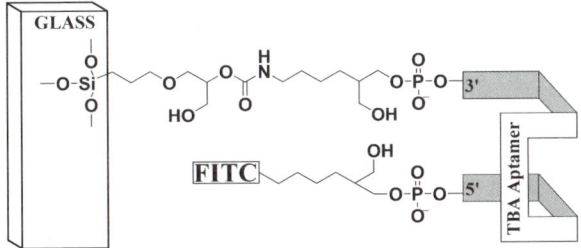

Figure 33 *A fluorescein modified DNA oligonucleotide that functions as biosensor for thrombin*
(See ref 61).

fluorescence anisotropy. Likewise, serine proteases other than thrombin did not bind to the fluorescein-labeled TBA. This TBA biosensor was sensitive and rapid, as it could detect as little as 0.7 amol of thrombin over a dynamic range of three orders of magnitude (from nanomolar to micromolar) in less than 10 min.

Lee and Walt[62] used a related strategy to build a thrombin biosensor by covalent attachment of the TBA sequence to silica microspheres. They then used a fiber optic device to detect the binding of fluorescein-labeled thrombin to these glass beads. They also developed a more practical assay that involved the competitive binding and displacement of fluorescein-labeled thrombin by unlabeled protein. Despite the need for specialized equipment this paper described an assay for thrombin in solution that was highly selective, rapid, and reproducible.

In 2001, Stanton and colleagues[63] described the use of "aptamer beacons" for the direct detection of thrombin binding. They chemically synthesized an oligonucleotide that contained the TBA sequence embedded within a longer DNA strand that was designed to form a stem-loop structure in the absence of thrombin. This DNA oligonucleotide contained a fluorescein chromophore at its 5'-end and a quencher group at its 3'-end. Thus, when the oligonucleotide was in its stem-loop conformation the 5'-fluorescein was quenched by the nearby 3'-DABCYL unit. Addition of thrombin shifted the DNA's conformational equilibrium from the stem-loop structure to a folded G-quadruplex, causing an increase in the chromophore-quencher distance and a fluorescence enhancement (Figure 34). The authors stressed that this method for thrombin detection could, in principle, be applied to other nucleic acid aptamers by simply embedding the protein binding sequence within an unproductive stem-loop structure that contained juxtaposed fluorescent label and quencher. Binding of the target protein should shift the conformational equilibrium and stabilize the aptamer's structure, resulting in fluorescence enhancement as the fluorophore-quencher separation changes. They envisioned using this strategy to make biosensors for proteomics applications using high-throughput, automated selection techniques.

Tan and colleagues also used the aptamer beacon strategy to develop real-time sensing of thrombin.[64,65] In addition to using fluorescence quenching, they also used both FRET and excimer strategies that allowed for significant

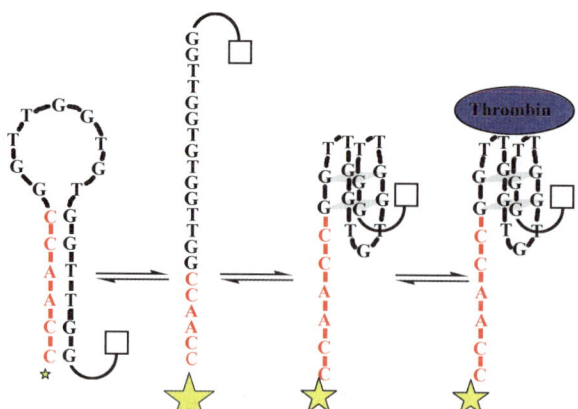

Figure 34   *A protein biosensor based on the "aptamer beacon" strategy. Thrombin shifts the DNA's conformational equilibrium to G-quadruplex and produces an increase in fluorescence as the donor-quencher groups get farther apart, compared to the stem-loop structure*
  (See ref 63).

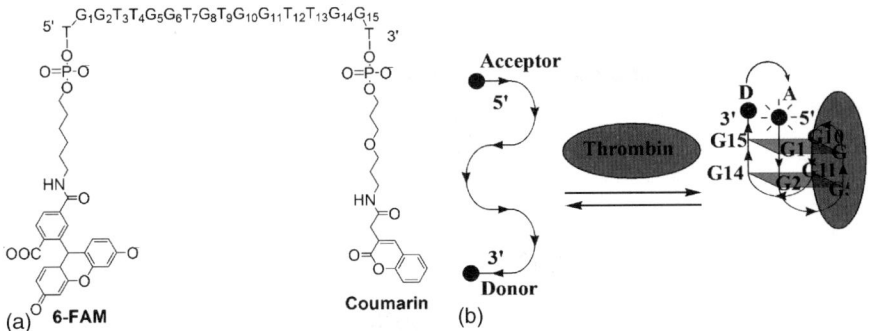

Figure 35 *Structure of (a) acceptor-donor TBA and (b) schematic showing FRET upon binding of thrombin to TBA*
(See ref 64).

fluorescence enhancement upon formation of a DNA–thrombin complex. Their aptamer beacon design involved labeling the 5′-end of a 15-mer with an energy acceptor, 6-FAM, and the 3′-end with a coumarin group as an energy donor (Figure 35). This modified 15-mer tended to favor a random coil conformation in low salt and the absence of thrombin, whereas the equilibrium was shifted to the folded TBA G-quadruplex in the presence of thrombin. This conformational change resulted in a significant enhancement in the fluorescence signal for 6-FAM as the chromophores came closer together in the folded state. These assays were highly sensitive giving a detection limit of 112 ± 9 pM for thrombin.

In 2003, Nutiu and Li[66,67] described a strategy for the preparation of fluorescent sensors based on their use of the so-called "Structure-Switching Signaling Aptamers." These DNA aptamers work by undergoing a major structural change from duplex DNA to a DNA-target complex. The starting duplex is formed between a DNA strand that contains the aptamer sequence and two shorter oligonucleotides; one of the shorter oligonucleotides contains a fluorophore and the other short strand contains a quencher. In the absence of the thrombin target, the aptamer strand binds to the short oligonucleotide containing the quencher, bringing it into proximity to the fluorophore and causing maximum quenching. Upon addition of the thrombin protein, the aptamer sequence releases the short oligonucleotide containing the bound quencher, resulting in a strong fluorescence enhancement (Figure 36).

In 2005, Heyduk and Heyduk[68] took advantage of the fact that thrombin has two different DNA binding epitopes to facilitate the simultaneous co-association of two different aptamers. Each aptamer was outfitted with a flexible linker region and a DNA sequence that would allow DNA duplex formation and enable simultaneous FRET enhancement. In the absence of the thrombin analyte, the two DNA strands do not associate because the complementary binding region is too short. However, when both sequences are bound to thrombin the increased entropy favors duplex formation and subsequent FRET enhancement (Figure 37).

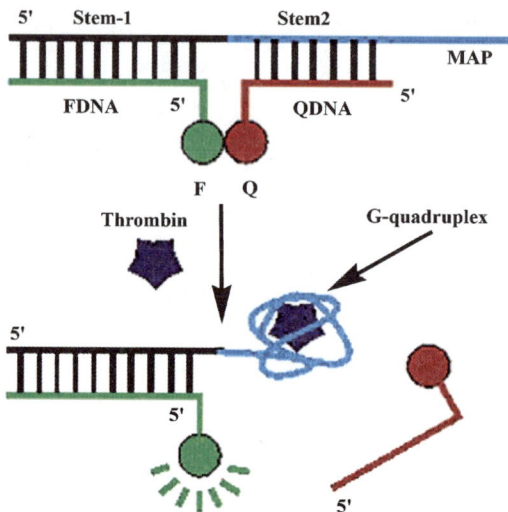

Figure 36 *The structure-switching signaling aptamer. A DNA duplex composed of three strands of DNA places a fluorophore (F) close to a quencher group (Q). Upon addition of thrombin, the QDNA piece is released, and the fluorescence increases*
(See ref 67).

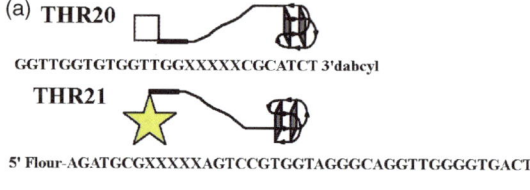

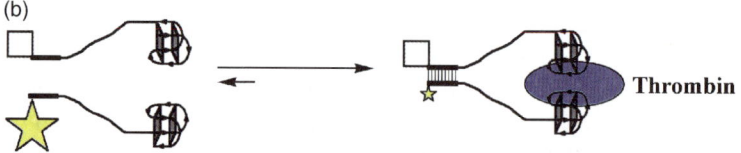

Figure 37 *(a) Detection of thrombin by binding two different G-quadruplexes at different epitope binding sites. (b) Association of the two strands of DNA on the thrombin surface leads to fluorescence quenching*
(See ref 68).

Willner and colleagues[69] used thrombin's two binding epitopes to design an ingenuous method for the optical detection of thrombin. Willner's group used gold nanoparticles functionalized with thiolated aptamers to enable the amplified detection of thrombin both in solution and on glass surfaces (Figure 38).[70] Reaction of the functionalized Au nanoparticles with thrombin in solution led to significant aggregation, since thrombin's two binding epitopes enabled

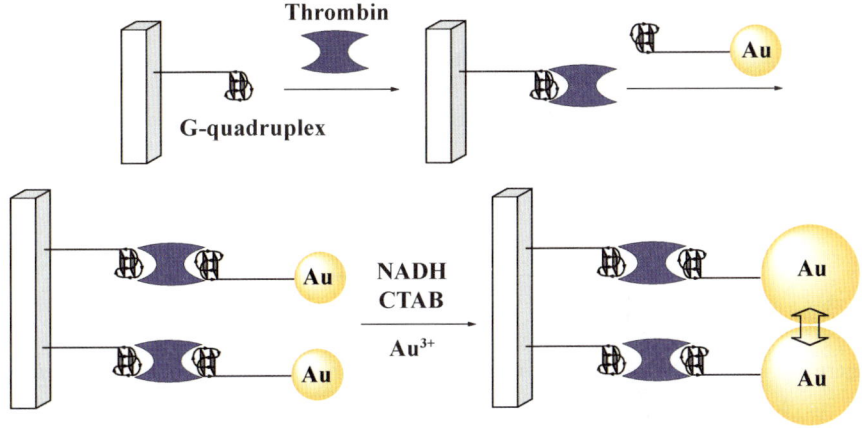

Figure 38 *Amplified detection of thrombin based on enlargement of Au nanoparticles* (See ref 69).

crosslinking of the Au nanoparticles. Addition of thrombin led to a significant decrease in the plasmon absorbance for the Au nanoparticles. The isolated precipitates were resuspended in solution containing a CTAB surfactant and then used to seed nanoparticle growth using $HAuCl_4$ and NADH. This catalytic growth of the nanoparticles was monitored by the gold's increased plasmon absorbance at 530 nm. Furthermore, the enlarged nanoparticles showed a red-shifted absorbance at 650 nm that was proposed to originate from a coupled plasmon exciton due to contacts between enlarged Au nanoparticles. These solution protocols for Au nanoparticle growth were also adapted to enable the optical sensing of thrombin on glass. A TBA oligonucleotide containing a siloxane unit was covalently attached to a glass surface and thrombin was bound to the resulting monolayer. The Au nanoparticles containing the thiolated TBA were then allowed to bind to thrombin through the second epitope site. Catalytic growth of the bound Au nanoparticles was then carried out in the presence of $HAuCl_4$, CTAB, and NADH. Both absorbance spectra and QCM measurements confirmed that the thrombin could be detected in a concentration dependent fashion. SEM images also showed that the Au nanoparticles came in contact with each other, entirely consistent with the presence of the interparticle absorbance band at 650 nm.

10.2.4 G-Quadruplexes in the Electrochemical Detection of Proteins

In the past few years a new direction in TBA-based biosensors has been the development of methods for the electrochemical detection of thrombin. Some of the reported advantages of these electrochemical biosensors are their potential to provide high sensitivity, fast response times, low costs, easy fabrication, and the possibility for miniaturization. Ikebukoro and colleagues[71] were the first to report on a TBA-based electrochemical sensor. Like others,

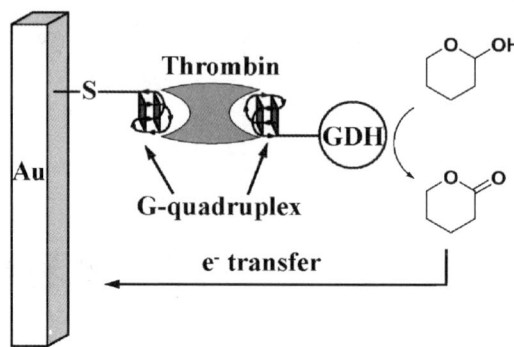

Figure 39   *Electrochemical detection of thrombin through the coupled oxidation of glucose
by glucose dehydrogenase*
(See ref 71).

they took advantage of thrombin's two separate binding sites. Fabrication of
the device involved immobilizing a thiolated TBA sequence onto a gold
electrode. A second oligonucleotide that can fold into a G-quadruplex structure
was covalently modified at its 3'-end with the enzyme glucose dehydrogenase
(GDH). Addition of thrombin to this solution resulted in formation of a
sandwich structure wherein the GDH was brought close to the gold electrode.
Oxidation of glucose by the immobilized GDH enzyme resulted in a measurable electrical current (Figure 39). No current was detected in the absence of
thrombin, demonstrating that the GDH needs to be close to the Au electrode.
Using this electrochemical detection device, thrombin at concentrations as low
as 1 µM could be detected.

In the last year a flurry of papers from the groups of Hianik, Plaxco, Lee, and
O'Sullivan have appeared describing a variety of approaches for the electrochemical detection of thrombin.[72–77] Plaxco and colleagues[73] used a 5'-
thiolated DNA oligonucleotide containing both the TBA sequence and an
electrochemically active group (methylene blue) attached to the 3'-end of the
DNA. This modified DNA oligonucleotide was attached *via* its thiol tether to
the gold electrode. In the absence of thrombin, the DNA adopts a conformation such that the electroactive methylene blue label can bind to the gold
surface and enable electron transfer with the electrode. However, binding of
thrombin by the folded TBA sequence results in a conformational change that
turns off electron transfer between the 3'-methylene blue label and the gold
electrode. Presumably the aptamer's conformational change significantly increases the electron-tunneling distance between the electrode and the electroactive label. This particular sensor, used to measure thrombin in blood serum,
demonstrated excellent dynamic range of 10–700 nM and outstanding sensitivity, such that thrombin at 10–100 nM concentrations could be measured
from blood plasma [Figure 40(a)].

Radi and O'Sullivan[77] recently described a similar approach wherein they
attached a thiolated TBA sequence containing a redox-active ferrocene group

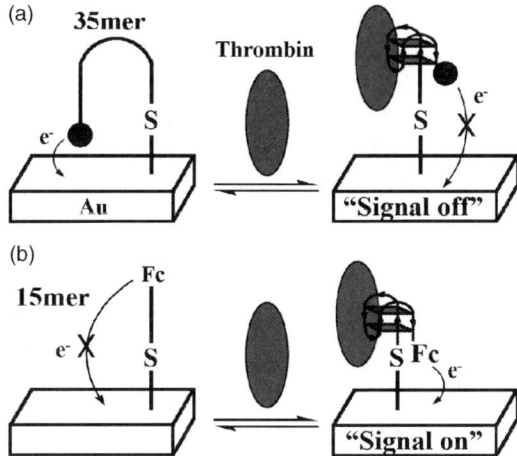

Figure 40 *Electrochemical biosensors for the detection of thrombin (a) "signal-off system" described for a 35mer oligonucleotide containing the TBA sequence in ref 73; (b) "signal-on" system described in ref 77*

to a gold electrode. A bifunctional 15-base TBA derivative with a ferrocene group and a thiol at its respective 5′ and 3′ termini was prepared. After anchoring this electroactive aptamer to a gold electrode the rest of the gold surface was coated with 2-mercaptoethanol to form a mixed monolayer. Cyclic voltammetry (CV), differential pulse voltammetry (DPV), and electrochemical impedance spectroscopy (EIS) were used to characterize this DNA-modified electrode. The modified electrode gave a voltammetric signal due to the redox reaction of the TBA's ferrocene group. The increase in signal intensity upon binding thrombin to the TBA sensor was attributed to a conformational transition from random coil to the folded G-quadruplex [Figure 40(b)]. In this "signal-on" system, which contrasts to Plaxco's "signal-off" system,[73] the authors noted an increased electrochemical signal upon binding thrombin. They suggested that the short length of their DNA tether resulted in a conformational change that brings the ferrocene label closer to the electrode surface and increases electron transfer. This "signal-on" electrochemical biosensor was used for the detection of thrombin without the need for any special reagents. The sensor had nanomolar detection limit for its target and showed little interference from nonspecific proteins. The aptasensor could be easily regenerated and reused 25 times without any loss in detection sensitivity.[74]

10.2.5 Biosensors for Nucleic Acids

DNA can also be optically detected using TBA–thrombin interactions.[78–80] Fan and colleagues[78] used an electrochemical version of the "molecular beacon" approach to detect DNA hybridization by measuring the electrochemical signal that accompanied a conformational change in the sensor. Their

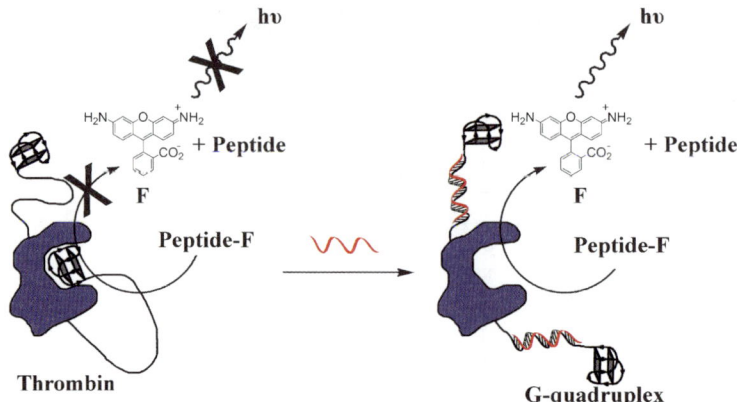

Figure 41   *Optical detection of DNA by catalytic activation of thrombin upon dissociation of an intramolecular thrombin–TBA complex*
                 (See ref 80).

strategy involved attaching a ferrocene tag to a thiolated TBA sequence within a stem-loop DNA structure, followed by subsequent attachment of the labeled DNA to a gold electrode. Hybridization of this sensor with a complementary DNA sequence then triggered a conformational change in this surface-confined TBA sensor, which led to a corresponding change in the electron-tunneling distance between the Au electrode and the ferrocene label. Using CV, target DNA concentrations as low as 10 pM could be measured using this sensor.

In an elegant approach toward DNA detection, Willner and colleagues introduced the use of "catalytic beacons."[79,80] Their method is illustrated in Figure 41. The thrombin protein was covalently modified with an oligonucleotide containing the TBA sequence.[80] In the absence of a complementary DNA strand the appended TBA sequence folds into a G-quadruplex and blocks the enzyme's active site. Addition of a complementary DNA strand unfolds the G-quadruplex, resulting in substrate access to the thrombin's active site. Hydrolysis of a fluorophore-labeled peptide then results in a readily detectable optical signal.

10.2.6   The Use of G-Quadruplexes in Building Nanomachines

The TBA sequence has also been used as the basis for single molecule systems that have been coined "nanomachines" or nanomotors.[81–83] Li and Tan[81] first demonstrated that conformational switching of a DNA oligonucleotide between its duplex and its folded G-quadruplex forms resulted in a flexing motion. They used FRET to follow this shrinking and expansion motion in real time. In a similar fashion, Alberti and Mergny[82] reported that the conformational equilibrium between DNA duplex and quadruplex defines a nanomolecular machine. Thus, the conformational states of a 21-mer DNA oligonucleotide, modified with 5′-fluorescein donor and 3′-rhodamine acceptor groups, could be readily detected by using FRET techniques. Switching between the folded unimolecular G-quadruplex and a duplex conformation caused a 5–6 nm displacement along the length of the oligonucleotide. This nanomachine could be cycled between its closed G-quadruplex state and open duplex state by sequential addition of other DNA strands, the so-called "C-fuel" and a "G-fuel." The "C-fuel" unfolded the unimolecular G-quadruplex to generate a duplex, while the "G-fuel" strand was used to liberate the labeled 21-mer so that it could refold into a G-quadruplex structure.

Simmel and co-workers[83] recently described a nanomachine that can bind and release thrombin as it undergoes conformational switching (Figure 42). In this DNA-based machine, the TBA sequence was fused to another DNA sequence that can partially bind another DNA sequence (Q). Upon addition of the Q DNA to the TBA–thrombin complex, the G-quadruplex region unfolds and releases the bound thrombin protein. Addition of another DNA strand (R) that is complimentary to Q frees up the TBA sequence and allows it to refold. Thus, this nanomachine represents a new way to control the reversible binding of thrombin in solution.

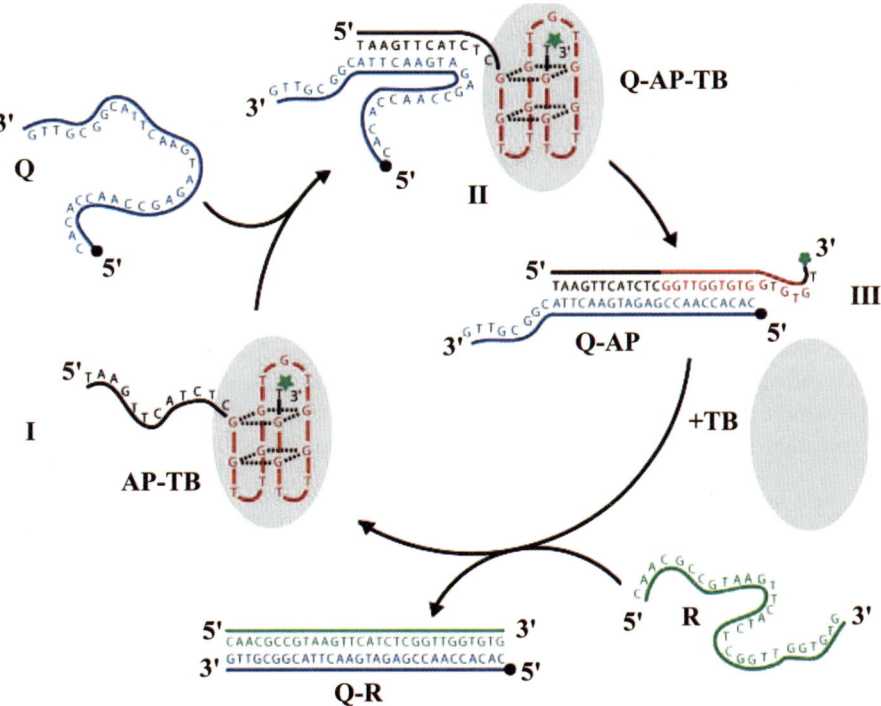

Figure 42 *A DNA-based nanomachine that binds and releases thrombin. Binding of DNA strand Q to TBA–protein complex release thrombin, and addition of complementary DNA strand R removes Q and shifts equilibrium back to the TBA–thrombin complex*
(Adapted from ref 83 with permission).

10.2.7 New G-Quadruplex Structures from Synthetic DNA Analogs

Polymers other than canonical DNA and RNA oligonucleotides can also form G-quadruplexes. The ability to alter the polymer backbone may result in G-quadruplexes with a variety of potential applications in supramolecular chemistry, biotechnology, and nanotechnology. In addition, studies on nucleic acid analogs may lead to insights into the structural factors that control fundamental issues about the thermodynamics and kinetics of the G-quadruplex motif in the parent DNA and RNA nucleic acids.

For example, locked nucleic acids (LNA) have conformationally constrained ribose units that are fixed in a C3'-endo conformation by a methylene bridge between the 2'-O and 4'-C atoms (Figure 43).[84,85] This RNA-like C3'-endo sugar pucker reduces backbone flexibility and helps drive the attached nucleobase to adopt an *anti* conformation about the glycosidic bond. Dominick and Jarstfer[86] recently showed that replacement of individual dG residues with LNA nucleotides in the Oxy28 telomeric sequence d(G4T4G4T4G4T4G4) dramatically alter the topology of the resulting G-quadruplex. Oligonucleotides with four G-rich tracts can adopt either parallel or antiparallel intramolecular G-quadruplexes. For example, the human telomeric sequence d(AGGG (TTAGGG)₃) forms a unimolecular propeller structure whose phosphate backbone sections are all parallel to one another. On the other hand, the Oxy28 sequence forms an antiparallel crossover basket, with the G residues alternating in a syn-anti-syn-anti fashion along the individual G4 tracts. Because 3'-endo nucleotides prefer to adopt an *anti* glycosidic bond, the authors postulated that incorporation of LNA residues into Oxy28 might drive the formation of a parallel G-quadruplex.

Dominick and Jarstfer inserted LNA into specific positions of the Oxy28 sequence and used CD spectroscopy to determine both the folding topology and thermodynamic stability of a family of modified oligonucleotides. In all cases, substitution of an LNA residue led to G-quadruplexes that were destabilized relative to the parent Oxy28 sequence. However, in some cases, even single nucleotide changes shifted the G-quadruplex from an antiparallel to a parallel propeller structure in the presence of K^+ (Figure 44). This

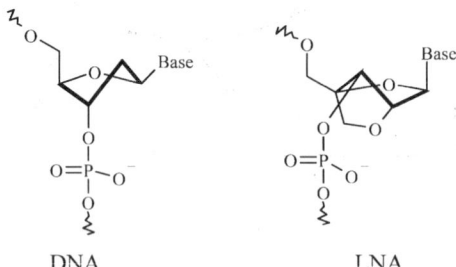

DNA   LNA

Figure 43   *Structures of DNA and LNA showing (a) DNA in the C2'-endo conformation and (b) LNA locked into the C3'-endo conformation*

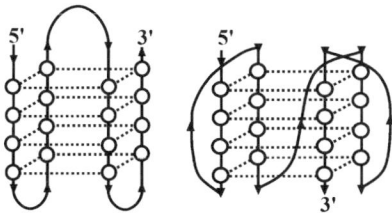

Figure 44 *Schematic showing (a) antiparallel DNA G-quadruplex and (b) parallel DNA*
G-quadruplex. Substitution of a single DNA monomer with an LNA analog
results in conformational switching between the two structures
(See ref 86).

remarkable finding drives home the point that even single internal modifications within the oligonucleotide backbone can dramatically influence the
structure of the resulting G-quadruplex.

In another informative study, Mayol and colleagues[87] demonstrated the
significant impact that LNA can have on both the thermodynamics and kinetics
of G-quadruplex folding. They used ^{1}H NMR and CD measurements to
characterize a well-defined G-quadruplex [tgggt]$_4$ with three stacked G-quartets. Like the analogous d[TGGGT]$_4$, this LNA G-quadruplex formed a
symmetric structure with all four strands parallel to one another. The LNA
G-quadruplex [tgggt]$_4$ was more stable, with a higher melting temperature, than
the corresponding DNA and RNA quadruplexes. Importantly, these CD
melting and annealing measurements also revealed that the LNA strands had
a much faster association rate than do DNA and RNA at micromolar
concentrations. Mayol's study indicates that the significant preorganization
of the LNA backbone, coupled with the stabilization of *anti* glycosidic bonds,
provides an entropy gain that leads to faster kinetics for G-quadruplex
formation.[88]

Peptide nucleic acids (PNA), nucleobase oligomers wherein the anionic
phosphate backbone is replaced by neutral *N*-(2-aminoethyl) glycine linkages,
also form a variety of G-quadruplex structures. Both DNA and RNA G-
quadruplexes can be invaded by a homologous PNA strand to give hybrid
PNA$_2$–DNA$_2$ G-quadruplexes.[89,90] Armitage and colleagues showed that the
PNA H-G$_4$T$_4$G$_4$-Lys-NH$_2$ hybridizes with its homologous DNA d(G$_4$T$_4$G$_4$) to
give a G-quadruplex consisting of two strands of DNA and two strands of
PNA. FRET measurements using labeled polymers indicated that strands were
organized such that the two DNA strands are parallel with each other and the
5′ ends of the DNA point in the same direction as the N-termini of the PNA
strands (Figure 45). Of the two possible structures envisioned for such a PNA$_2$–
DNA$_2$ hybrid Armitage favored the "alternating" structure (A) over the
"adjacent" structure (B) for two reasons: electrostatic repulsion would be
minimized by separating the two anionic DNA strands and FRET experiments
indicated that the donor and acceptor were closer to one another when both the
PNA and DNA strands were labeled. Armitage also made some important
observations about G-quadruplex kinetics in comparing the CD melting

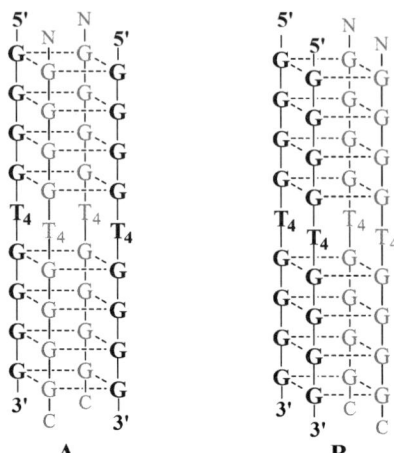

A   **B**

Figure 45 *Possible structures of hybrid 1:1 PNA₂–DNA₂ quadruplexes (bold=DNA, gray=PNA) where the PNA strands are (A) diagonally opposite or (B) adjacent to each other*
(Adapted from ref 89 with permission).

profiles for this hybrid PNA_2–DNA_2 G-quadruplex with that for the hairpin dimer formed by the homologous DNA. The DNA hairpin dimer showed significant hysteresis upon cooling, indicating that the rate of association of two strands to make the hairpin dimer is relatively slow. In contrast, the hybrid PNA_2–DNA_2 showed little hysteresis in the melting and annealing process, indicating that the kinetics for strand association are much faster for the four-stranded PNA G-quadruplex. Armitage suggested that this faster hybridization kinetics was due to the lack of negative charges along the PNA backbone and, possibly, due to electrostatic attraction of the PNA's positively charged N-terminus with the anionic DNA. Armitage concluded his paper by noting that, in principle, PNA_4 G-quadruplexes should be possible.

Indeed, shortly after Armitage's paper on hybrid PNA_2–DNA_2 G-quadruplexes, Balasubramanian and colleagues[91] reported formation of intermolecular G-quadruplexes composed solely of four PNA strands. Based on the combined ESI-MS, UV, and CD data they identified a four-stranded PNA quadruplex $(Lys$-TG_3-$NH_2)_4$ that aligned in an antiparallel fashion. This PNA sequence, which contains only one chiral center at its terminal Lys residue, exhibited an induced CD spectrum characteristic for stacked G-quartet chromophores, with a negative CD band at 270 nm and a positive band at 288 nm. UV melting experiments revealed that this particular PNA G-quadruplex was not nearly as stable, nor as cooperative in its formation, as the corresponding DNA quadruplex $(TG_3)_4$. Subsequent to Balasubramanian's report, Armitage and colleagues showed that another PNA sequence also form intermolecular G-quadruplexes. Thus, depending on the conditions, the PNA $(H$-$G_4T_4G_4$-Lys-$NH_2)$ forms either a four-stranded quadruplex or a two-stranded hairpin dimer (Figure 46).[92] Unlike the $(Lys$-TG_3-$NH2)_4$ G-quadruplex studied by

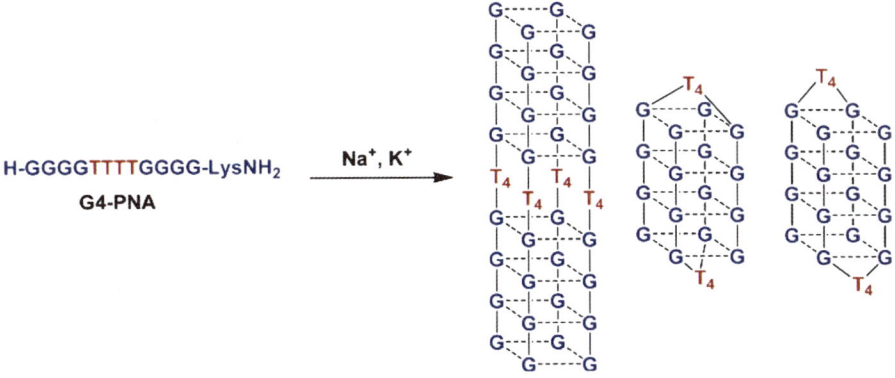

Figure 46 *Possible G-Quadruplex structures formed by PNA* (See ref 92).

Balasubramanian, this $(H-G_4T_4G_4-Lys-NH_2)_4$ PNA quadruplex was stabilized by the presence of Na^+ and K^+. Since their backbones are neutral, Armitage noted that a PNA G-quadruplex might be an excellent candidate for transporting cations across cell membranes.

Most recently, Giancola and colleagues[93] reported on the thermodynamic and kinetic properties of G-quadruplexes formed from chimeras 5'-tGGGT-3' and 5'-TGGG-3'-t, sequences that contain a single PNA residue at the ends of a DNA sequence. Using CD spectroscopy and calorimetry, they found that these chimeric PNA–DNA quadruplexes were thermodynamically more stable than the corresponding DNA G-quadruplex. Furthermore, the kinetics of quadruplex formation, as measured by melting and annealing experiments, indicated a reaction order of 4.0 in strand concentration. Both chimeric G-quadruplexes assembled more slowly than the corresponding DNA, as the rate constants at 20°C were $(3.0 \pm 0.2 \times 10^7)$ for [5'-TGGGGT-3']$_4$ and $(2.1 \pm 0.2 \times 10^7)$ for the chimeras. Giancola et al. also identified a kinetically stable intermediate, suggested to be a dimer during the process of G-quadruplex formation. Their data agreed with a mechanism for G-quadruplex formation, first put forth by Wyatt for DNA,[94] wherein single and double strand species are in an equilibrium favoring single strand, and the step going from dimer to quadruplex is rate limiting. Such a mechanism is consistent with the fourth-order dependence of the association rate on single-strand concentration, but does not require the unlikely event of a four-body collision. Studies on such nucleic acid analogs, showing that incorporation of a single PNA residue into a DNA strand can influence biophysical properties, may well help guide the design of new biopolymer conjugates with improved molecular recognition properties.

Finally, it is important to recognize that folded oligonucleotides can also be functional. For example, Sen's[95–100] group has described a series of DNA oligonucleotide aptamers that are catalysts. Aptamers, selected with a transition state analog *N*-methylmesoporphyrin, catalyzed the Cu^{2+} and Zn^{2+} metallation of porphyrins.[95] This catalytic DNA, which requires K^+ for its activity, may either bind the porphyrins by external stacking or by intercalation between G-quartets. Li and Sen[96] concluded that these DNA chelatases used substrate binding energy to distort the porphyrin's planar conformation, making the porphyrin more basic and easier to metallate. They suggested that the G-quartet is sufficiently rigid to enable this substrate distortion. Sen's group also identified G-quadruplex DNA aptamers with peroxidase activity.[98,99] Their DNA–hemin complexes had enhanced peroxidase activity, when compared to the heme cofactor alone. Again, they concluded that the folded DNA activates the bound heme and enhances peroxidase activity.[99] Willner's group used this hemin binding aptamer as the basis for the clever development of a DNA sensor.[79]

Most recently Sen and Chinnapen[100] used *in vitro* selection to discover a DNA aptamer that can catalyze the photoreactivation of thymine–thymine cyclobutane dimers in DNA. Thus, a 42-mer nucleotide repaired a thymine–thymine dimer substrate with 305 nm light, showing an efficiency that rivaled the native photolyase enzyme. A G-quadruplex unit, formed by specific guanine bases within this 42-mer deoxyribozyme, was proposed to function as a light-harvesting antenna, with photoreactivation of the thymine–thymine dimer proceeding *via* electron donation from an excited guanine base within the G-quadruplex. These studies by Sen underscore the potential for using G-quadruplexes to function as catalysts.

10.3   Summary

In this chapter, we have described just some of the supramolecular structures that have been built using guanine self-assembly. These synthetic G-quartet systems, in addition to providing models for understanding assembly in DNA and RNA, also have potential impact on sensor development, materials science, and nanoscience. Combining this inspiration from Nature with the fertile imagination of supramolecular chemists will undoubtedly provide more discoveries and advances. The use of nucleobase self-assembly to form self-assembled ionophores, synthetic ion channels, dynamic liquid crystals, hydrogels, noncovalent polymers, nanomachines, biosensors, therapeutic aptamer, and catalysts highlights the many functions that can arise from these interesting supramolecular assemblies.

References

1. W. Guschlbauer, J.F. Chantot and D. Thiele, *J. Biomol. Struct. Dyn.*, 1990, **8**, 491.
2. G.P. Spada and G. Gottarelli, *Synlett*, 2004, 596.
3. J.T. Davis, *Angew. Chem. Int. Ed.*, 2004, **43**, 668.
4. M. Gellert, M. Lipsett and D. Davies, *Proc. Natl. Acad. Sci. USA*, 1962, **48**, 2013.
5. T. Pinnavaia, C. Marshall, C. Mettler, C. Fisk, H. Miles and E. Becker, *J. Am. Chem. Soc.*, 1978, **100**, 3625.

6. A. Wong, R. Ida, L. Spindler and G. Wu, *J. Am. Chem. Soc.*, 2005, **127**, 6990.

7. N. Sreenivasachary and J.M. Lehn, *Proc. Natl. Acad. Sci. USA*, 2005, **102**, 5938.

8. G. Gottarelli, S. Masiero and G.P. Spada, J. Chem. Soc., *Chem. Commun.*, 1995, 2555.

9. T. Giorgi, F. Grepioni, I. Manet, P. Mariani, S. Masiero, E. Mezzina, S. Pieraccini, L. Saturni, G.P. Spada and G. Gottarelli, *Chem. Eur. J.*, 2002, **8**, 2143.

10. R. Rinaldi, G. Maruccio, A. Biasco, V. Arima, R. Cingolani, T. Giorgi, S. Masiero, G.P. Spada and G. Gottarelli, *Nanotechnology*, 2002, **13**, 398.

11. T. Giorgi, S. Lena, P. Mariani, M.A. Cremonini, S. Masiero, S. Pieraccini, J.P. Rabe, P. Samori, G.P. Spada and G. Gottarelli, *J. Am. Chem. Soc.*, 2003, **125**, 14741.

12. A.L. Marlow, E. Mezzina, G.P. Spada, S. Masiero, J.T. Davis and G. Gottarelli, *J. Org. Chem.*, 1999, **64**, 5116.

13. S.L. Forman, J.C. Fettinger, S. Pieraccini, G. Gottarelli and J.T. Davis, *J. Am. Chem. Soc.*, 2000, **122**, 4060.

14. M.S. Kaucher, Y.F. Lam, S. Pierracini, G. Gottarelli and J.T. Davis, *Chem. Eur. J.*, 2005, **11**, 164.

15. A. Wong, S.L. Forman, J.C. Fettinger, J.T. Davis and G. Wu, *J. Am. Chem. Soc.*, 2002, **124**, 742.

16. G. Wu, A. Wong, Z. Gan and J.T. Davis, *J. Am. Chem. Soc.*, 2003, **125**, 7182.

17. F.W. Kotch, J.C. Fettinger and J.T. Davis, *Org. Lett.*, 2000, **2**, 3277.

18. X.D. Shi, K.M. Mullaugh, F.C. Fettinger, Y. Jiang, S.A. Hofstadler and J.T. Davis, *J. Am. Chem. Soc.*, 2003, **125**, 10830.

19. M.M. Cai, V. Sidorov, Y.F. Lam, R.A. Flowers and J.T. Davis, *Org. Lett.*, 2000, **2**, 1665.

20. M.M. Cai, A.L. Marlow, J.C. Fettinger, D. Fabris, T.J. Haverlock, B.A. Moyer and J.T. Davis, *Angew. Chem. Int. Ed.*, 2000, **39**, 1283.

21. X.D. Shi, J.C. Fettinger, M.M. Cai and J.T. Davis, *Angew. Chem. Int. Ed.*, 2000, **39**, 3124.

22. M.M. Cai, X.D. Shi, V. Sidorov, D. Fabris, Y.F. Lam and J.T. Davis, *Tetrahedron*, 2002, **58**, 661.

23. X.D. Shi, J.C. Fettinger and J.T. Davis, *J. Am. Chem. Soc.*, 2001, **123**, 6738.

24. V. Andrisano, G. Gottarelli, S. Masiero, E.H. Heijne, S. Pieraccini and G.P. Spada, *Angew. Chem. Int. Ed.*, 1999, **38**, 2386.

25. S. Tirumala, A.L. Marlow and J.T. Davis, *J. Am. Chem. Soc.*, 1997, **119**, 5271.

26. F. van Leeuwen, W. Verboom, X.D. Shi, J.T. Davis and D.N. Reinhoudt, *J. Am. Chem. Soc.*, 2004, **126**, 16575.

27. F. van Leeuwen, C.J.H. Miermans, H. Beijleveld, T. Tomasberger, J.T. Davis, W. Verboom and D.N. Reinhoudt, *Environ. Sci. Technol.*, 2005, **39**, 5455.

28. J.L. Sessler, M. Sathiosatham, K. Doerr, V. Lynch and K.A. Abboud, *Angew. Chem. Int. Ed.*, 2000, **39**, 1300.

29. F.W. Kotch, V. Sidorov, K. Kayser, Y.F. Lam, M.S. Kaucher, H. Li and J.T. Davis, *J. Am. Chem. Soc.*, 2003, **125**, 15140.

30. R. Otero, M. Schock, L.M. Molina, E. Laegsgaard, I. Stensgaard, B. Hammer and F. Besenbacher, *Angew. Chem. Int. Ed.*, 2005, **44**, 2270.

31. V. Gubala, J.E. Betancourt and J.M. Rivera, *Org. Lett.*, 2004, **6**, 4735.

32. J.L. Sessler, J. Jayawickramarajah, M. Sathiosatham, C.L. Sherman and J.S. Brodbelt, *Org. Lett.*, 2003, **5**, 2627.

33. H. Fenniri, M. Packiarajan, K.L. Vidale, D.M. Sherman, K. Hallenga, K.V. Wood and J.G. Stowell, *J. Am. Chem. Soc.*, 2001, **123**, 3854.

34. H. Fenniri, B.L. Deng, A.E. Ribbe, K. Hallenga, J. Jacob and P. Thiyagarajan, *Proc. Natl. Acad. Sci. USA*, 2002, **99**, 6487.

35. H. Fenniri, B.L. Deng and A.E. Ribbe, *J. Am. Chem. Soc.*, 2002, **124**, 11064.

36. J.G. Moralez, J. Raez, T. Yamazaki, R.K. Motkuri, A. Kovalenko and H. Fenniri, *J. Am. Chem. Soc.*, 2005, **127**, 8307.

37. (*a*) S.J. Rowan, S.J. Cantrill, G.R.L. Cousins, J.K.M. Sanders and J.F. Stoddart, *Angew. Chem.*, 2002, **114**, 1528; (*b*) Angew. Chem. Int. Ed., 2002, 41, 898.

38. S. Otto, R.L.E. Furlan and J.K.M. Sanders, *Science*, 2002, **297**, 590.

39. S.M. Gowan, J.R. Harrison, L. Patterson, M. Valenti, M.A. Read, S. Neidle and L.R. Kelland, *Mol. Pharmacol.*, 2002, **61**, 1154.

40. G.R. Clark, P.D. Pytel, C.J. Squire and S. Neidle, *J. Am. Chem. Soc.*, 2003, **125**, 4066.

41. S.M. Haider, G.N. Parkinson and S.J. Neidle, *J. Mol. Biol.*, 2003, **326**, 117.

42. J.A. Schouten, S. Ladame, S.J. Mason, M.A. Cooper and S. Balasubramanian, *J. Am. Chem. Soc.*, 2003, **125**, 5594.

43. A.M. Whitney, S. Ladame and S. Balasubramanian, *Angew. Chem. Int. Ed.*, 2004, **43**, 1143.

44. S. Ladame, A.M. Whitney and S. Balasubramanian, *Angew. Chem. Int. Ed.*, 2005, **44**, 5736.

45. Y. Krishan-Ghosh, A.M. Whitney and S. Balasubramanian, *Chem. Commun.*, 2005, 3068.

46. A. Ghoussoub and J.-M. Lehn, *Chem. Commun.*, 2005, 5763.

47. M.S. Kaucher, W.A. Harrell Jr. and J.T. Davis, *J. Am. Chem. Soc.*, 2006, **128**, 38.

48. N. Sakai, Y. Kamikawa, M. Nishii, T. Matsuoka, T. Kato and S. Matile, *J. Am. Chem. Soc.*, 2006, **128**, 2218.

49. L.C. Bock, L.C. Griffen, J.A. Latham, E.H. Vermaas and J.J. Toole, *Nature*, 1992, **355**, 564.

50. K.Y. Wang, S. McCurdy, R.G. Shea, S. Swaminathan and P.H. Bolton, *Biochemistry*, 1993, **32**, 1899.

51. R.F. Macaya, P. Schultze, F.W. Smith, J.A. Roe and J. Feigon, *Proc. Natl. Acad. Sci. USA*, 1993, **90**, 3745.

52. V.M. Marathios and P.H. Bolton, *Biochemistry*, 1999, **38**, 4355.
53. M. Vairamani and M.L. Gross, *J. Am. Chem. Soc.*, 2003, **125**, 42.
54. K. Padmanabhan, K.P. Padmanabhan, J.D. Ferrara, J.E. Sadler and A.J. Tulinsky, *Biol. Chem.*, 1993, **268**, 17651.
55. D.M. Tasset, M.F. Kubik and W. Steiner, *J. Mol. Biol.*, 1997, **272**, 688.
56. S. Nagatoishi, T. Nojima, B. Juskowiak and S. Takenaka, *Angew. Chem. Int. Ed.*, 2005, **44**, 5067.
57. H. Ueyama, M. Takagi and S. Takenaka, *J. Am. Chem. Soc.*, 2002, **124**, 14286.
58. H.-A. Ho and M. Leclerc, *J. Am. Chem. Soc.*, 2004, **126**, 1384.
59. H.-A. Ho, M. Béra-Abérem and M. Leclerc, *Chem. Eur. J.*, 2005, **11**, 1718.
60. F. He, Y. Tang, S. Wang, Y. Li and D. Zhu, *J. Am. Chem. Soc.*, 2005, **127**, 12343.
61. R.A. Potyrailo, R.C. Conrad, A. Ellington and G.M. Hieftje, *Anal. Chem.*, 1998, **70**, 3419.
62. M. Lee and D.R. Walt, *Anal. Biochem.*, 2000, **282**, 142.
63. N. Hamaguchi, A. Ellington and M. Stanton, *Anal. Biochem.*, 2001, **294**, 126.
64. J.W.J. Li, X.H. Fang and W.H. Tan, *Biochem. Biophys. Res. Commun.*, 2002, **292**, 31.
65. C.J. Yang, S. Jockusch, M. Vicens, N.J. Turro and W. Tan, *Proc. Natl. Acad. Sci. USA*, 2005, **102**, 17278.
66. R. Nutiu and Y.F. Li, *J. Am. Chem. Soc.*, 2003, **125**, 4771.
67. Y. Li and R. Nutiu, *Chem. Eur. J.*, 2004, **10**, 1868.
68. E. Heyduk and T. Heyduk, *Anal. Chem.*, 2005, **77**, 1147.
69. V. Pavlov, Y. Xiao, B. Shlyahovsky and I. Willner, *J. Am. Chem. Soc.*, 2004, **126**, 11768.
70. For 2 other examples of nanoparticle assemblies that incorporate G-quartets, see: (a) F. Seela, A.M. Jawalekar, L. Chi and D. Zhong, *Chem. Biodiversity*, 2005, 2, 84; (b) Z. Li and C. Mirkin, *J. Am. Chem. Soc.*, 2005, 127, 11568.
71. K. Ikebukoro, C. Kiyohara and K. Sode, *Anal. Lett.*, 2004, **37**, 2901.
72. T. Hianik, V. Ostatna, Z. Zajacova, E. Stoikova and G. Evtugyn, *Bioorg. Med. Chem. Lett.*, 2005, **15**, 291.
73. Y. Xiao, A.A. Lubin, A.J. Heeger and K.W. Plaxco, *Angew. Chem. Int. Ed.*, 2005, **44**, 5456.
74. Y. Xiao, B.D. Piorek, K.W. Plaxco and A.J. Heeger, *J. Am. Chem Soc.*, 2005, **127**, 17990.
75. H.-M. So, K. Won, Y.H. Kim, B.H. Ryu, P.S. Na, H. Kim and J.-O. Lee, *J. Am. Chem. Soc.*, 2005, **127**, 11906.
76. A.E. Radi, J.L.A. Sanchez, E. Baldrich and C.K. O'Sullivan, *Anal. Chem.*, 2005, **77**, 6320.
77. A.E. Radi, J.L.A. Sanchez, E. Baldrich and C.K. O'Sullivan, *J. Am. Chem. Soc.*, 2006, **128**, 117.

78. C.H. Fan, K.W. Plaxco and A.J. Heeger, *Proc. Natl. Acad. Sci. USA*, 2003, **100**, 9134.
79. Y. Xiao, V. Pavlov, T. Niazov, A. Dishon, M. Kotler and I. Willner, *J. Am. Chem. Soc.*, 2004, **126**, 7430.
80. V. Pavlov, B. Shlyahovsky and I. Willner, *J. Am. Chem. Soc.*, 2005, **127**, 6522.
81. J.W.J. Li and W.H. Tan, *Nano Lett.*, 2002, **2**, 315.
82. P. Alberti and J.-L. Mergny, *Proc. Natl. Acad. Sci. USA*, 2003, **100**, 1569.
83. W.U. Dittmer, A. Reuter and F.C. Simmel, *Angew. Chem. Int. Ed.*, 2004, **43**, 3350.
84. A.A. Koshkin, V.K. Rajwanshi and J. Wengel, *Tetrahedron Lett.*, 1998, **39**, 4381.
85. S. Obika, D. Nanbu, Y. Hari, J. Andoh, K. Morio, T. Doi and T. Imanishi, *Tetrahedron Lett.*, 1998, **39**, 5401.
86. P.K. Dominick and M.B. Jarstfer, *J. Am. Chem. Soc.*, 2004, **126**, 5050.
87. A. Randazzo, V. Esposito, O. Ohlenschläger, R. Ramachandran and L. Mayol, *Nucleic Acids Res.*, 2004, **32**, 3083.
88. J.-L. Mergny, A. De Cian, A. Ghelab, B. Sacca and L. Lacroix, A comprehensive study of the kinetics of quadruplex association, *Nucleic Acids Res.*, 2005, **33**, 81.
89. B. Datta, C. Schmitt and B.A. Armitage, *J. Am. Chem. Soc.*, 2003, **125**, 4111.
90. V.L. Marin and B.A. Armitage, *J. Am. Chem. Soc.*, 2005, **127**, 8032.
91. Y. Krishman-Ghosh, E. Stephens and S. Balasubramanian, *J. Am. Chem. Soc.*, 2004, **126**, 5944.
92. B. Datta, M.E. Bier, S. Roy and B.A. Armitage, *J. Am. Chem. Soc.*, 2005, **127**, 4199.
93. L. Petraccone, B. Pagano, V. Esposito, A. Randazzo, G. Piccialli, G. Barone, A. Mattia and C. Giancola, *J. Am. Chem. Soc.*, 2005, **127**, 16215.
94. J.R. Wyatt, P.W. Davis and S.M. Freier, *Biochemistry*, 1996, **35**, 8002.
95. Y.F. Li and D. Sen, *Biochemistry*, 1997, **36**, 5589–5599.
96. Y.F. Li and D. Sen, *Chem. Biol.*, 1998, **5**, 1.
97. C.R. Geyer and D. Sen, *J. Mol. Biol.*, 2000, **299**, 1387.
98. P. Travascio, Y.F. Li and D. Sen, *Chem. Biol.*, 1998, **5**, 505–517.
99. P. Travascio, P.K. Witting, A.G. Mauk and D. Sen, *J. Am. Chem. Soc.*, 2001, **123**, 1337.
100. D.J.-F. Chinnapen and D. Sen, *Proc. Natl. Acad. Sci. USA*, 2004, **101**, 65.

Subject Index